Colloidal Active Matter

What do bird flocks, bacterial swarms, cell tissues, and cytoskeletal fluids have in common? They are all examples of active matter. This book explores how scientists in various disciplines, from physics to biology, have collated a solid corpus of experimental designs and theories during the last two decades to decipher active systems.

The book addresses, from a multidisciplinary viewpoint, the field of active matter at a colloidal scale. Concepts, experiments, and theoretical models are put side by side to fully illuminate the subtilities of active systems. A large variety of subjects, from microswimmers or driven colloids to self-organized active fluids, are analysed within a unified perspective. Generic collective effects of self-propelled or driven colloids, such as motility-induced flocking, and new paradigms, such as the celebrated concept of active nematics in reconstituted protein-based fluids, are discussed using well-known experimental scenarios and recognized theories. Topics are covered with rigor and in a self-consistent way, reaching both practitioners and newcomers to the field.

The diversity of topics and conceptual challenges in active matter have long hampered the chance to explore the field with a general perspective. This monograph, the first single-authored title on active matter, is intended to fill this gap by bridging disparate experimental and theoretical interests from colloidal soft matter to cell biophysics.

Francesc Sagués Mestre, Universitat de Barcelona.

Advances in Biochemistry and Biophysics

This monograph series offers expert summaries of cutting edge topics across all areas of biological physics. Individual titles address such topics as molecular biophysics, statistical biophysics, molecular modeling, single-molecule biophysics, and chemical biophysics. The goal of the series is to facilitate interdisciplinary research by training biologists and biochemists in quantitative aspects of modern biomedical research and to teach key biological principles to advanced students in physical sciences and engineering.

Pulling Rabbits Out of Hats
Using Mathematical Modeling in the Material, Biophysical, Fluid Mechanical, and Chemical Sciences
David Wollkind, Bonni J. Dichone

Colloidal active matter
Concepts, experimental realizations and models
Francesc Sagués Mestre

Colloidal Active Matter

Concepts, Experimental Realizations, and Models

Francesc Sagués Mestre

Universitat de Barcelona

CRC Press
Taylor & Francis Group
Boca Raton London New York

CRC Press is an imprint of the
Taylor & Francis Group, an **informa** business

First edition published 2023
by CRC Press
6000 Broken Sound Parkway NW, Suite 300, Boca Raton, FL 33487-2742

and by CRC Press
4 Park Square, Milton Park, Abingdon, Oxon, OX14 4RN

CRC Press is an imprint of Taylor & Francis Group, LLC

ISBN: 978-1-032-28894-9 (hbk)
ISBN: 978-1-032-29840-5 (pbk)
ISBN: 978-1-003-30229-2 (ebk)

DOI: 10.1201/9781003302292

Typeset in font CMR10
by KnowledgeWorks Global Ltd.

Publisher's note: This book has been prepared from camera-ready copy provided by the authors.

To Margot, Carles and Gerard

Contents

Preface

Over the last two decades, the field of active matter has been modish from many different perspectives and responding to a huge variety of interests. I'm glad to having had the opportunity to modestly participate in this advent and, at this point, to share with the readers of this monograph my views on this fascinating topic.

The vastness of active matter, even when restricted to a colloidal scale, might have reasonably dissuaded me to undertake the preparation of this text. But I have attempted to overcome this challenge trying to benefit from my personal attitude when facing the realm of active systems. In this sense, I confess that during the last ten years of research in the field, I have been as captivated by experiments, as absorbed in understanding the theoretical pieces proposed to model them. In short, the ultimate purpose when I started to prepare this book was, thus, to provide a sort of "big picture" of colloidal-based active systems, encompassing comprehensively experimental scenarios along with theoretical developments. I am confident that readers, rather than feeling distracted with this choice, will appreciate such a holistic view of the discipline.

As the book subtitle indicates, *Concepts, Experimental Realizations*, and *Models*, referring to colloidal active preparations, are presented in a balanced way throughout the text. The three practically equal length and more important chapters are dedicated, respectively, to *Particle-based Systems* (microswimmers and driven colloids) in Chapter 3, *Protein-based Active Fluids* (experimental realizations) in Chapter 4, and *Modeling Active Fluids* (theoretical modeling) in Chapter 6. Meanwhile, Chapter 2 introduces the most fundamental and classical concepts of isotropic and anisotropic colloidal suspensions. Chapter 5 somehow summarizes, with a similar spirit, those concepts that have originally emerged from the study of their active counterparts. For the sake of completeness, I comment in Chapter 7 on the most relevant coarse-grained hydrodynamic approaches as they apply to dry active systems. The monograph is completed with four Appendices, a couple of them related still to microswimming, respectively in constrained and complex fluids, and the remaining pair dealing with the application of concepts borrowed from active fluids to motility assays and cell tissues. A special effort has supposed to face the disparity in notation commonly employed in the different contexts. From the very beginning, I abandoned the idea to unify it. I admit that this choice is debatable, but doing otherwise it would have been an exhausting and, even, detrimental task, given the range of dealt with subjects. This has been, hopefully, remedied in two ways. First, by preparing a lot of specially set up footnotes that connect notation when differently used in different contexts. Second, by compiling an extended six page glossary that lists the more frequent symbols, with indication of their dimensions.

Concerning the style of the monograph, the text was conceived neither to be an authoritative specialized text on some specific subject, nor as a simply compiled review of the many developments that have grounded the domain of active matter. The primary intention has been to organize the manuscript in a way it can be useful to readers with different backgrounds and sympathies in relation to active matter. First, I ambition this monograph can be fruitful to practitioners, or, at least, to those already familiar, one way or another, with the field, and who are willing to delve into some poorly recognized, or simply unnoticed, aspects of this endless field. People in this group, hopefully, will find the contents of the monograph worthy and inspiring enough. In relation to newcomers, my highest reward would be to have provided a general overview that might persuade them to further digging into specific references or general treatises, both of them widely quoted among the text. In some cases, an option has been clearly made to retain original papers, while leaving apart follow-up publications, just as a tribute to the originality and freshness of first coming reports. In any case, readers will find at the end sections of Chapters 3 and 6, rather exhaustive lists of briefly commented review papers, pertaining to the respective ambit of microswiming and active fluids. Needless to say, I apologize for the overlooked material.

The level of presentation has been aimed such that experimentalists can learn from the brief sketches of the theories at hand, while the theoreticians can perceive the essentials of the succinctly reported experimental scenarios. This means that details in both contexts have been restricted as much as possible. Practically all the main experimental scenarios I am aware of are commented, whether they are dedicated to self-moving or driven colloids in the first part, or to active fluids in the second. I have deliberately kept to a minimum the topic of biological microswimming in simple or complex fluids that, to my understanding, merits a specific treatment on its own. On the other hand, the criterion to choose the exposed theoretical materials obeys a couple of reasons. Either since they constitute the basic theoretical grounds in the development of active fluids, thinking, for instance, of the Leslie-Ericksen and Beris-Edwards approaches, that are presented in depth and accompanied of some illustrative applications. Alternatively, and specially pertinent in the context of self-propelling or driven colloids, because these modeling bits refer to very concrete experimental scenarios that can be best apprehended when discussed side-by-side with a minimum of theory. More precisely, the text as it stands does not require a specific education, beyond what is granted for a graduate in Physics, although, for most of its contents, it should also be possible to be easily followed from parallel backgrounds in disciplines like Physical Chemistry or Engineering.

I dedicate the last words of this preface to express my gratitude to many senior colleagues, together with the lively group of postdocs and doctorate students, who have accompanied my journey through the world of active systems. Discussions with all of them have strengthened my understanding of the subtleties of this subject. First, my warmest recognition is to acknowledge the cordiality offered by Jordi Ignés-Mullol. This text would have never come to be possible without his sustained collaboration over more than two decades. I've learned from him practically everything I know on experimental Soft Condensed Matter Physics, but more importantly, I have come to deeply appreciate his "easy way of doing". He also largely revised parts of the manuscript, and helped greatly in the preparation of the artwork.

I entered the classical word of colloids following a collaboration with Pietro Tierno that started more than fifteen years ago. This interest turned more recently into the project of driven colloids when dispersed in Liquid Crystals, again as a joint research program with Jordi Ignés-Mullol. In the context of driven colloids, I acknowledge conversations with Ignacio Pagonabarraga and Carles Calero. No doubt, researching in active fluids has been the most fascinating part of my work during the last decade. In this respect, I'm singularly indebted to Pau Guillamat who bravely started this experimental research in the group with a fruitful and careful dedication. I have gone deeper into this particular subject by discussing the theory of active gels with Jaume Casademunt and Ricard Alert. A little bit more occasionally, I've enjoyed discussions with Alberto Fernandez-Nieves and Teresa Lopez-Leon, mainly from the perspective of confined active fluids. Special words should go to Maria Angeles Serrano for an encouraging conversation during a nice walk in Berlin at the beginning of the adventure to write this book, and to Berta Martinez-Prat for the design of the cover.

Finally, many colleagues from abroad have contributed, probably in an unnoticed way to deepen my interest for active systems. The list is practically endless, but some names come first to my mind. So, many thanks to I. Aranson, M. Bär, A. Baskaran, C. Calderer, H. Chaté, R. Goldstein, G. Gompper, J. F. Joanny, O. Lavrentovich, H. Löwen, M. C. Marchetti, M. Polin, M. Shelley, L. Schimansky-Geier, I. Sokolov, A. Snezhko, H. Stark, J. Toner, A. Vilfan, J. Viñals and J. Yeomans. Additionally, some of them also contributed generously with original figures: R. Adkins, D. Bartolo, J. Baudry, A. R. Bausch, C. Bechinger, E. Clément, A. Doostmohammadi, A. Estevez-Torres, A. Fernandez-Nieves, S. Fraden, L. Giomi, R. Golestanian, W. Irvine, J. Palacci, S. Ramaswamy, T. Ross, P. Silberzan, T. Speck, H. Stark, Y. Sumino, S. Thutupalli, and K. Wu. Finally, but very specially, my recognition goes to Zvonimir Dogic for introducing me, during a 2012 short visit to Brandeis, into the field of reconstituted, microtubule-based, active fluids. And to all that should appear listed here as well, but I have sadly missed.

Barcelona, June 2022

List of Figures

Symbols

Symbol Description

a Defect core radius, [m], Sect. 6.2.2.[1]

a, b   Radii of spheres in doublet roller, [m], Sect. 3.2.1.1.

$\hat{a}$ Unit director in an ensemble of rods, Sect. 2.2.1.

c Concentration field, [m^{-3}], Sect. 3.1.1.

c_i Molar concentration of ionic species, [mol dm^{-3}],[2] Sect. 2.1.2.

D Colloid (translational) diffusion coefficient, [m^2 s^{-1}], Sects. 2.1.1 and 3.1.1.4.

D_r Colloid (rotational) coefficient, [s^{-1}], Sect. 3.1.1.4.

D   Distance between surfaces (in relation to depletion forces), [m], Sect. 2.1.3.

$\mathcal{D}$ Chemicals diffusion coefficient in the context of self-phoretic swimmers, [m^2 s^{-1}], Sect. 3.1.1.

D   Electric displacement, [C m^{-2}], Sect. 3.2.4.

E   Electric field, [V m^{-1}], Sect. 3.2.3; Strain rate tensor, [s^{-1}], Sect. 6.2.1.

$\mathcal{E}$ Kinetic energy per unit mass density, [m^4 s^{-2} in d=2], Sect. 6.3.

$E(q)$ Spectral density of kinetic energy (per unit mass density),[3] [m^3 s^{-2} in d=2], Sect. 6.1.4.

$F(q)$ Spectral density of orientational energy in nematics (used in energy balances), [J m^{-1}≡ N in d=2], Sect. 6.1.4.

$F_{elastic}$ Elastic contribution to the nematic free energy density, (denoted F_{FO} with the subindex Frank-Oseen when expressed simply in terms of distortions of the nematic

---

[1]Units, preferably in S.I. system, are given between brackets [ ]. In practice, submultiples in the micro range, specially in length, are used to adapt to the colloidal scale. In most cases, rather than the fundamental, i.e. **kilogram** (kg) for mass, the **meter** (m) for length, the **second** (s) for time, the **kelvin** (K) for temperature, the **ampére** (A) for electrical quantities, and the **mole** (mol) for quantity of mass, derived S.I. units, more conventional, are employed.

[2]In the definition of ionic strength, ionic concentrations are calculated in molar units, but without considering these units explicitly in the corresponding expression to render a dimensionless quantity.

[3]Spectral densities are defined considering isotropic systems after angular integration and incorporating a q factor.

director) [J m^{-2} in d=2], Sect. 2.2.3.

F_{phase} Bulk, phase, i.e. temperature dependent, contribution to the nematic free energy density, [J m^{-2} in d=2], Sect. 2.2.

$\mathcal{F}/F$ Nematic free energy/Nematic free energy density, without explicit consideration of kinetic contributions, [J] ([J m^{-2} in d=2]), Sects. 2.2.1 and 6.2.1.

$\mathcal{F}_t$ Total free energy in nematics, including kinetic contributions, [J], Sect. 6.1.1.

$\mathcal{F}_p$ Polarization free energy, [J], Sects. 6.1.1 and 7.1.1.

$\mathcal{F}_Q$ Free energy for particles interacting nematically on a substrate, [J], Sect. 7.1.2.

$\mathcal{F}_{pQ}$ Free energy for self-propelled rods interacting nematically, [J], Sect. 7.1.3.

f_d Colloidal driving force, [N], Sect. 2.1.1.

$f_{\hat{a}}$ Angular distribution of oriented rods, Sect. 2.2.1.

$\mathbf{G}(\mathbf{r}_a, \mathbf{r}_b)$ Hydrodynamic mobility matrix in a colloidal doublet, [m N^{-1} s^{-1}], Sect. 3.2.1.1. A similar notation designates the Green function or Green operator, $G(\mathbf{r} - \mathbf{r}')$, in the discussion of the effects of hydrodynamic coupling betwwen a bulk fluid and an active monolayer, Sect. 6.3.

$\mathbf{g}$ Momentum density in the theory of active gels, [kg m^{-2} s^{-1}], Sect. 6.1.1.

g Gravitational field, [N kg^{-1}], Sect. 2.1.1.

$\mathbf{H}$ Magnetic field strength in magnetic driving of colloids,

[A m^{-1}], Sect. 3.2.1; Molecular (orientational) field in Beris-Edwards formulation, [J m^{-2} in d=2], Sect. 6.2.1.

$\mathbf{h}$ Molecular (orientational) field in Leslie-Ericksen formulation, [J m^{-2} in d=2], Sect. 6.1.1.

I Inertial momentum or momentum of inertia, [kg m^2], Sect. 3.1.1.4. Used also as the momentum of inertia density in Sect. 3.2.1.3.

$\mathbf{I}$ Unit matrix or unit tensor

I Ionic strength, Sect. 2.1.2.

$\mathbf{J}$ Diffusional flux, [m^{-2} s^{-1}], Sects. 2.1.1 and 3.1.1.4.

$\mathbf{J}_r$ Rotational flux, [m^{-3} s^{-1}], Sect. 3.1.1.4.

K_i Frank-Oseen elastic constants for nematic Liquid Crystals, [N in d=3, J in d=2], Sects. 2.2.3 and 6.1.1. Without subindices, K is conventionally used under the single constant approximation.

k_B Boltzmann constant, [J K^{-1}], Sect. 2.1.1.

L Single elastic contant approximation in $\mathbf{Q}$ formulation (with dimensions as K), Sect. 2.2.3.

l Bond length in a polymer, [m], Sect. 2.1.3.

l_a Active length scale, [m]. Extensively employed in different contexts in Chapters 4 and 6.

l_{nem} Molecular scale in a nematic material, [m], Sect. 2.3.

$\mathbf{M}$ Magnetization (i.e magnetic dipole moment per unit volume), [A m^{-1}], Sect. 2.2.2.

m	Mass of a colloid, [kg], Sect. 2.1.1.	R	Colloid radius, [m], Sect. 3.1.1.
n_i	Number of ions per unit volume for ionic species, [m^{-3}], Sect. 2.1.2.	R_g	Radius of gyration, [m], Sect. 2.1.3.
$\hat{\mathbf{n}}$	Nematic director, Sects. 2.2.2, and 6.1.4.	$\mathcal{R}$	Gradient-like operator corresponding to rotations on the unit sphere (dimensionless), Sect. 3.1.1.4. The symbol without bold, and with subindices $_a$ and $_c$ is used in Sect. 4.3.1 to denote annihilation, respectively creation, rates per unit area of topological defects in active nematic preparations, [m^{-2} s^{-1}].
$P(x;t)$	Distribution function for a one-dimensional random walk, Sect. 2.1.1.		
$\mathbf{P} = \frac{D\mathbf{p}}{Dt}$	Comoving and corotational derivative of the polarization vector, [s^{-1}], Sect. 6.1.1.		
$\mathbf{p}$	Orientation (dimensionless) variable in ABP models, Sect. 3.1.1.4. The same notation is employed to denote the polarization vector, Sects. 7.1 and 6.1.1. Sometimes used capitalized with units [C m] as in Sect. 3.2.2 to refer to the electric polarization of the (Quincke) rolling sphere.	S	Scalar (dimensionless) nematic order parameter, Sect. 2.2.2. This quantity is denoted q in Sect. 6.2.1.
		T	Absolute temperature, [K], used throughout the whole text.
		$\mathbf{T}_m$	Magnetic torque (doublet roller), [N m], Sect. 3.2.1.1.
p	Pressure variable in the context of the Navier-Stokes equation, [N m^{-2}]. It may appear capitalized in some contexts, for instance in Sect. 6.1.1, and with different subindices in the discussion of the different contributions to the swim pressure, Sect. 5.4.2.	U	Interaction potential between colloids, [J], Sect. 7.1.
		V	Magnitude of colloid velocity (eventually denoted V_0 as terminal velocity), [m s^{-1}], Sect. 2.1.1).
		$\mathbf{v}^s$	Slip velocity, [m s^{-1}], Sect. 3.1.1.
Q	Tensorial (dimensionless) order parameter in nematics, Sects. 2.2.2, 6.2, and 7.1.	$\mathbf{v}$	Fluid flow velocity in the context of the Navier-Stokes equation, [m s^{-1}].
$\mathbf{q}$, or q	Employed, depending on dimensionality, as a vectorial [m^{-2} in d=2] or, respectively, scalar notation [m^{-1} in d=1], to denote wave number in Fourier decomposition.	W_a	Anchoring energy for a nematic liquid crystal, [J m^{-2}], Sect. 2.3.
		W_i	Excess energy to build the ionic boundary layer in relation to an ionic species, [J], Sect. 2.1.2.
		w	Channel width in confined active fluids, [m], Sect. 4.4.2.

z_i Electric charge of an ionic species, [C], Sect. 2.1.2.

α Generic notation to refer to an activity parameter in an active nematic system, [N m^{-1} in d=2], Sect. 4.3.1. This notation may also appear, with different meanings, as a coefficient in the expansion of the polarization free energy, as for instance in Sect. 6.2, or in Sect. 7.1. Dimensions in this latter case can be easily inferred from the terms where these symbols appear.

$\tilde{\alpha}$ Apparent activity coefficient, Sect. 3.1.1. In the context of self-diffusiophoresis has units of m^{-2} s^{-1}.

β Boltzmann factor, $\beta = \frac{1}{k_B T}$, [J^{-1}], Sect. 3.1.1.4. This notation may also appear, with different meanings, as a coefficient in the expansion of the polarization free energy, as for instance in Sect. 6.2, or in Sect. 7.1. Dimensions in this latter case can be easily inferred from the terms where these symbols appear.

$\tilde{\beta}$ Squirmer (dimensionless) parameter, $\tilde{\beta} > 0$ for pullers, $\tilde{\beta} < 0$ for pushers, Sect. 3.2.4.

$\beta(\hat{a}, \hat{a}')$ Coupling coefficient in the calculation of the excluded volume contribution to the free energy for a system of aligned rods, [m^3], Sect. 2.2.1.

γ Colloidal frictional (translational) coefficient, [kg s^{-1}], Sect. 2.1.1; Referred to fluid (translational) friction in Navier-Stokes equation, [kg m^{-3} s^{-1}], or [kg m^{-2} s^{-1} in d=2].

γ_r Colloidal frictional (rotational) coefficient, [kg m^2 s^{-1}], Sect. 3.1.1.4. I use instead γ_1 referring to rotational viscosities of Liquid Crystals, with units [kg m^{-1} s^{-1}] (Sect. 3.2.4.2), or of active fluids, [kg s^{-1} in d=2], in the Leslie-Ericksen theory, Sect. 6.1, and in the Beris-Edwards formulation for active nematics, Sect. 6.2. Notice, however, that in this latter context this quantity is sometimes denoted Γ.

$\gamma_\rho, \gamma_p, \gamma_Q$ Kinetic coefficients in the dynamical equations of the corresponding quantities indicated as subindices, used in the formulation of hydrodynamic-like theories for dry active systems, [J m s], [J m^{-3} s], [J m^{-3} s], Sect. 7.1.

δ Sedimentation length, [m], Sect. 3.1.1.3.

δ_{roller} Disparity in size for particles in doublet roller, Sect. 3.2.1.1.

$\tilde{\delta}$ Hydrodynamic screening length, [m], Sects. 3.2.1.3 and Chapter 7 (Introductory paragraphs).

$\epsilon, \epsilon_0, \epsilon_r$ Absolute, vacuum ($4\pi\epsilon_0 = 1.11 \ 10^{-10}$ C V^{-1} m^{-1}), and relative dielectric permittivities, Sect. 2.1.2. ϵ is also used in some specific contexts as a generic perturbative parameter.

$\epsilon_\parallel \epsilon_\perp$ Parallel and perpendicular components of the dielectric tensor in nematics (with or without dimensions depending on whether they refer to absolute or relative values), Sect. 3.2.4.

ϵ_{ij} 2d antisymmetric Levi-Civitta tensor, Sect. 3.2.1.3.

ζ ζ-potential, [V], Sect. 3.2.3. Also employed to denote a scaling exponent, Sect. 7.1.

$\zeta\Delta\mu$ Active contribution to the stress, [N m^{-1} in d=2]. Used in presenting the general theory of active polar fluids, Sect. 6.1.1, to respect the notation employed in the original articles where this theory was introduced. In some applications, for instance in Sect. 6.1.4, or in the Beris-Edwards presentation of the theory of active nematics, Sect. 6.2.1, a shorthand notation is sometimes introduced into the single parameter ζ.

η Shear (dynamic) fluid viscosity, [kg m^{-1} s^{-1}], Sect. 2.1.1. In some contexts, for instance in Sect. 6.3, subindices are added to distinguish the shear viscosity of the active nematic (2d) layer with dimensions [kg s^{-1}] from (bulk, i.e. three dimensional) viscosities of other fluids in contact.

η_r Rotational fluid viscosity, used in Sect. 3.2.1.3 to describe a d=2 fluid of spinners with units kg s^{-1}.

θ Angular variable employed mostly to describe in-plane alignment of nematic, either passive or active, materials.

κ Employed as $\kappa^{-1} = \lambda_D$, to define the Debye screening length, [m], Sect. 2.1.2.

$\mathcal{K}$ Gaussian curvature, [m^{-2}], Sect. 4.4.1.

λ, λ_1 Phenomenological coefficients introduced in the early theory of active gels, Sect. 6.1.1, to account for active contributions to the dynamics of the polarization field. λ is also used in the context of active nematics formulated in the **Q** representation, Sect. 6.2.1 to denote the flow alignment parameter. Different parameters related to the notation λ, with dimensions of velocity, are also employed in the context of the formulation of the theory of hydrodynamic-like theories for dry active systems, Sect. 7.1.

λ_c Wavelength associated to the instability of an aligned active nematic, [m], Sect. 6.1.4.

λ_{dGK} Ratio between elastic and anchoring energies known as de Gennes-Kleman length, [m], Sect. 2.3.

μ Chemical potential, Sect. 2.1.1; $\tilde{\mu}$ corresponds to the generalized form including gravitation effects. This symbol is also used as $\mu = \frac{D}{k_B T}$, [kg^{-1} s], or variants indicated with subindices, either generic as μ_{ph}, or specific for electrophoresis μ_e, [m^2 s^{-1} V^{-1}], and it applies to different variants of the colloid mobility.

μ_0 Vacuum magnetic permeability, $\mu_0 = 4\pi \ 10^{-7}$ V s A^{-1} m^{-1}, Sect. 3.2.1.1.

ν_1 Flow (dimensionless) alignment parameter in the theory of active gels, Sect. 6.1.1.

ξ Employed to denote a scaling exponent, Sect. 7.1.

$\boldsymbol{\xi}, \boldsymbol{\xi}_r$ Translational and rotational noise forces in Langevin-like equations of the ABP model [s$^{-1/2}$ in Eqs. 3.11 and 3.12], Sects. 3.1.1.4 and 5.2. They can be renormalized for convenience as in the force equations for sedimenting colloids (see Eq. 3.15).

$\boldsymbol{\xi}_\rho, \boldsymbol{\xi}_p, \boldsymbol{\xi}_Q$ Noise terms [m^{-2} s^{-1}, s^{-1}, s^{-1}]. Employed in the dynamical equations of the corresponding quantities indicated in subindices that are used in the formulation of hydrodynamic-like theories for dry active systems, Sect. 7.1.

ρ Colloid number density, [m^{-3}], Sect. 2.1.1. Also it may indicate mass density of a fluid system, [kg m^{-3}].

ϱ Electric charge density, [C m^{-3}], Sect. 3.2.3.

σ Electric conductivity, [S m^{-1}= kg^{-1} m^{-3} s^3 A^2], denoted $\boldsymbol{\sigma}$ as a (conductivity) tensor for nematics, Sect. 3.2.4.2. The latter symbol also stands for the stress tensor (mostly in two dimensions) throughout the text [N m^{-1} in d=2]. The symbol may be decorated with different superindices

(referring to reactive, dissipative, total), or subindices (for elastic, viscous, active) to denote specific contributions to the stress.

ς Cross-section for defect annihilation, [m in d=2], Sect. 4.3.1.

τ This symbol indicates, with different specific subindices, corresponding characteristic times, i.e., τ_c or τ_e, charging and electrode blocking times, respectively; τ_a an active time scale, [s].

τ_r Persistence time in colloid motion, [s], Sect. 3.1.1.3.

Φ Generic phoretic potential field with dimensions depending on the context, Sect. 3.1.1.

ϕ Electrostatic potential in the context of boundary layers, [V], Sect. 2.1.2. It is used also as a generic angular variable, a volume fraction of colloids in solution, or a phase-field variable in specific contexts.

ϕ_{pol} Volume fraction (dimensionless) of polymers in solution, Sect. 2.1.3.

χ Magnetic susceptibility (dimensionless) used in different contexts throughout the text.

$\tilde{\chi}$ Euler (dimensionless) characteristic, Sect. 4.4.1.

Ψ, ψ One-particle (complete), or (reduced) distribution function in a system of colloids, [m^{-3}], Sect. 3.1.1.4. ψ is also used as a generic angular variable, or a stream function, $[\text{m}^2 \text{ s}^{-1}]$, in specific contexts.

Ω Vorticity tensor, $[\text{s}^{-1}]$, Sect. 6.2.1.

Ω Used as solid angle, angular velocity for colloid rotation, $[\text{s}^{-1}]$, or enstrophy, $[\text{m}^2 \text{ s}^{-2}$ in integral form for d=2$]$, in the corresponding specific contexts.

ω Used as an angular velocity, $[\text{s}^{-1}]$, frequency, $[\text{s}^{-1}]$, and as phenomenological coefficients in the expression for the polarization free energy used in the formulation of hydrodynamic-like theories for dry active systems, Sect. 7.1. Dimensions in this latter case can be easily inferred from the terms these symbols appear.

Dimensionless Numbers

Re **Reynolds number**: ratio between inertial and viscous forces, Sect. 3.1.1.

Pe **Péclet number**: ratio between advection and diffusion fluxes, Sect. 3.1.1.

Er **Ericksen number**: ratio between viscous and elastic forces, Sect. 3.2.4.

De **Deborah number**: ratio between the time a material takes to adjust to applied stress, and a characteristic experimental time, Sect. 9.1.

Wi **Weissenberg number**: product of the fluid relaxation time and the liquid rate of deformation, Sect. 9.2.

Abbreviations

c.a.	approximately; from the Latin *circa*
et al.	and others; from the Latin *et alia*
i.e.	that is, that is to say; from the Latin *id est*
vs.	against; from the Latin *versus*
Ch.	Chapter
Sect.	Section
l.h.s.	left hand side
r.h.s.	right hand side
Tr	Trace of a matrix
ABL	Active Boundary Layer
ABP	Active Brownian Particle
AN	Active Nematic
ATP	Adenosine Triphosphate
DADAM	Dry aligning Dilute Active Matter
ICEO	Induced-Charge Elestroosmosis
ICEP	Induced-Charge Electrophoresis
ICEK	Induced-Charge Electrokinetics
LC	Liquid Crystal
LCEEK	Liquid Crystal-Enabled Electrokinetics
LCEEP	Liquid Crystal-Enabled Electrophoresis
MIPS	Motility-Induced Phase Separation
MSD	Mean Squared Displacement
MT	Microtubule
PDMS	Polydimethylsiloxane
PEG	Polyethylene Glycol
TFCD	Toroidal Focal Conic Domains

1

Introduction

From a generic point of view, **Active Matter** deals with systems composed by units that convert internal or externally supplied energy into characteristic patterns of motion. They do so by assembling collectivities that, without the need for any leadership, are able to self-organize in exquisite ways with the unique concourse of the self-interactions established from the composing units.

Active Matter has become a very trendy discipline during the last two decades. Presumably, there is no single answer to explain this sustained interest, but one may anticipate a few arguments that might justify the attraction that this field has spurred among different scientific communities.

For some of the practitioners in the field, most likely those more inclined to appreciate the formalisms that the study of active systems has brought about mainly within the field of Physics, the study of Active Matter supposes in a certain way to revitalizing well-established paradigms in Condensed Matter and Statistical Physics. In this respect, sooner or later, we recognize that there is no need to invoke any new Physics to deal with active systems. In other words, what is new is not the Physics underlying their behavior, but their original trends we aim at describing, and the way we employ classical concepts of non-equilibrium Physics to address these new scenarios in an innovative manner.

On the other hand, the phenomena involved when referring to active systems admit experimental realizations that are, in general, modestly difficult but strikingly rewarding. In simple words, Active Matter is an arena where theories can be easily challenged through dedicated experiments, or, the other way around, where experiments appear more provocative to theoreticians. This spirit is, I believe, clearly captured in this text, where experiments and modeling are treated side-by-side on equal basis and depth.

Another aspect that makes active systems particularly motivating is that the commonly invoked realizations span a huge range of spatial length scales, from the microns to the meters and above. As a matter of fact, it has become practically a convention to start any academic presentation in the field with the classical slide spotting admirably performing bird swarms, observed at naked eye, abreast of some colloidal swimmers assembling flocks when observed through a microscope. My choice, clearly indicated in the chosen title, **Colloidal Active Matter**, is to restrict to active systems at a colloidal scale, i.e. behavioral features at the micron-size scale, performed by active units of the most disparate nature, are those I will be mostly concerned with in this monograph.

Last but not least, another key reason to explain the interest for Active Matter is that it is essentially a multifaceted topic. Physicists might claim their ownership since many different disciplines from Physics contribute to the study of active systems, from Statistical Physics to Hydrodynamics or Soft Matter, to mention a few. But this would be certainly unfair. Chemists might as well allege that experimental realizations are based in chemical processes fueling activity one way or another. More importantly, Biophysicists have also been attracted toward Active Matter, as they admit that some of the complex behaviors at the subcell or cell scales might be disentangled with artificial replicas or modeling schemes typical of active systems.

Active matter is thus a conceptually elegant scientific discipline, easily demonstrated through captivating experimental realizations and placed at the crossroad of complementary disciplines. Everything seems accomplished, yet this is not. Certainly, we can congratulate ourselves of the reached milestones, but there is plenty of room that remains to be discovered, as Active Matter is still a blooming area in Science. To my understanding, the greatest venture of Active Matter is twofold, depending on whether we want to test it from the point of view of Material Sciences or from the perspective of Life Sciences. In the first ambit, we need to make Active Matter, not only beautiful but practical. In other words, Active Matter will turn mature when its specialists are going to be faced to engineer active devices that might exploit activity in still unexplored ways. From the perspective of Life Sciences, one needs to translate our deeper and deeper knowledge of *in-vitro* replicas to approach closely the complexity of biological systems at an organism level. Ultimately, this would allow to gain useful information on the functionality that some active biocomponents might carry out in cells and tissues.

Whether we still keep fascinated by the present developments of Active Matter or, rather, we cannot refrain from looking forward imaging their future potentialities, no doubt we will benefit from an accompanied tour through its main realms. This is exactly what this text aims at providing to the interested reader.

2

Fundamental Concepts: Isotropic and Anisotropic Colloidal Suspensions

Colloidal science encompasses a broad range of seemingly very different systems, from paints to biological fluids, just to mention a couple of disparate examples. The underlying concept that unifies colloidal systems is their inherent heterogeneous nature. By this we mean that they are composed of a discontinuous (minority) substance dispersed in a (majority) continuous medium. Although, in principle, each of the two components may appear in any particular phase state (gas, liquid, or solid), colloidal systems in the context analyzed in this monograph will be mostly restricted to two cathegories. I will first refer to colloidal preparations consisting of a continuous liquid phase embedding solid dispersed particles (*colloidal sols* or, simply, *sols*), as commented in Chapter 3. A similar attention will be paid to systems (nominally) close to the reverse scenario, i.e. aqueous dispersions permeating an elastic network when referring to aqueous-based (active) gels in Chapter 4.

The aim of this introductory chapter is to briefly present the most classical concepts of colloidal suspensions. However, since the main interest is on active matter, only those aspects that will turn to be central to the discussion of active (colloidal) realizations will be explicitly considered.

A notorious characteristics of colloidal dispersions that will be most relevant in the context of this book, although it is often quite overlooked in the standard textbooks dedicated to colloidal science, is the shape of the colloidal inclusion. More precisely, and in addition to standard isotropic dispersions, I will pay a distinctive attention to those systems for which the anisometric nature of the dispersed entities endows the colloidal system with clear anisotropic properties. As a matter of fact, in the next chapter dealing with both autonomous swimmers and specimens driven to move, the colloid shape, albeit potentially important, is subsumed into the principles that govern their self-sustained or driven motion. In other words, I will assume there that the specific colloid shape does not impart any significant (macroscopic) difference with respect to conventional dispersions (i.e. such dispersions are going to be mostly considered, even for anisometric colloids, as isotropic in space). Contrarily, when referring later on to protein-based active fluids, the filamentary nature of these biological components, and in turn the *liquid crystal* nature of their assemblies, will be much emphasized, as it is crucial to

DOI: 10.1201/9781003302292-2

contextualize their behavioral trends and to envision, as well, an appropriate level of modeling.

With this idea in mind, I have decided to split the contents of this chapter into two main and at first sight rather disparate sections. The first one is dedicated to classic isotropic (dilute) colloidal suspensions, while the second introduces the basic concepts of Liquid Crystals. A final section closes this chapter, while introducing a sort of mixed category that bridges colloid and liquid crystals concepts. In the first part, I will revisit the most classical concepts of the (equilibrium) microscopic colloidal behavior (Brownian motion, diffusion, etc.). I will add a taste of other macroscopic features, in special those rooted in the established boundary layers that turn out to be fundamental in the (non-equilibrium) electrokinetic scenarios that will be examined in the following chapter.

The second section is justified by the previously announced interest in *dense anisotropic colloidal systems*, specifically referred to here as *Lyotropic liquid crystals*. Although in standard textbooks lyotropic liquid crystals are normally presented from the self-assembly of amphiphilic compounds, I use here this term as equivalent to *colloidal liquid crystals*, i.e. liquid crystalline phases obtained from the assembly of anisometric (commonly rod-like) colloids. This approach will allow us to introduce in this section generic concepts of **Liquid Crystals** (LCs), regardless of being referred to either lyotropic systems, as just mentioned, or to the most common thermotropic materials. *Thermotropic liquid crystals* correspond to pure, or mixed, oily preparations, whose phase behavior is essentially controlled by the temperature, as fundamental control parameter. Conversely, in (aqueous) lyotropics, apart from temperature, the variable concentration plays, as expected, a crucial role in phase space. Among the most typical LC concepts, I will emphasize those referring to symmetries (*nematic, smectic*), long- and short-range order characteristics, topological defects, etc. All of them will be of much use later on in relation to biologically inspired active matter made of assembled proteins (microtubules and molecular motors).

The final section somehow bridges these two contexts, i.e. it combines concepts from standard colloids with those of thermotropic liquid crystals. More precisely, the contents of this final section refer to a particular category of composite-like systems, the so-called nematic colloids, higher in the hierarchy of Soft Matter, prepared after dispersing colloidal inclusions in an anisotropic (thermotropic) liquid crystal solvent.

I choose to adapt the level of the presentation in each section to what is standard in well-recognized textbooks. As primary references I mention, respectively, R. J. Hunter textbook, *Introduction to Modern Colloidal Science* (Oxford University Press, 2003) for colloidal suspensions, and P G. de Gennes and J. Prost monograph on Liquid Crystals, *The Physics of Liquid Crystals* (Oxford University Press, 1993). Readers will enjoy consulting them to acquire a solid overview on either discipline, beyond their use in relation to active systems that is the declared motivation in this book.

2.1 Isotropic Dilute Suspensions

I start with what is, probably, the most universal feature of colloidal suspensions, i.e. the endless and erratic motion of their dispersed constituents. *Brownian motion* refers to these aimless displacements that colloidal particles undergo when free of external interventions and under equilibrium conditions. As a well-known historical note, the name comes from the English botanist R. Brown who first detected it nearly two centuries ago when observing through an optical microscope a suspension of pollen grains dispersed in water. Since then the concept of Brownian motion has become central to many areas of Physics, singularly in Statistical Mechanics and related disciplines.

Brownian motion underlies many macroscopic phenomena in colloidal science, and more in particular, all the aspects related to diffusion and sedimentation. Since both aspects turn out to be specially relevant in relation to swimmers as presented in next chapter, I briefly introduce these two basic scenarios in what follows.

2.1.1 Microscopic Colloidal Behavior: Diffusion, Sedimentation and Random Walk Models

In standard textbooks in Thermodynamics, we all learned that in the absence of external fields, all the components dispersed in a phase in thermodynamic equilibrium must be distributed homogeneously. A more precise way to express the same principle using more elaborated terms is to say that the *chemical potential*, μ, of all and every component of an isolated system at equilibrium must be distributed uniformly in space. If, otherwise, a spatial gradient of chemical potential of any of its constituents is established, this will elicit a flux of the unbalanced component in order to ultimately reestablish the equilibrium (homogeneous) conditions. In the most familiar situations, the unbalance comes from a gradient of concentration or, more precisely, of (chemical) activity, that drives a *material (diffusional) flux* from the more to the less concentrated parcels of the suspension.

The simplest one-dimensional description of a diffusion process assumes a situation not far from thermodynamic equilibrium. In this case, the situation is easily rationalized in terms of a simple (differential) equation relating the *diffusional flux* and the corresponding *driving force*,

$$f_d = -\frac{d\mu}{dx}, \tag{2.1}$$

and by reference to the (number) concentration variable (i.e. the number density) ρ (assuming diluted enough solutions),

$$f_d = -\frac{d}{dx}(\mu^0 + k_B T \ln \rho) = -\frac{k_B T}{\rho}\frac{d\rho}{dx}. \tag{2.2}$$

In this last equation μ^0 stands for the characteristic (temperature-dependent) chemical potential, and k_B denotes the Boltzmann constant. This driving force is compensated with a *drag force* of viscous origin due to the surrounding fluid, whose magnitude defines a *frictional coefficient* γ. The terminal colloid velocity V_0 thus fulfills,

$$f_d = \gamma V_0. \tag{2.3}$$

The material flux, defined in the usual way $J = V_0 \rho$, is itself expressed in terms of *Fick's first law* i.e. $J = -D\frac{d\rho}{dx}$, where D stands for the (translational) *diffusion coefficient*. Using the relations above, one readily arrives at the Einstein's relation connecting thermal fluctuations and frictional dissipation (i.e. a *fluctuation-dissipation relation*),

$$D = \frac{k_B T}{\gamma}. \tag{2.4}$$

This last equation can be easily transformed into the well-known *Stokes-Einstein* formula by using the expression of the friction coefficient for a spherical colloid of radius R in a medium of (dynamic) viscosity η,

$$D = \frac{k_B T}{6\pi\eta R}. \tag{2.5}$$

A similar scheme of equations can be employed to describe the equilibrium distribution of sedimenting colloids under a gravitational field $\phi_{gr} = g$. As mentioned earlier, the starting piece of the argument when referring to thermodynamic equilibrium is to invoke the spatial uniformity of chemical potentials. This later notion needs here to be generalized to include the effect of the gravitational effect, $\tilde{\mu} = \mu + m\phi$, for m the buoyant mass. In this situation, the applicable differential equation is,[1]

$$d\ln\rho = -\frac{mgdh}{k_B T}. \tag{2.6}$$

The integrated form of this differential equation is the well-known *barometric profile* for the distribution in the number of sedimenting particles,

$$N = N_0 \exp\left(-\frac{mg(h - h_0)}{k_B T}\right). \tag{2.7}$$

As a matter of fact, this was the handle that J. Perrin used in 1909 to estimate the Avogadro number from careful analysis of the distribution of gum droplets in a monodisperse suspension, and further relate the Boltzmann constant with the at that time known value for the universal constant R.

---

[1]Notice that one can equally write this equation multiplying by the Avogadro number both numerator and denominator that would render the expression written in terms of, respectively, the molar mass and the universal constant R.

Notice that this distribution complies with Boltzmann prescription for the distribution of colloids under gravity.

Underlying the just delineated scenarios of diffusion and sedimentation there is the notion of *one-dimensional random walk*, as the simplest conceptual basis for a Brownian colloid. The latter model describes the motion of a particle that takes steps of length l at regular time intervals and completely at random while sitting on a line (its generalization to higher dimensionality is straightforward). If a step is taken at intervals τ, the probability for the particle to get to a position denoted x from its original starting location after t/τ steps is given by,[2]

$$P(x;t) = \left(\frac{2\tau}{\pi t}\right)^{1/2} \exp\left(-\frac{x^2\tau}{2tl^2}\right), \tag{2.8}$$

rendering a *normal* or *Gaussian distribution*.

It results instructive to compare this distribution with the solution of the one-dimensional diffusion problem of an initial layer spread uniformly at the origin along a y/z plane of a container,

$$\rho(x;t) = \rho_0 \left(\frac{1}{4\pi Dt}\right)^{1/2} \exp\left(-\frac{x^2}{4Dt}\right). \tag{2.9}$$

By direct comparison, it turns out that $\tau = l^2/2D$. Alternatively, one may be interested in measuring how far the substance spreads under diffusion. This is nicely quantified in terms of the *(root) mean square displacement*. The latter is simply the standard deviation of the normal distribution given above, i.e. in one-dimension,

$$\langle x^2 \rangle^{1/2} = (2Dt)^{1/2}. \tag{2.10}$$

Notice that under isotropic conditions, this expression is easily generalized to 3d situations as $\langle r^2 \rangle = 6Dt$.

All these formal results will be compared in next chapter with their corresponding counterparts for autonomous or forced to move colloids (see Sect. 3.1.1.3).

2.1.2 The Boundary Layer Concept: Electrically Charged Interfaces

In the study of flow processes in heterogeneous (multiphase) systems, the interface itself has often been considered to play a quite circumstantial role. In most of the instances, boundary conditions imposing the continuity of velocity

---

[2]Notice that this corresponds to a probability for a discrete variable. To obtain from it a (normalized) distribution function on the continuous real axis we should introduce a dividing factor $2l$.

and stress are normally invoked, this permitting to fix the appropriate flow conditions. At most, when the interface is allowed to play a more determinant role, this latter effect is commonly subsumed in a single parameter, the *surface tension*, responsible to elicit *surface stresses*. However, all these perspectives really ignore the structure of the boundary layer itself.

In the context of colloidal science, the true message that needs to be conveyed from the very beginning, though, is that such *boundary layers* may play in particular situations a crucial active role, albeit being of very small spatial extend compared to bulk length scales. This is particularly true in the context of hydrodynamic flows, singularly thinking on *electrokinetic phenomena*, and, more in general, in relation to different phoretic scenarios, as I will much emphasize in the ambit of microswimming analyzed in the following chapter.

Electrophoresis is a particularly rewarding example. Colloids normally bear surface charges that are balanced by a diffuse cloud of counterions that fade away from the colloid surface to restore the (neutral) bulk ionic distribution within the dispersing solvent. Considered together the surface charge and the diffuse layer (the so-called *double layer* construction) confers the colloid a electroneutral nature. Yet, the colloid may move electrophoretically under the driving action of an externally applied electric field.

The rationale behind this observation is that the wrapped body is indeed neutral electrically but does not behave as rigid. As a matter of fact, the diffuse layer moves in a direction opposite to that of the charged particle. In other words, a *slip velocity* is established that sets the difference between the fluid velocity at the outer edge of the boundary layer and the particle velocity.

Let's stay within the context of electrophoresis and briefly discuss the issue of the electrically charged interface established between a solid colloid and its dispersing solvent.

The description I am looking for relates the electrostatic potential originated near the colloid surface with the amount of electrical charge accumulated in the diffuse layer built around it. The latter is assumed to be composed by ions described as point charges which (self-consistently) adapt to the locally created electrostatic potential. At the beginning of last century, L. G. Gouy first and later D. L. Chapman developed a treatment describing in simple terms such a situation.

The model assumes that an excess of ions of charge opposite to that of the colloid surface spreads from it, such that planes parallel to that surface are equipotential.[3] Neglecting fluctuations, the distribution of point charges is accounted for in terms of a *(volume) charge density* $\varrho(x)$ that satisfies the fundamental equation of electrostatics, i.e. the *Poisson equation*, in terms of

---

[3]One implicitly assumes that the colloid radius is much larger than the width of the boundary layer itself, in such a way that the interface can be viewed as planar. Sometimes this is called the *Smoluchowski limit*.

the *electrostatic potential* $\phi(x)$, measured at a distance x from the particle's contour. For a flat surface,

$$\frac{d^2\phi}{dx^2} = -\frac{\varrho}{\epsilon_0\epsilon_r}, \tag{2.11}$$

in terms of the product of vacuum and relative electric permittivities. Ions themselves are assumed to fulfill *Boltzmann distribution*, i.e.,

$$n_i = n_i^0 \exp(-W_i/k_BT), \tag{2.12}$$

where n_i stands for the number of ions per unit volume for each ionic species. The notation W_i is assumed to refer to the excess energy involved in getting the ionic species of charge z_i at their precise location within the diffuse layer, i.e. $W_i = z_ie\phi$, with the zero potential taken in the bulk region far from the colloid surface. The charge density is then,

$$\varrho = \sum_i n_iz_ie = \sum_i n_i^0z_ie\exp(-z_ie\phi/k_BT). \tag{2.13}$$

The resulting differential equation for the electrostatic potential adopts then the form of the classical *Poisson-Boltzmann equation*,

$$\frac{d^2\phi}{dx^2} = -\frac{1}{\epsilon_0\epsilon_r}\sum_i n_i^0z_ie\exp(-z_ie\phi/k_BT). \tag{2.14}$$

This second-order differential equation can be solved quite straightforwardly. However, an approximate solution is even more instructive. It turns out to be obtained through a linear approximation of the exponential form and it is known under the name of *Debye-Hückel* limit, valid for small electrostatic potentials. In this case,

$$\frac{d^2\phi}{dx^2} = -\frac{1}{\epsilon_0\epsilon_r}\left(\sum_i n_i^0z_ie - \sum_i n_i^0z_i^2e^2\phi/k_BT\right). \tag{2.15}$$

Clearly, the first term is null due to the bulk electroneutrality, and the equation reduces to the particularly simple form,

$$\frac{d^2\phi}{dx^2} = \left(\frac{e^2\sum_i n_i^0z_i^2}{\epsilon_0\epsilon_rk_BT}\right)\phi = \kappa^2\phi, \tag{2.16}$$

that defines one of the central concepts in colloidal science, i.e. the so-called *Debye (screening) length* as $\kappa^{-1} = \lambda_D$.[4] The solution of this equation is an exponential law whose spatial scale is naturally set by the Debye length (see Sect. 3.2.3).

---

[4]In water at 25^oC, $\kappa = 3.288I^{1/2}$ (nm^{-1}), where I is the so-called *ionic strength*, i.e. $I = \frac{1}{2}\sum_i c_iz_i^2$, where c_i refer to molar concentrations.

2.1.3   Effects of Polymers on Colloidal Stability

A polymer is a macromolecule composed of many monomeric units or segments. If all the monomers are the same, the macromolecule is referred to as a *homopolymer*. Otherwise, we use the generic term *heteropolymer*. A common, and for sure the simplest, way to characterize polymers in solution describes polymers as expanded *(ideal) chains*, assuming there are no interactions between distant monomers in the chain, even if they come close in space. At this level, the most basic model to describe polymer chains is the so-called *freely jointed chain* model that assumes no correlations between bond directions.

A useful mental picture of such a polymer configuration is that of a *random coil* with a radius of gyration R_g. For ideal conditions, the latter is given in terms of the bond length l and the number of repeating units n as $R_g^2 = nl^2/6$.[5] Ideal conditions refer here to the absence of exclude volume effects taking into account all the interactions between the segments within the solvent. In real solvents, this ideal situation is nothing but a useful standard, since the effective size of the colloid is commonly larger or smaller than the above given determination. In *good solvents*, there is net repulsion between segments and the polymer swells when in solution. Conversely, when considering *bad* or *poor solvents*, segments attract each other and the coil shrinks with respect to the random coil configuration. A bad solvent can be "improved", i.e. transformed into a good solvent principally by raising the temperature above a critical value known as *theta temperature*.

Polymers may adsorb easily on surfaces and this applies as well to colloidal surfaces when colloids and polymeric components are mixed in solution. When two polymer-covered surfaces approach each other, one may reasonably expect that they experience an interaction as soon as the outer segments start to overlap (singularly thinking of good solvents). In the simplest situations, this leads to an osmotic repulsive force due to the negative entropy balance associated to the compression of the chains between the two surfaces. In colloidal science, this is a well-known scenario of steric or overlap repulsion that may play a central role in relation to colloid stability.

However, polymers may have a completely different effect regarding colloid stability. As a matter of fact, co-dissolved polymers may originate as well (non-specific) attractive interactions between colloids, this being the situation we are more interested in examining on what follows. This occurs in relation to polymers that are neither attracted nor repelled from surfaces, and it is known as a polymer-originated *depletion effect*. Such an interaction is weak and attractive, yet may be dominant in biocolloid systems. I mention it here because depletion interactions play a central role in the engineering of active microtubule-based active gels, as it will become clear in Chapter 4.

Depletion forces have again an entropic nature and were first recognized by Asakura et al. [12]. The simplest way to understand them is to think of

---

[5]In textbooks of Polymer Science such as M. Rubinstein and R. H. Colby, *Polymer Physics* (Oxford University Press, 2003), this result is named after P. Debye.

an osmotic pressure between the bulk solution that contains a polymer at (number) concentration ρ and the polymer-depleted zone between two close placed surfaces. The result is an effective flux of water out from the inner-depleted region into the outer solute-concentrated area. This effect leads in turn to an effective attraction between the two enclosing surfaces.

On what follows, I provide a simple analytical basis of this depletion effect that illustrates the basic parameters that control it.[6] In the case of two flat surfaces separated a distance $\mathcal{D}$, the elicited interaction energy (per unit area) is evaluated from the accumulated osmotic pressure expressed in terms of the number density ρ, i.e. $P = -\rho k_B T$, as,

$$W(\mathcal{D}) = -\int_{R_g}^{\mathcal{D}} P(\mathcal{D})d\mathcal{D} = \int_{R_g}^{\mathcal{D}} \rho k_B T d\mathcal{D} = -\rho R_g k_B T(1 - \mathcal{D}/R_g), \quad (2.17)$$

valid for $\mathcal{D} < R_g$, with the latter denoting the radius of gyration of the polymer. That is, the energy is zero for $\mathcal{D} \geq R_g$, and decreases linearly to $W(0) = -\rho R_g k_B T$ at contact. One can apply the same argument to curved surfaces, as correspond typically to colloids of radii R, to evaluate the depletion force and corresponding interaction energy,

$$F_{col}(\mathcal{D}) = \pi R W(\mathcal{D}) = -\pi R \rho R_g k_B T(1 - \mathcal{D}/R_g), \quad (2.18)$$

and,

$$W_{col}(\mathcal{D}) = -\int_{R_g}^{\mathcal{D}} F_{col}(\mathcal{D})d\mathcal{D} = -\frac{1}{2}\pi R R_g^2 \rho k_B T(1 - \mathcal{D}/R_g)^2. \quad (2.19)$$

This equation can be similarly written down in terms of the volume fraction of polymer particles in solution $\phi_{pol} \approx \rho R_g^3$,

$$W_{col}(\mathcal{D}) = -\frac{1}{2}\pi (R/R_g)\phi_{pol} k_B T(1 - \mathcal{D}/R_g)^2. \quad (2.20)$$

This expression simply tell us that to have depletion forces intense enough one would need to consider high volume concentrations and large polymer gyration radii. The two conditions are, however, incompatible given the limit of close packing coils. In practice, this means that depletion forces are enhanced when using rather concentrated solutions of middle or low molecular weight polymers.

In summary, the basic idea underlying the depletion effect is that the number density of depleting agents being much greater than that of colloids, the entropic cost associated with clustering a few large colloids is more than offset by the increase in free volume (entropy) accessible to depletant molecules.

---

[6]In this particular respect, I follow J. N. Israelachvili's textbook *Intermolecular and molecular forces* (Elsevier, 3rd. edition 2011).

Obviously, the previous ideas can be similarly exploited in reference to colloidal systems made of rod-like entities. As mentioned in the introductory paragraphs of this chapter, rod-based systems are the essential ingredient behind the concept of colloidal liquid crystals. Thus, the bottom line idea of employing depletion forces in the latter context is that adding non adsorbing polymers to colloidal rods significantly alters their liquid crystalline state. This is a feature much exploited in relation to the preparation of active fluids, as discussed in Chapter 4.

After having referred to them several times, I can not postpone any longer to introduce the basic ingredients to describe and characterize liquid crystal phases.

2.2   Anisotropic Dense Suspensions: Colloidal Liquid Crystals

This second section is built around the concept of "liquid crystallinity". Its importance is justified in different ways within the context of this book. First and more distinctive, because, as already stated, the biological inspired systems I am going to refer to later on in relation to active fluids (see Chapter 4) show more or less prominently the characteristics of liquid crystal phases, singularly in their active nematic preparations. Secondly, because in the discussion of scenarios of colloidal swimming that will be analyzed in subsection 3.2.4 of Chapter 3 we will discover new, non standard, mechanisms of colloid motion when the latter are dispersed in anisotropic thermotropic liquid crystals, i.e. with reference to nematic colloids.

Following the spirit of the previous section, I first provide simple arguments justifying the need to invoke liquid crystal concepts when referring to concentrated (lyotropic) dispersions of anisotropic colloids setting the basis for L. Onsager's study of colloidal liiquid crystals, and in particular his analysis of the isotropic-nematic transition. The remaining couple of sections present a basic overview of generic properties of liquid crystal phases, irrespective of their thermotropic or lyotropic nature.

2.2.1   The Role of Colloid Shape and Concentration

Colloidal liquid crystals combine the unique features of colloids and molecular liquid crystals. In fact and from an historical perspective, it should be worth remembered that interest in liquid crystals surged from the research devoted to concentrated aqueous dispersions of the *tobacco mosaic virus*. It was precisely in this system where the first indications of an isotropic-nematic transition were revealed near ninety years ago. These discoveries motivated Onsager to formulate his seminal theory describing such a phase transition around 1950, now a common place in most of the treatises dedicated to Liquid Crystals. I briefly outline on what follows the essentials of *Onsager's theory* for a system of non-interacting hard rods. The ultimate goal is to unable the reader to become familiar with free energy-based concepts that will be central later on for the development of theoretical models applied to active fluids.

Onsager's theory is a mean-field construction based in three basic postulates: the system is assumed to be composed of slender rods, further dissolved at small volume fractions, and, finally, rods are considered athermal, i.e. one neglects any sort of self-interaction between them, apart from the obvious steric (hard core) effects that are interpreted as dominant. In this respect Onsager assumed that rods behave as completely non penetrable objects. Under these assumptions the free energy minimization shows multiplicity of solutions, i.e. an isotropic solution always exists, but such trivial solution is replaced by an aligned nematic phase when increasing the volume fraction as single-control parameter in phase space.

We can turn these arguments a little bit more quantitative by expanding the free energy in powers of the number density, i.e. proposing a *virial expansion*,

$$\mathcal{F} = \mathcal{F}_0 + k_B T \left(\int f_{\hat{\mathbf{a}}} \log(4\pi f_{\hat{\mathbf{a}}} \rho) d\Omega + \frac{1}{2}\rho \int \int f_{\hat{\mathbf{a}}} f_{\hat{\mathbf{a}}'} \beta(\hat{\mathbf{a}}, \hat{\mathbf{a}}') d\Omega d\Omega' \right), \quad (2.21)$$

written specifically for anisotropic systems, as exemplified by the use of the distribution function $f_{\hat{\mathbf{a}}}$.[7] The latter denotes the angular distribution of oriented rods, i.e. $\rho f_{\hat{\mathbf{a}}} d\Omega$ stands for the number of rods per unit volume pointing within a small solid angle $d\Omega$ around a prescribed direction specified by unit director $\hat{\mathbf{a}}$.

The first term is a reference additive constant. The logarithmic contribution favors rotationally invariant isotropic phases, whereas the next term in the expansion introduces the notion of *excluded volume* through the coupling constant $\beta(\hat{\mathbf{a}}, \hat{\mathbf{a}}')$. This latter term favors aligned configurations and dominates at sufficiently high densities. In other words, minimizing excluded volume increases packing entropy at expenses of reducing the mixing counterpart. The balance between these two terms terms sets the location of the (first-order) isotropic-nematic transition along the number density axis (for fixed rod aspect ratio). Although Onsager's theory is mainly qualitative, the transition is easily parametrized by solving a self-consistent equation for the distribution function $f_{\hat{\mathbf{a}}}$. Although elegant and quite straightforward, I do not pursue here this formal analysis since it would take us far away from our true interests in presenting basic concepts of Liquid Crystals in relation to active matter.

---

[7]Although in this last expression $\mathcal{F}_0$ remains undetermined at this point, I prefer, respecting the spirit of its derivation, to consider this free energy form as arising from density effects of the material. In this sense, I assume it does no contain either hydrodynamic contributions, or those associated to electric or magnetic effects.

Any colloid that significantly deviates from a spherical shape undergoes an isotropic-nematic phase transition. Consequently, numerous experimental systems exhibit a nematic phase, including biological polymers such as DNA, actin, and various filamentous viruses [85].

2.2.2   Basic Concepts of Liquid Crystals: Phases and Order Parameter

Liquid crystalline states are fluid mesophases (meso for intermediate) displaying middle symmetries between the proper of an ordered crystal and those intrinsic to isotropic liquids. In other words, Liquid Crystals are systems in which a liquid-like order exists at least in one direction (i.e. the density auto-correlation function displays in this particular direction no sign of periodicity), and in which some degree of anisotropy is present.

These two conditions can be simultaneously satisfied in the simplest examples of liquid crystal phases. With reference to anisotropic substances, one can think of imposing a reduced positional order, while introducing an alternative orientational order of the constituting non-spherical units. This argument readily leads to identify the two most common phases of liquid crystals: the *nematic phase*, referred to above in relation to Onsager's treatment, supposes the total absence of positional order, but bringing about an identifiable long range orientational order of elongated objects. In turn, the *smectic phase* generically refers to layered (fluid) configurations, thus retaining the minimum one-dimensional degree of positional order for, otherwise, aligned units.

With reference to the contents of this book the nematic phase, more specifically its uniaxial version, will be by far the most dealt with, while I will make a single comment on smectics in relation to interfaced active and passive liquid crystals in Sect. 4.3.2.

Since nematics will play a fundamental role in this book, it is necessary to start by introducing properly the order parameter corresponding to this particular symmetry. The *nematic order parameter* must be non zero in the nematic state but should vanish in the isotropic phase. There are two ways to define it, depending on whether one adopts a microscopic or a macroscopic picture. The first approach renders an scalar, whereas the second a order parameter of tensorial nature.

The microscopic approach will refer to a collection of rigid rods with cylindrical symmetry (uniaxial nematics). Axes of the rods will be denoted by a three-dimensional unit vector $\hat{a}$. Relative to a reference axis, taken z for simplicity, the state of alignment is characterized by a distribution function $f_{\hat{a}}(\theta, \phi)$ as indicated above in the presentation of Onsager's theory. Rather than working with this distribution itself, we choose a numerical observable (scalar) derived from it. The lowest order non null moment of this distribution defines the **scalar nematic order parameter** as,[8]

$$S = \frac{1}{2}\langle 3\cos^2\theta - 1\rangle = \int f_{\hat{a}}(\theta, \phi)\frac{1}{2}(3\cos^2\theta - 1)d\Omega. \qquad (2.22)$$

With this definition in terms of the tilt angle θ, it is clear that S varies between $S = 0$ in an isotropic configuration of rods ($f_{\hat{a}}(\theta)$ independent of θ), and $S = 1$ for perfect parallel alignment ($f_{\hat{a}}(\theta)$ peaked around 0 or π).

From a macroscopic point of view one could employ any macroscopic property whose value for an liquid crystalline material would naturally depend on whether it is measured in its isotropic (liquid) or nematic phase. For instance, considering the magnetic moment one would write,

$$M_\beta = \chi_{\alpha\beta}H_\beta. \qquad (2.23)$$

In isotropic conditions $\chi_{\alpha\beta} = \chi\delta_{\alpha\beta}$, while for uniaxial nematics,

$$\chi_{\alpha\beta} = \begin{bmatrix} \chi_\perp & 0 & 0 \\ 0 & \chi_\perp & 0 \\ 0 & 0 & \chi_\| \end{bmatrix}$$

A convenient tensorial magnitude to express the nematic order is defined extracting the anisotropic part of the magnetic sysceptibility matrix,

$$Q_{\alpha\beta} = G\Big(\chi_{\alpha\beta} - \frac{1}{3}\delta_{\alpha\beta}\sum_\gamma \chi_{\gamma\gamma}\Big). \qquad (2.24)$$

Notice that this tensor is real, symmetric and traceless. G is a normalization constant commonly chosen so that $Q_{zz} = 1$ in a fully oriented configuration. This tensor can be always diagonalized in the form,

$$\mathbf{Q} = \begin{bmatrix} -q/2 & 0 & 0 \\ 0 & -q/2 & 0 \\ 0 & 0 & q \end{bmatrix}$$

with $tr(\mathbf{Q})^2 = 3q^2/2$ as the second-order invariant. Next order invariants allow to construct a (density) free energy expansion, encoding in a coarse-grained way the phase state of the anisotropic material,

$$F_{phase} = F_0 + \frac{3}{4}Aq^2 + \frac{B}{4}q^3 + \frac{9}{16}Cq^4 + O(q^5), \qquad (2.25)$$

with the coefficient of the quadratic term A being a function of the temperature.[9] As a matter of fact, this dependence is assumed linear close enough

---

[8]The following expression refers strictly to a three-dimensional nematic configuration. One needs to adapt it when the space of angular configurations spanned by the rods is restricted. In general, one would prescribe $S = \frac{1}{d-1}\langle d(\mathbf{a}\cdot\mathbf{n})^2 - 1\rangle$, where $\mathbf{n}$ denotes the common direction of alignment, and d the spatial dimension of the space of orientations.

[9]This expression is written for a free energy volume density, and denoted correspondingly, following [323]. I do so to better connect with the notation employed later on in Sect. 6.2.1.

to a threshold defined as the lowest temperature down to which one could supercool the isotropic phase (not to be confused with the *clearing point* or transition temperature where the nematic phase starts to melt in a temperature ramp experiment). The presence of a non-vanishing cubic contribution guarantees that the isotropic-nematic transition is first order, as argued in de Gennes et al.'s book.

In an ideal nematic phase, molecules are (on average) aligned along a common direction $\hat{n}$, we refer to as the **(nematic) director**. There is an important symmetry to be recognized here, since directions $\hat{n}$ and $-\hat{n}$ are considered totally equivalent.[10] The orientational fluctuations of the molecules around the director are captured through the (scalar) order parameter S introduced above. Thus, the **tensorial nematic order parameter** can be rewritten as,[11]

$$Q_{\alpha\beta} = Q(T)\left(n_\alpha n_\beta - \frac{1}{3}\delta_{\alpha\beta}\right), \qquad (2.26)$$

or formally in terms of the scalar order parameter S,[12]

$$Q_{\alpha\beta} = \frac{3}{2}S\left(n_\alpha n_\beta - \frac{1}{3}\delta_{\alpha\beta}\right). \qquad (2.27)$$

Defined in this way the tensorial order parameter, has the largest eigenvalue given by the scalar value S and a couple of degenerated values given by $-S/2$. Since $tr(\mathbf{Q})^2 = 3S^2/2$, the above free energy functional is written as,[13]

$$F_{phase} = F_0 + \frac{3}{4}a(T - T_{NI}^*)S^2 + \frac{B}{4}S^3 + \frac{9}{16}CS^4 + O(S^5), \qquad (2.28)$$

where T_{NI}^* denotes the temperature below which the isotropic phase is absolutely unstable with respect to the nematic ordering.

2.2.3   Long- and Short-Range Order: Orientational Distortions and Defects

Nematic materials accumulate elastic energy if their alignment is subject to spatial gradients. In other words, the basic state of a nematic liquid crystal

---

[10]Notice that this invariance has nothing to do with any eventual symmetry at the level of individual molecules. As a matter of fact, if the latter carry a permanent dipole, we are admitting implicitly that there are as many dipoles "up" as dipoles "down".

[11]In writing the director in terms of explicit components I avoid to use the notation $\hat{n}$ that identifies its unit norm.

[12]The following expression refers again to the three-dimensional case. Its obvious generalization would introduce a dimensionality factor $\frac{d}{d-1}$, although in some of the applications discussed later on in Chapter 6, the reduced form $Q_{\alpha\beta} = S(n_\alpha n_\beta - \frac{1}{2}\delta_{\alpha\beta})$, i.e. $tr(\mathbf{Q})^2 = S^2/2$, is chosen for two-dimensional realizations.

[13]In two-dimensional systems, the cubic term is taken null.

in equilibrium maximally preserves a uniform long range orientational order compatible with the imposed boundary conditions at the sample boundaries. Deviations in the form of spatial inhomogeneities are penalized in terms of an (elastic) energy cost.

Phenomenologically one can construct a free energy functional accounting for such distortion energy, similarly to what has been postulated earlier for the phase (bulk) contribution. Obviously we can envisage two ways to construct such a functional. Either one employs a director-based description, admitting that the "nematicity" is fixed, or, alternatively, one chooses a form based on the full tensor order parameter that eventually incorporates inhomogeneities in the scalar order parameter itself.

The first choice is elaborated in de Gennes's book and it is known as the Frank or **Frank-Oseen (distortion) free energy** density. Invoking basic symmetry requirements, it is written combining the allowed second-order spatial variations of the (nematic) director field in terms of a triad of elastic constants, each one representing the elementary distortion modes of a nematic configuration, i.e. *splay* (K_1), *twist* (K_2) and *bend* (K_3),[14]

$$F_{elastic} = \frac{1}{2}K_1(\boldsymbol{\nabla} \cdot \hat{\mathbf{n}})^2 + \frac{1}{2}K_2[\hat{\mathbf{n}} \cdot (\boldsymbol{\nabla} \times \hat{\mathbf{n}})]^2 + \frac{1}{2}K_3[\hat{\mathbf{n}} \times (\boldsymbol{\nabla} \times \hat{\mathbf{n}})]^2. \quad (2.29)$$

As easily checked, the elastic constants have units of energy per unit length, with typical values for conventional thermotropic materials of the order of tens of piconewtons [77],[15]. Often, a one-constant approximation is invoked to simplify the use of the above expression. In this case,[16]

$$F_{elastic} = \frac{1}{2}K[(\boldsymbol{\nabla} \cdot \hat{\mathbf{n}})^2 + (\boldsymbol{\nabla} \times \hat{\mathbf{n}})^2] = \frac{1}{2}K(\boldsymbol{\nabla}\hat{\mathbf{n}})^2 = \frac{1}{2}K\partial_\alpha n_\beta \partial_\alpha n_\beta. \quad (2.30)$$

On the other hand, the distortion energy expressed in term of the tensor **Q** reads (under the single constant approximation),

$$F_{elastic} = \frac{1}{2}L\frac{\partial Q_{\alpha\beta}}{\partial x_k}\frac{\partial Q_{\alpha\beta}}{\partial x_k}, \quad (2.31)$$

---

[14]Other contributions as for example a saddle-splay term are disregarded [197] on all what follows since I am not aware of any eventual influence they could have in the context of active materials covered in this book. Notice also that the twist contribution vanishes in two-dimensional nematic layers.

[15]I have explicitly decided to keep the subindices "phase" and "elastic" in these introductory notes. Later on as the corresponding expressions for these (density) free energy functionals are going to be used, either combined or separately, in the specific context of active fluids I am not going to use these subindices anymore since their respective meaning is fully understandable.

[16]Here and throughout the book the Einstein convention for summing over repeated indexes is adopted.

and adds to the previous contribution the corresponding one arising in the nematic order parameter S. In three-dimensional nematics,

$$F_{elastic} = \frac{1}{2}K(\boldsymbol{\nabla}\hat{\mathbf{n}})^2 + \frac{3}{4}L(\nabla S)^2, \qquad (2.32)$$

with $K = 9LS^2/2$.

Up to now I was mostly concerned with long range distortions in the nematic alignment, yet the director was supposed to be well-defined at any spatial location. Nevertheless, there are situations where the field $\hat{\mathbf{n}}$ displays singularities in specific points, i.e. it does not show smooth spatial variations around them. This idea introduces the concepts of *defects*, most commonly referred to on all what follows as **topological defects**. The presence of defects is quite natural in Liquid Crystals. As a matter of fact, the name nematics comes precisely from the observation of flexible dark threads in thick nematic samples, later referred to as *disclinations*. In this book, I will be more interested, though, in point-defects rather than in their three-dimensional extended counterparts.

The arrangement of the molecules around such point defects is commonly observed by polarization microscopy. This allows to classify defects with an index or **topological charge** whose values can be positive or negative integer or semi-integers. For most of the contents of the chapters devoted to active fluids, and in particular active nematics, positive and negative seminteger defects of value $\pm 1/2$ will play a crucial role.

This index is readily identified fro direct observations. Assume that the director around the defects can be simply described by a planar structure. One would place a two-dimensional coordinate reference system centered at the defect location. Denoting $\phi(\mathbf{r})$ the polar coordinate of a point around the singularity and $\theta(\mathbf{r})$ the corresponding angle spanned by the director with respect to the x-axis, then the simplest relationship holds,

$$\theta(\mathbf{r}) = m\phi(\mathbf{r}) + const., \qquad (2.33)$$

equivalently to say that if one follows a closed circuit around the singularity, and make a full turn $\Delta\phi = 2\pi$, then $\Delta\theta = 2\pi m$. This is acceptable only if m is an integer or seminteger index, since $\hat{\mathbf{n}}$ and $-\hat{\mathbf{n}}$ are indistinguishable.

2.3 A Composite System: Nematic Colloids

In a particular context of next chapter, that of electrokinetic phenomena as driving mechanisms (see subsection 3.2.4), I will refer to a composite system made of colloids dispersed in thermotropic liquid crystals. These preparations are commonly referred to in the literature as **nematic colloids** [213, 268, 353, 441]. I find opportune to briefly comment on them in this last section as the best way to bridge the contents of the two previous sections of this introductory chapter. Nevertheless, before going directly into this specific topic we need to comment on the issue of anchoring of liquid crystals, and, in particular, on the role of anchoring in establishing the state of alignment of liquid crystals around inclusions.

The contact surfaces of Liquid Crystals impart to the director a direction of preferred orientation, that is referred to as an *easy axis*. Deviations of the local director from this prescribed condition involves an energy cost, that, when assumed restricted to the plane normal to the surface, is expressed by the Rapini-Papoular expression,

$$W = \frac{1}{2}W_a \sin^2(\theta - \theta_0). \tag{2.34}$$

W_a is called the *anchoring coefficient*, with units of energy per unit surface, θ_0 accounts for the angle between the easy axis and the normal to the interface, and θ is the actual tilt of the director. For a typical thermotropic (5CB) aligned tangentially (*planar anchoring*) at a passivated (for instance polyimide-coated) surface, $W_a \sim 10^{-4}$ Jm^{-2}. One or two orders of magnitude smaller values are measured for nematics anchored perpendicularly to a surfactant-functionalized surface (*homeotropic anchoring*). In general, slightly smaller values apply to lyotropic LCs.

An important ratio with dimensions of length, called the *de Gennes-Kleman length*, measures the relative importance of the elastic and anchoring energies $\lambda_{dGK} \equiv K/W_a$. For the typical values given previously, λ_{dGK} is of the order of 0.1μm, orders of magnitude larger that the nanometer molecular scale of the nematic l_{nem}, and of the order or, more typically, smaller than the size of colloids typically dispersed in nematic dispersions. This will have important consequences as commented next.

The balance between elastic and anchoring energies will determine the local configuration of the director in the neighborhood of a colloid dispersed in a (nematic) LC matrix. The argument is simply illustrated for a (spherical) colloid of radius R with normal (radial) homeotropic anchoring that is placed in a uniformly aligned LC configuration [217]. It is easy to see in this case that the far field director of the nematic is non-compatible locally with the prescribed boundary condition at the colloidal surface. Either the elastic energy prevails, and one pays the anchoring energy penalty given by $F_{anch} \sim W_a R^2$, or, the other way around, normal anchoring is enforced at the prize of distorting elastically the liquid crystal matrix by an energy amount $F_{elastic} \sim K(\frac{\theta - \theta_0}{R})^2 R^3 \sim KR$, where K accounts for an averaged elastic constant.

Clearly, the two scalings correspond to different orders in the colloid radius R, the de Gennes-Kleman length λ_{dGK} connecting them in a precise way. In particular, one can expect that for small particles $R \ll \lambda_{dGK}$ the linear (elastic) contribution penalizes more than its anchoring counterpart, and the director would be barely affected by the immersed colloid (see Fig. 2.1(a)).

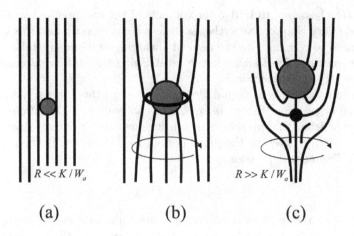

FIGURE 2.1
Main defect configuration around a colloid under homeotropic anchoring. Different configurations correspond to (a) small particle; (b) Saturn ring (quadrupolar); (c) hyperbolic hedgehog (dipolar). Image extracted from Lavrentovich [217].

In the opposite situation for large colloids, $R \gg \lambda_{dGK}$, it is essentially the anchoring condition that is satisfied locally in the colloid neighborhood at the expenses of the bulk elastic energy. These two scenarios can be respectively referred to as *weak* and *strong* anchoring.

On what follows I restrict to the case of strong anchoring, and since the figured condition just discussed was that of a colloid with homeotropic anchoring we stay first in this situation. The actual distortions of the nematic LC around a colloid may depend on different factors (cell dimensions, eventual electric forcing, etc), but singularly on the orientation of the easy axis and the colloid shape and size.

From a topological point of view, the simplest matching solution for the director under strong homeotropic anchoring corresponds to a disclination ring of strength $-1/2$ that embraces the colloid through its poles and known as *Saturn ring* (see Fig. 2.1(b)). This structure has a quadrupolar symmetry. Its stability is normally an issue except for thin cells of the order of twice the colloid radius. Otherwise, it may easily collapse into a point defect structure, named *hyperbolic hedgehog* of dipolar nature and topological charge -1 (see Fig. 2.1(c)). The pair hedgehog-sphere (with accumulated zero topological charge) is represented by an elastic dipole **p**, conventionally directed from the satellite-defect towards the particle. For planar anchoring, the less distorted configuration around the colloid is satisfied with a pair of defects symmetrically located at the poles (equatorial with respect to the far field) and configuring a *(quadrupolar) double-boojum*.

Elastic distortions in the nematic matrix brought about by dispersed colloids nematic are at the origin of the multiple possibilities for equilibrium self-assembling that have been analyzed both with basic and applied interests in a vast list of references. For an compilation one can consult the recent reviews by Blanck and Muševič [35, 267].

A rough estimate of the order of magnitude of the energies involved in liquid crystal-based colloid dispersions can be obtained by computing the cost in elastic energy associated to a strongly anchored colloid in an, otherwise, uniformly aligned nematic. A simple scaling argument goes as follows $F_{elastic} \sim KR \sim (U_{LC}/l_{nem})R \sim k_B T_{NI}(R/l_{nem}) \sim 10^3 \, k_B T_{NI}$, where an averaged elastic constant is estimated as $K \sim U_{LC}/l_{nem}$ with U_{LC} [77], being the energy of the molecular chemical component

3

Particle-based Active Systems

Motion is the most essential aspect of **active systems**, at least in the way we understand this concept here. In this chapter, I will concentrate on particle-based systems, i.e. systems composed of discrete entities that show activity in the form of sustained motion. I leave for next chapter the consideration of continuous active media, that we qualify as active fluids. Even though the latter could be considered as multicomponent systems as well, the individual characteristics of the dispersed active specimens in this case will not be in general a major concern, neither for the interpretation of their behavior, nor for most of the modeling approaches they have motivated in the past.

More precisely, I will refer in this chapter to systems with characteristics of classical colloidal dispersions that are additionally endowed with distinctive individual and collective features that are directly rooted in the motility of their dispersed units. The latter can **self-propel**, i.e. behave as **microswimmers**, or, alternatively, be **externally actuated** as **colloids driven to swim**. In other words, motility is here understood in a very generic way, considered to be originated either from an intrinsic activity of the dispersed entities, or, alternatively, be forced externally following a purposed actuation. As a general principle, swimming at the microscale implies hydrodynamic conditions corresponding to very low *Reynolds numbers*, when inertia is overwhelmed by viscous and friction effects. This imparts specific properties to the flow patterns of individual particles and collective assemblies as well.

During these last two decades, a large variety of laboratory-tailored developments based on different propulsion mechanisms have been proposed referring either to microswimmers or to driven colloids. Some of them were clearly inspired by living realizations, while others were designed to achieve particularly aimed performances. Definitively, the present challenge is to downscale these or other prototypes towards the nanoscale, fostered by their potential utility in health care or for environmental applications, as reviewed by several groups [297, 125].

Swimming at the microscale will be considered in this chapter as taking place essentially in unconstrained aqueous media, apart from a brief consideration of anisotropic, Liquid Crystals-based, dispersions of driven colloids. I refer the reader to specific appendices to find very brief discussions that extend this framework. In particular, Appendix 1 (see Chapter 8) considers microswimming in disordered and constrained environments, while Appendix 2

DOI: 10.1201/9781003302292-3

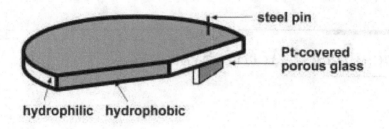

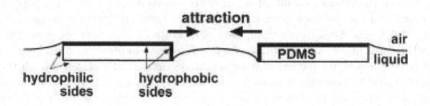

FIGURE 3.1
Autonomous movement of a chemically functionalized plate. (Top)
Schematics of the self-propelling object. A thin PDMS plate (1 to 2 mm thick
and 9 mm in diameter) with specified faces rendered hydrophilic. Additionally, a piece of porous glass filter (covered with platinum by an electron beam
evaporation) was mounted on the PDMS piece with a stainless steel pin. (Bottom) A diagram illustrating self-assembly by capillary interactions. Image and
caption text adapted from Ismagilov et al. [177].

(see Chapter 9) is devoted to microswimming in complex, i.e. non-Newtonian,
fluids.

To better contextualize the contents of this chapter it results particularly
illustrative to close these short introductory remarks by referring to a pair
of pioneering examples dedicated to specific realizations of, respectively, autonomous and driven motion.

A seminal work that opened the field of **(chemically powered) artificial swimming** was reported by Ismagilov et al. [177] (see Fig. 3.1). These
authors pioneered the use of interfacial reactions to support the autonomous
propulsion of floating blocks that are chemically heterogeneous. These blocks
consisted of millimeter-size, hemicylindrical PDMS plates with a small area of
platinum on one surface. Plates were observed to move under the impulse of
the bubbles generated by the chemical decomposition of the hydrogen peroxide catalyzed by Pt. Propulsion speeds achieve values of a few millimeters per
second and last for several hours. Looking for dynamical self-assembly possibilities, the plates were also patterned to have hydrophilic and hydrophobic
regions to promote attraction driven by capillary forces originated at the hydrophobic edges

The second representative example corresponds to an external (magnetic)
actuation as proposed by Dreyfus et al. [91] to model an artificial flagellum

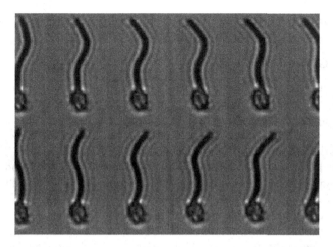

FIGURE 3.2
Beating of a magnetic flexible filament attached to a red blood cell.
The filament length is $L = 24$ μm. The magnetic field is applied along the vertical direction in the shown image. Time interval between images is 5 ms. Image and caption text adapted from Dreyfus et al. [91].

(see Fig. 3.2).[1] In this case, the prototype was based on a red blood cell with an attached colloidal chain of DNA-linked magnetic beads. The whole assembly was forced to oscillate following an external magnetic field. By breaking time-reversal invariance, this forcing induces a beating pattern that results from the coupling of magnetic forces, filament flexibility, and viscous drag from the solvent.

3.1 Self-propelled Swimmers

In this section, I will consider swimmers that are able to self-propel. The principal attention will be devoted to swimmers based on **self-phoretic effects**, to which I dedicate most of the first subsection. In the second one, I will refer to **Marangoni-based swimmers**. For the sake of completeness, I reserve a last subsection to briefly tackle the issue of biological microswimmers.

---

[1]In a parallel development, bio-hybrid swimmers consisting of a polymeric filament with a rigid head and a slender tail decorated with cultivated heart-muscle cells, as designed by Williams et al. [432], supposes a new twist on the idea of flagella replication, and a closer realization of an autonomous sperm-like cell.

3.1.1 Self-phoretic Swimmers and their Active Brownian Particle (ABP) Models

3.1.1.1 General Concepts

It is well-known from colloidal science that external fields are commonly used to achieve transport (*phoresis*) at the microscale. One of the most paradigmatic examples is **electrophoresis**. In this case, the electrical charge accumulated at the particle surface is separated from a diffuse cloud of counterions, whose spatial extent is set by the Debye screening length, as revisited in the previous chapter (see Sect. 2.1.2). Considered together, the solid body and the accompanying diffuse ionic layer constitute a non-rigid (electrically) neutral body. Under an externally applied electric field, the charged solid body and the diffuse cloud move in opposite directions, while building a slip velocity at the outer edge of the diffuse layer (see also Sect. 3.2.3 for a more detailed treatment). Yet, neutrality of the combined particle and counterions layer means that there is no net force (neither a net torque) on the system.

This characteristic fully distinguishes this situation from, for instance, the sedimentation of such a colloid under the action of gravity. In short, what is generic in any phoretic process is the existence of a *gradient of a field-variable* that induces a slip velocity. In the case of colloidal electrophoresis, the field-variable is the electrostatic potential, whose gradient is associated with the separation of electric charges. Alternative scenarios correspond to gradients of other fields, such as chemical concentration (**diffusiophoresis**) or temperature (**thermophoresis**).

To achieve autonomous, i.e. **self-phoresis**, the gradient in the field-variable must be created by the colloid itself. This is commonly realized by introducing some degree of (spatial) asymmetry at the level of the particle. This permits to break up its fore-aft symmetry and enables to gain partial control on the direction of movement. This latter characteristics is best illustrated by the use of the so-called **Janus-like** particles, i.e. two-sided particles with distinctive different properties on each side. In the most conventional case, the particle heterogeneity is of chemical nature, and when it gets dispersed in an appropriate solvent a sided distribution of solutes is unavoidably created around the inclusion. In turn, this leads to a spatial gradient of concentration that supports, at least partially, the particle motion. This situation is the prototypical example of **self-diffusiophoresis** and was theoretically addressed by Golestanian et al. in a pioneering paper [140].

A comprehensive theoretical treatment of the role of interfacial forces on colloidal transport can be found in the classic review by Anderson [7]. I adapt it here as it is presented by Illien et al. [172]. The first and most important result to be retained is that, independently of its origin, *the slip velocity is proportional to the gradient of a field-variable*, the proportionality factor being characteristic of the material properties at the interface. This linear relation translates further into a similar dependence for the colloid speed in terms of a generic local activity at its surface.

Anderson's introductory general treatment [7] on phoretic effects refers to planar interfaces, but the arguments can be easily translated to the boundary layer of a spherical particle. In this latter case, one usually considers extremely thin boundary layers compared to the sphere radius (i.e. the Smoluchowsli limit as mentioned in the previous introductory chapter). In particular, this supposes to disregard any eventual advection effect that might affect the solute. As a matter of fact, this simplifying assumption has been relaxed by some authors under particular circumstances. As representative of early and more recent references to this issue, I mention a couple of papers published, respectively, by Michelin et al. [257] and by De Corato et al. [76]. On what follows I present a simplified treatment that captures the essentials of self-diffusiophoresis.

The interaction force between the solute particles in the fluid and the nearby particle surface can be represented by a potential energy $\phi(\mathbf{r})$, referred to a coordinate system linked to the colloid. The forces applied to individual molecules are transmitted to the fluid and aggregate into a body force $-c\nabla\phi$, written in terms of the concentration field c.[2] One assumes, similarly to the analysis of the diffuse double layer in Sect. 2.1.2, that the concentration field follows the Boltzmann distribution,

$$c(\mathbf{r}) = c_o(\mathbf{r}_s)exp(-\phi(\mathbf{r})/k_BT), \qquad (3.1)$$

where $\mathbf{r}_s$ specifies a point at the surface. The next step is to formulate the condition of momentum balance, expressed as the **Navier-Stokes** (NS) equation for steady flows,

$$-\eta\nabla^2\mathbf{v} = -\nabla p - c\nabla\phi. \qquad (3.2)$$

This equation introduces the hydrostatic pressure p and the fluid viscosity η. Projecting the NS equation into the radial direction, such a balance translates into a corresponding one between the hydrostatic and osmotic pressures, i.e. $p - k_BTc = const$. Using an appropriate boundary condition for the fluid velocity along the tangential axis, one easily arrives at an explicit expression for the slip velocity,

$$\mathbf{v}^s(\mathbf{r}_s) = -\frac{k_BT}{\eta}\left(\int_o^\infty r\left(exp(-\phi(\mathbf{r})/k_BT) - 1\right)dr\right)\nabla_{\|}c(\mathbf{r}_s), \qquad (3.3)$$

evaluated along the outer edge of the boundary layer.

This linear dependence of the slip velocity on the gradient of a field-variable results similarly from the treatment of equivalent phoretic scenarios. This permits to write the general form,

$$\mathbf{v}^s(\mathbf{r}_s) = \mu_{ph}(\mathbf{r}_s)\nabla_{\|}\Phi(\mathbf{r}_s), \qquad (3.4)$$

---

[2] In the previous chapter, this variable was named as number concentration and denoted differently by ρ. I prefer here to change to c for concentration as it is more standard in the context of diffusiophoresis.

in terms of a generic phoretic potential field Φ, and a characteristic *local phoretic mobility* μ_{ph} that only depends on the colloidal surface properties. In terms of the local normal to the surface $\mathbf{n}$ and the unit tensor $\mathbf{I}$, this result can be more formally written as,[3]

$$\mathbf{v}^s(\mathbf{r}_s) = \mu_{ph}(\mathbf{r}_s)(\mathbf{I} - \mathbf{nn}) \cdot \boldsymbol{\nabla}\Phi(\mathbf{r}_s). \tag{3.5}$$

The second step in the derivation of a phoretic mobility is to take the reciprocal view (more precisely we need to invoke the reciprocal theorem for low Reynolds numbers [154]) to transform the slip velocity into a colloid speed.[4] This central result establishes that the components of the colloid velocity $\mathbf{V}_0$ in a basis $\hat{e}_i$ are given by,

$$\mathbf{V}_0 \cdot \hat{f}_i = -\int_S \mathbf{n} \cdot \sigma_i \cdot \mathbf{v}^s d\mathbf{r}_s, \tag{3.6}$$

where σ_i denotes the components of the stress tensor applied to to the surface of the colloid as it is dragged by an applied unit force $\hat{f}_i = \hat{e}_i$ in absence of slip.

Again I summarize Anderson's treatment in the context of self-diffusiophoresis [141]. Since colloids are micron-size and their velocities are in the scale of microns/second, the **Péclet number**, $Pe = \frac{V_0 R}{\mathcal{D}}$ is very small (R denotes the colloid radius, and $\mathcal{D}$ the diffusion parameter of the chemical species involved in the diffusiophoresis mechanism). This justifies to assume a *Laplace equation* for the concentration field, i.e. $\nabla^2 c = 0$. Finally, a boundary condition at the colloid surface is imposed on the concentration field $\mathcal{D}\mathbf{n} \cdot \boldsymbol{\nabla} c(\mathbf{r}_s) = \tilde{\alpha}(\mathbf{r}_s)$. This last relationship introduces an *apparent activity coefficient* $\tilde{\alpha}$ that clearly depends on the reactive properties of the colloid surface. Since for a sphere of radius R, $\mathbf{n} \cdot \sigma_i = (1/4\pi R^2)\hat{e}_i$ the final result for the colloid speed reads,[5]

$$\mathbf{V}_0 = -\frac{1}{4\pi R^2}\int_S \mu_{diffph}(\mathbf{r}_s)(\mathbf{I} - \mathbf{nn}) \cdot \boldsymbol{\nabla} c(\mathbf{r}_s)d\mathbf{r}_s. \tag{3.7}$$

The generalization of this scheme to other self-phoretic effects, proceeds similarly. For self-electrophoresis, one would replace c and $\mathcal{D}$, respectivey, by the electrostatic potential and the electric conductivity (a simple dimensional analysis of $\tilde{\alpha}$ characterizes it as an injected current density). For thermophoresis, respective changes involve the temperature field and the thermal conductivity.

---

[3]To avoid any possible confusion with the nematic director much used in the context of Liquid Crystals, I simply denote the normal vector $\mathbf{n}$ without any reference to its unit norm.

[4]The Reynolds number is the fundamental dimensionless parameter in hydrodynamics directly rooted in the Navier-Stokes equation. It is given as a ratio between inertial and viscous forces, i.e. $Re = \rho v/\eta$, in terms of the mass density, flow velocity and viscosity.

[5]Notice that torque-free conditions would introduce as well an angular velocity on the colloid $\Omega \propto \frac{1}{R}\int_S(\mathbf{v}^s \times \mathbf{n})$, which is null for axisymmetric bodies.

Even without solving the problem in full, we can already get a valuable estimation of the order of magnitude and functionality of the computed colloid velocity, given by,

$$V_0 \propto \frac{\tilde{\alpha}\mu_{diffph}}{\mathcal{D}}. \tag{3.8}$$

The importance of this result is that, at least under the approximations it has been derived, the colloid speed appears not to depend on particle size, but it is simply given by a bilinear dependence on two surface properties, a mobility, related to the scalar field responsible of the phoresis, and an activity parameter (eventually shape- and surface-dependent). At the same time, the colloid speed is inversely proportional to an *effective medium conductivity* (here expressed as a medium diffusivity, by no means to be confused with the colloid self-diffusion coefficient itself).

3.1.1.2 Experimental Realizations of Phoretic Swimmers

The previously announced three main types of self-phoresis have been all realized experimentally. On what follows, I briefly comment on the most distinctive features of such designs (see Fig. 3.3).

Electrophoretic-based self-propulsion of colloids was first demonstrated in a couple of papers published by Paxton et al. [299, 300].[6] The proposed microswimmers consisted in striped bimetallic gold/platinum nanorods dispersed in a peroxide solution (see Fig. 3.3 (a)). This prototype works as an elementary electrochemical cell at the microscale;: the hydrogen peroxide is oxidized by the platinum and produces electrons that are used for the reduction of the solvent at the gold end.[7] Exchanged electrons in this redox process, do not get into the solvent but simply circulate through the rod. Concurrently, a proton flux is established along the rod/solution interface. Rods preferentially move along their axis in the direction of the Pt end, at speeds as large as ten body lengths per second. Complementary experiments replacing water by water/ethanol mixtures demonstrated that the velocity is proportional to the area-normalized oxygen evolution rate, indicating the catalytic (chemical) origin of the propulsion.

Collective effects and the possibility to assemble and drive close-packed rafts of passive tracer particles was reported from the same group [429]. This system also provided the first experimental demonstration of **chemotaxis** outside the biological world [165]. The specific observation was that the platinum-gold nanorods exhibited direct motion following an externally

---

[6]Synthetic self-propelled nanorotors based on the same mechanism were published nearly simultaneously by Fournier et al. [108].

[7]Notice that this mechanism has nothing to do with a propulsion mechanism induced by bubble ejection. Bubble formation and further spontaneous symmetry-breaking ejection has been observed, however, to propel hollow tubular structures, whose internal surface catalyzes the decomposition of the hydrogen peroxide, as reported by Solovev et al. [370] (a review from the same group is [254]).

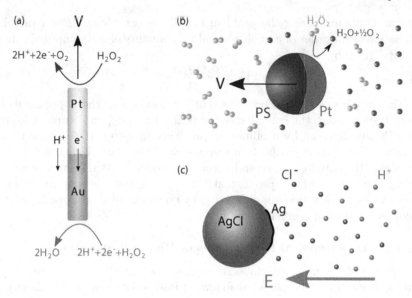

FIGURE 3.3
Schematics of three basic realizations of swimmers based on self-phoretic mechanisms. (a) Bimetallic rod (self-electrophoresis). (b) Janus particle (self-diffusiophoresis). (c) Self-propulsion of AgCl particles under UV illumination. In this latter case, the electric field is self-created due to the different diffusivities of the ionic species H^+ and Cl^-. Image and caption text adapted from Illien et al. [172].

imposed hydrogen peroxide gradient (towards higher concentration of the aqueous reactant).

Howse et al. [166] were the first to report a self-diffusiophoretic swimmer based on a prototype that has become very popular in the recent years. It is realized through a Janus polystyrene particle of micrometer size half-coated with platinum that performs the electrochemical oxidation mentioned above but limited to the metallic face (see Fig. 3.3 (b)). This generates a fuel gradient in solution. As a matter of fact, recent studies by Ibrahim et al. [170] suggest that the mechanism is a bit more complex, since motion is driven by a multiphoretic (ionic-diffusio and electrophoretic) concourse.

A similar interplay appears to explain the self-propulsion of AgCl particles under UV illumination as reported by Ibele et al. [169] (see Fig. 3.3 (c)). In this particular system, schooling behavior was evidenced as a distinctive collective effect of microswimming. Another variant of microswimming under illumination was proposed by Palacci et al. a few years later [292]. These authors reported a design still based on Janus particles, but this time bearing a hematite cube that catalytically decomposes hydrogen peroxide under blue

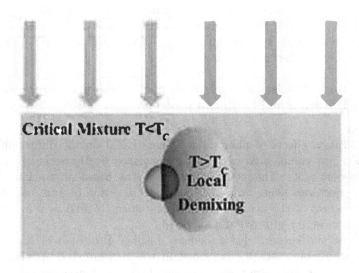

FIGURE 3.4
Active Brownian microswimmer in a critical binary mixture.
Schematic for the self-propulsion mechanism: a Janus particle is illuminated and the cap is heated above T_c inducing a local demixing that eventually propels the particle. Image and caption text adapted from Buttinoni et al. [49].

illumination. This design opened the appealing possibility to work with swimmers under a switch on/switch off mode. Again, striking collective effects were explored by the authors who coined the concept of *"living crystals"* to refer to dynamic assemblies of such surfers.

Self-thermophoresis has been realized according to a couple of different experimental schemes. The first and simplest one refers to Janus silica or polystyrene particles half-coated with a gold layer under defocused laser beam, as reported by Jiang at el. [181]. Specific absorption at the metal-coated side of the particle creates the necessary temperature gradient. The latter, in addition to viscosity inhomogeneities, is responsible for the colloid transport.[8]

The second realization is a little bit more sophisticated, having the advantage with respect to the previous one that it requires much less power injection, i.e. lower light intensities. The basic principle consists in thermally forcing a local phase separation process through a similar Janus particle but this time placed in a binary liquid mixture close to its critical temperature. This realization was proposed by Volpe et al. [425, 49] (see Fig. 3.4).

---

[8]Heating of Janus colloids adsorbed at water-oil interfaces to realize active micrometric **Marangoni surfers** was recently reported by Dietrich et al. [83].

It turns out that the mechanism is again multiphoretic, the diffusiophoretic part overweighting the thermophoretic contribution. Theoretical analysis of this phenomenon were reported in a couple of successively papers published respectively by Wurger and Samin et al. [445, 336].

3.1.1.3 Basic Statistical Properties of Self-phoretic Swimmers: Diffusion and Sedimentation

Diffusion

Self-diffusion effects of phoretic swimmers based on the diffusiophoresis mechanism just commented were already analyzed by Howse et al. in their original report [166]. This standard analysis was based on tracking Janus particle trajectories using as a benchmark a dispersion of pure polystyrene beads. Measures of the mean squared displacement (MSD) as a function of elapsed times were registered and analyzed.

As it is well-known, classic colloids display Brownian characteristics provided a reliable statistics is gained in the performed experiments (see Sect. 2.1). The fundamental result in conventional colloidal dispersions is that MSD grows linearly with time in steady and non-biased trajectories, the proportionality factor providing the translational diffusion parameter of the colloid (see Sect. 2.1.1). According to the classical Stokes-Einstein relation $D = \frac{k_B T}{6\pi\eta R}$ for a spherical particle of radius R. At the same time a *rotational diffusion* coefficient could be similarly defined as $D_r = \frac{k_B T}{8\pi\eta R^3}$.

For passive particles it is intuitively clear that translational and rotational modes are separated. However, this is no longer true for swimmers whose rotational symmetry is intimately broken by their intrinsic displacement direction. Elementary stochastic effects at the level of the rotation mode result unavoidably in equivalent translational displacement fluctuations. Quite intuitively, the key parameter governing this coupling is a characteristic time scale defined as the swimmer **persistence time** τ_r, whose inverse is precisely a measure of D_r.[9] Alternatively one can refer to a swimmer **persistence length**, $l_p = V_0 \tau_r$.

Results from the paper by Howse et al. for the two-dimensionally projected MSD are,

$$\langle \mathbf{r}^2(t) \rangle = 4Dt + \frac{V_0^2 \tau_r^2}{2} \left[\frac{2t}{\tau_r} + e^{-2t/\tau_r} - 1 \right]. \tag{3.9}$$

What is particularly illustrative of this result is to look at its asymptotic forms. For short times, i.e. $t << \tau_r$, $\langle \mathbf{r}^2(t) \rangle = 4Dt + V_0^2 t^2$, in an expression that combines diffusive and ballistic contributions. For very long times, i.e. $t >> \tau_r$, the MSD displays diffusive characteristics with an effective diffusion coefficient $D_{eff} = D + \frac{1}{4} V_0^2 \tau_r$.

---

[9]In general $\tau_r = \frac{1}{(d-1)D_r}$ for a d-dimensional trajectory.

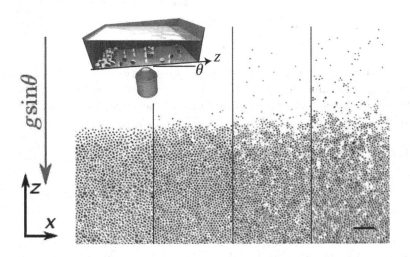

FIGURE 3.5
Sedimentation of phoretic gold-platinum Janus colloids. From left to right, panels correspond to increasing activity (leftmost panel corresponds to zero activity). Image and caption text adapted from Ginot et al. [131].

Sedimentation

Suspensions of active colloids under gravity provide an interesting scenario to probe not only individual kinematic characteristics, for instance the colloid persistence time, as follows from the expression just quoted for the effective diffusion coefficient, but also global properties of the suspension, with the perspective to extract **(active) thermodynamic parameters**. Interesting theoretical concepts and experimental results along this line can be found in a couple of papers published by Palacci et al. [291] and Ginot et al. [131] (see Fig. 3.5), both devoted to Janus-like microswimmers.

It is worth remembering that the classical result in this context is the celebrated distribution of settling particles of buoyant mass m under gravity at thermal equilibrium. As commented in Sect. 2.1.1, this (barometric) profile fulfills a Boltzmann distribution that renders a vertical profile given by $\rho(z) = \rho_o \exp(-z/\delta)$ in terms of a sedimentation length $\delta = k_B T/mg$.

Stationary profiles reported in the quoted references were similarly fitted to an effective law $\rho(z) = \rho_o \exp(-z/\delta_{eff})$, with an **active sedimentation length**, δ_{eff}, that was found to depend strongly on the activity of the Janus-like particles, i.e. it increases with the increased speed of the colloids. Moreover, it was proved that active colloids satisfy Smoluchowski's relationship that relates the sedimentation length, the sedimenting terminal velocity V_0, and the diffusion coefficient, i.e. $\delta_{eff} = D_{eff}/V_0$. Notice that this last equality permits to establish a sort of **(active) fluctuation-dissipation relation** in terms of an effective **(active) temperature**,

$$k_B T_{eff} = D_{eff}/\mu = \delta_{eff} mg. \tag{3.10}$$

Effective temperatures were found to range between ambient values and extremely high values up to 10^3 K. Moreover, a relation between ambient and active temperatures was proposed in terms of the active version of the Péclet number.[10] More precisely, $k_B T_{eff} = k_B T (1 + \frac{2}{9} Pe^2)$.

3.1.1.4 The Active Brownian Model

Basic equations

The just commented experimental scenario perfectly motivates the discussion of one of the first and simplest theoretical models that have been vastly used in the context of active particle-based systems. I first introduce and later employ it precisely to obtain the distribution of sedimenting (non interacting) swimmers, following the analysis by Enculescu et al. [101]. This scheme is known under the name of **Active Brownian Particle** model, simplified routinely under the acronym ABP, and has been extensively reviewed by Schweitzer [351] and Romanczuk et al. [328].

The ABP model captures the basic three ingredients of active motion: i) overdamped dynamics of particle position **r** and orientation **p**; ii) motion performed with a fixed particle velocity V_0 along direction **p**, and iii) translational ξ and rotational noisy forces ξ_r, either of thermal origin or due to the activity itself.[11] Although particles are implicitly assumed spherical, rod-like (see [305, 18] for a couple of classical references), and active particles with more complex shapes [436, 278] have been studied as well, singularly looking for collective, i.e. clustering effects (see Sect. 5.1.2).[12]

The basic ABP equations are written as a set of *Langevin equations* to fully retain the stochastic nature of individual translational and rotational modes,[13]

---

[10]The active version of the Péclet number is defined as $Pe = V_0 R/D$ for a particle of radius R, with D referring to the colloid diffusion coefficient, see Sect. 3.1.1.4.

[11]A simpler variant of the ABP model is the **active Ornstein-Uhlenbeck particle model.** It refers purely to translational degrees of freedom with some kind of translational memory that establishes a persistence in the particle motion [107]. In one-dimension, this concept translates into an equation for the velocity in terms of a colored (Ornstein- Uhlenbeck) noise.

[12]Active Brownian motion of chiral objects has been addressed in theoretical approaches by Van Teeffelen et al. [419] (internal force plus an internal torque leading to circle swimming), and experimentally realized for L-shaped swimmers by Kummel et al. [212].

[13]Notice that I am here considering the equations for ABPs performing free of interactions with neighboring particles. This latter situation will be considered for instance later on in Sect. 7.2.

$$\dot{\mathbf{r}} = V_0 \mathbf{p} + \sqrt{2D}\boldsymbol{\xi}, , \tag{3.11}$$

$$\dot{\mathbf{p}} = \sqrt{2D_r}\boldsymbol{\xi}_r \times \mathbf{p}, \tag{3.12}$$

where V_0 is the (steady) particle swimming velocity, and D and D_r stand for its respective translational and rotational diffusion coefficients. For pure thermal fluctuations their explicit expressions were given in the precedent paragraph devoted to self-diffusion, although, strictly speaking, many other sources of noise could be potentially relevant in active systems. The stochastic, zero-mean and space-independent, forces satisfy a unit-variance and time uncorrelated (*white noise*) Gaussian statistics,

$$\langle \boldsymbol{\xi}(t) \rangle = 0 \qquad \qquad \langle \boldsymbol{\xi}(t)\boldsymbol{\xi}(t') \rangle = \mathbf{I}\delta(t - t'), \tag{3.13}$$

$$\langle \boldsymbol{\xi}_r(t) \rangle = 0 \qquad \qquad \langle \boldsymbol{\xi}_r(t)\boldsymbol{\xi}_r(t') \rangle = \mathbf{I}\delta(t - t'). \tag{3.14}$$

From the combination of directed motion and translational plus rotational noises, ABPs follow a variant of random walk, known as *persistent random walk* whose orientational time correlations satisfy $\langle \mathbf{p}(t) \cdot \mathbf{p}(0) \rangle = e^{-\frac{t}{\tau_r}}$.

Motion under an external field: Sedimentation

To apply these equations to a gas of sedimenting (non-interacting) particles in a viscous fluid, we need to slightly reformulate them in the form,[14]

$$\dot{\mathbf{r}} = \mathbf{V} \qquad \qquad \dot{\mathbf{V}} = -\frac{\gamma}{m}(\mathbf{V} - V_0\mathbf{p}) + V_0\boldsymbol{\Omega} \times \mathbf{p} + \mathbf{g} + \xi, \tag{3.15}$$

$$\dot{\mathbf{p}} = \boldsymbol{\Omega} \times \mathbf{p} \qquad \qquad \dot{\boldsymbol{\Omega}} = -\frac{\gamma_r}{I}\boldsymbol{\Omega} + \xi_r, \tag{3.16}$$

where I stands for the inertial momentum, γ and γ_r denote the, respectively, translational and rotational friction coefficients, and $\mathbf{g} = -g\hat{\mathbf{z}}$ stands for the gravitational field. The most general situation is here considered by allowing translational and rotational velocities as well. Note that, for the sake of convenience, I have kept explicit noise terms only at the level of the force equations. The latter actually express the absence of net acceleration terms under overdamped conditions. This actually explains the presence of the term $V_0\boldsymbol{\Omega} \times \mathbf{p}$ at the r.h.s. of the equation for the translational acceleration.

Without noise contributions, the swimming direction will be constant in time and the particle velocity would relax to $\mathbf{V} = V_0\mathbf{p} + m\mathbf{g}/\gamma$. Under noise

---

[14]Notice that the noise terms are slightly redefined in the next equations with respect to those just introduced with reference to the generic ABP model. I do so without any possibility of confusion to follow the original reference by Enculescu et al.

forcing, the velocity may change due to random forces and torques. The statistical prescriptions for the noise terms are the usual ones, so that stochastic translational and rotational accelerations will follow the corresponding Gaussian prescription,

$$\langle \boldsymbol{\xi}(t) \rangle = 0 \qquad \langle \boldsymbol{\xi}(t)\boldsymbol{\xi}(t') \rangle = \mathbf{I}\left(\frac{2\gamma k_B T}{m^2}\right)\delta(t - t'), \qquad (3.17)$$

$$\langle \boldsymbol{\xi}_r(t) \rangle = 0 \qquad \langle \boldsymbol{\xi}_r(t)\boldsymbol{\xi}_r(t') \rangle = \mathbf{I}\left(\frac{2\gamma_r k_B T}{I^2}\right)\delta(t - t'). \qquad (3.18)$$

From the Langevin description just proposed, one would write a *Fokker-Planck equation* for the *one-particle distribution function* $\Psi(\mathbf{r}, \mathbf{p}, \mathbf{V}, \boldsymbol{\Omega}; t)$. However, it is more convenient first to try to factorize such a distribution to get rid of the explicit dependences on both, translational and rotational, velocities. Using typical values for the parameters of the colloid γ, m, R and V_0, one would easily conclude that the relaxation time of both velocities (i.e. m/γ for the translational component and equivalently for the rotational velocity) are orders of magnitude smaller than the time scale associated to particle motion t_0. Thus, one is legitimated to invoke a *Maxwell-Boltzmann distribution* as an ansatz to obtain the reduced distribution function $\psi(\mathbf{r}, \mathbf{p}; t)$,[15]

$$\Psi(\mathbf{r},\mathbf{p}, \mathbf{V}, \boldsymbol{\Omega};t) = \psi(\mathbf{r}, \mathbf{p};t)\exp(-\beta m[\mathbf{V} - \langle\mathbf{V}\rangle(\mathbf{r}, \mathbf{p};t)]^2/2 - \beta I[\boldsymbol{\Omega} - \langle\boldsymbol{\Omega}\rangle(\mathbf{r}, \mathbf{p};t)]^2/2),$$
$$(3.19)$$

written in terms of the mean translational $\langle\mathbf{V}\rangle$ and rotational $\langle\boldsymbol{\Omega}\rangle$ velocities. To proceed further we obtain first the conservation equation for $\psi(\mathbf{r}, \mathbf{p}; t)$,

$$\frac{\partial\psi}{\partial t} + \boldsymbol{\nabla} \cdot (\psi\langle\mathbf{V}\rangle) + \boldsymbol{\mathcal{R}} \cdot (\psi\langle\boldsymbol{\Omega}\rangle) = 0, \qquad (3.20)$$

with $\boldsymbol{\mathcal{R}}$ denoting a gradient-like operator for rotations on the unit sphere, $\boldsymbol{\mathcal{R}} \equiv \mathbf{p}\times\boldsymbol{\nabla}_{\mathbf{p}}$. The previously introduced averaged quantities satisfy themselves the pair of equations,

$$\psi\frac{\mathcal{D}(\langle\mathbf{V}\rangle - V_0\mathbf{p})}{\mathcal{D}t} + (\gamma/m)\psi(\langle\mathbf{V}\rangle - V_0\mathbf{p}) = -(k_B T/m)\boldsymbol{\nabla}\psi + \mathbf{g}\psi, \qquad (3.21)$$

$$\psi\frac{\mathcal{D}\langle\boldsymbol{\Omega}\rangle}{\mathcal{D}t} + (\gamma_r/I)\psi\langle\boldsymbol{\Omega}\rangle = -(k_B T/I)\boldsymbol{\mathcal{R}}\psi, \qquad (3.22)$$

with $\mathcal{D}/\mathcal{D}t = (\partial/\partial t) + \langle\mathbf{V}\rangle\cdot\boldsymbol{\nabla} + \langle\boldsymbol{\Omega}\rangle\cdot\boldsymbol{\mathcal{R}}$, denoting the conventional (comoving and corotating) material derivative.[16]

---

[15]β is used in the following equation as a shorthand notation for the Boltzmann factor, i.e. $\beta = 1/k_B T$.

[16]Although I employ here the same notation for this derivative operator and for the diffusion coefficient of chemical species in the self-phoresis scenario commenter earlier in this chapter, it should not be any possibility of confusion.

Next, one invokes the usual limit of small Reynolds numbers, so that time derivatives are considered negligible with respect to damping times. In this way, one recovers deterministic-like equations for the averaged velocities,

$$\langle \mathbf{V} \rangle = V_0 \mathbf{p} + (m/\gamma)\mathbf{g} - (k_B T/\gamma) \boldsymbol{\nabla} \ln \psi, \tag{3.23}$$

$$\langle \boldsymbol{\Omega} \rangle = -(k_B T/\gamma_r) \boldsymbol{\mathcal{R}} \ln \psi, \tag{3.24}$$

each equation identifying the corresponding drift- and diffusion-like contributions. Using the latter expressions one transforms Eq. (3.20) above into a *Smoluchowski equation*,

$$\frac{\partial \psi}{\partial t} + \boldsymbol{\nabla} \cdot \mathbf{J} + \boldsymbol{\mathcal{R}} \cdot \mathbf{J}_r = 0, \tag{3.25}$$

with the corresponding translational and rotational fluxes, respectively given by $\mathbf{J} = -D\boldsymbol{\nabla}\psi + (V_0\mathbf{p} + (m/\gamma)\mathbf{g})\psi$ and $\mathbf{J}_r = -D_r\boldsymbol{\mathcal{R}}\psi$, written in terms of the corresponding diffusion coefficients ($D = k_B T/\gamma$, $D_r = k_B T/\gamma_r$).

More useful is an explicit and dimensionless form of this equation that is written in terms of the colloid Péclet number ($Pe = V_0 R/D$ for a particle of radius R) used together with an originally defined gravitational version of the latter denoted α in the original reference, $\alpha \equiv (mgR/k_B T)$. Furthermore, the analysis gets simpler if length is made dimensionless in terms of the particle size, whereas time is rescaled by means of the diffusion coefficient. The final form of the conservation equation reads,

$$\frac{\partial \psi}{\partial \tilde{t}} = \tilde{\nabla}^2 \psi + \frac{3}{4}\mathcal{R}^2 \psi - (Pe\mathbf{p} - \alpha\hat{\mathbf{z}}) \cdot \tilde{\boldsymbol{\nabla}}\psi. \tag{3.26}$$

Notice that for passive particles, a steady state solution reproduces the well-known barometric distribution commented above, $\psi \propto \exp(-\alpha\tilde{z})$, or, equivalently, in the original variables $\psi \propto \exp(-(mg/k_B T)z) = \exp(-z/\delta)$, in terms of the sedimentation length. The experiments by Palacci et al. mentioned above suggest to look for a steady-state solution of Eq. (3.26) of the form,

$$\psi(\mathbf{r}, \mathbf{p}) \propto \exp(-\alpha\tilde{z}/\zeta) \exp[Pe U_1(\cos\theta) + (Pe)^2 U_2(\cos\theta) + ...], \tag{3.27}$$

with $\zeta \equiv (\delta_{eff}/\delta)$ denoting the ratio of active and passive sedimentation lengths. In writing the solution in this way, one uses the axial symmetry of the problem around the z-axis and introduces the angle θ ($\cos\theta \equiv \mathbf{p} \cdot \hat{\mathbf{z}}$). Notice that with this prescription, a zero angle means that particles orient contrary to the gravitational field. According to [101], a perturbation analysis leads to $\zeta = 1 + (2/9)Pe^2$ at the lowest order, which agrees with experimental results (notice that in the experimental paper, the ratio T_{eff}/T rather than, but totally equivalent to, δ_{eff}/δ is plotted).

An interesting effect, already emphasized in the original Enculescu et al. paper, arises when working with the Smoluchowski equation written for the total density of active particles. It turns out that the established profile develops **polar order**, such that active particles partially align against gravity. More precisely, the result quoted in [101] reads $Pe\langle\cos\theta\rangle = \frac{\alpha(\zeta-1)}{\zeta} > 0$ for $\zeta > 1$. It is important to stress that this alignment effect results from the interplay between self-propulsion and a concentration profile without any need to account for particle interactions. Moreover, it appears without invoking a pure gravitational torque as it would be the case for particles with an asymmetric mass distribution. More details can be found in the original Enculescu et al. paper [101].

A parallel analysis would apply to the motion of active colloids, modeled as ABP, under external flows. The case of a *shear flow* has been reported by ten Hagen et al. in [395], whereas Zöttl et al. [460] considered *Poiseuille flows*. In any case, the simplest scenario corresponds to a colloid navigating in a fixed and unperturbed background flow field $\mathbf{v}(\mathbf{r}; t)$ at low Reynolds number. Its total velocity would contain self-propulsion plus the hydrodynamic drag, $\mathbf{V} = V_0\mathbf{p} + \mathbf{v}(\mathbf{r}(t))$. The phoretic speed is commonly assumed not to depend on the ambient flow, but its orientation may change continuously. A spherical particle would experiences a torque, proportional to the flow (local) vorticity, that results in an angular velocity $\mathbf{\Omega} = (1/2)(\mathbf{\nabla} \times \mathbf{v})$ (*Faxen's law*). Notice that for elongated particles, one would need to consider additional torques that would arise from the rate of strain tensor.

There are other important individual or collective aspects of colloidal swimmers that can be addressed with the use of Active Brownian Particle models. For instance, the description of several scenarios of taxis (i.e. directed motion under field gradients), or swimming near surfaces. I deliberately leave them aside in this monograph for the sake of brevity. Interested readers can find detailed discussions in the original references or in a few review papers, for instance that published by Zöttl and Stark. [462].

3.1.2   Swimmers Based on Marangoni Flows

A different class of self-propulsion at the microscale is based on *Marangoni stresses* associated to surface tension gradients localized at the colloid surface. This design was demonstrated for water-in-oil emulsions by Thutupalli et al. [403] (see Fig. 3.6). An important distinction with respect to the realizations reported in the previous subsection is that in this case the symmetry-breaking is spontaneously generated, rather than being built-in from the colloid design. Briefly described, propulsion in the Thutupalli's design arises from the spontaneous bromination of a surfactant with the bromine coming from inside the droplet. The result is a self-sustained bromination gradient along the drop surface that heterogeneously alters its surface tension and elicits flows which propel the droplet.

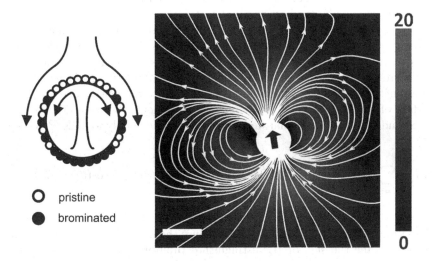

FIGURE 3.6
Microdroplet swimmer. Bromination of a surfactant induces a Marangoni stress that propels the droplet as a squirmer. The magnitude of the flow velocity (color code in original version) and streamlines along a horizontal section through the center of a squirmer droplet are shown. Scale bar: 100 μm; velocity scale (right) in units of μm/s. Image and caption text adapted from Thutupalli et al. [403].

A variant of this prototype was published by Izri et al. [178], and corresponds to the self-propulsion of pure water droplets in a biocompatible oil (squalane)-surfactant (monoolein) medium mediated by a concentration gradient of swollen reversed micelles. Still, a related design supposes changing the nature of the dispersed liquid phase from isotropic to anisotropic. In this case, motile, even curling, *(nematic) liquid crystal droplets* have been observed to propel in a surfactant solution with concentrations above the critical micelle concentration, while undergoing micellar solubilization [160, 205]. Rotational torques are generated by the interplay of surface flow and nematic order, such that the curling motion can be switched off by heating to the isotropic state and the droplet reverts to persistent swimming. In the particular context of swimming droplets, I address the reader to the review by Maass et al. [243], worth to mention since most of the published works on propelling colloids refer to solid particles rather than to droplet-based swimmers.

Surface tension effects can also be employed for propulsion at liquid-air interfaces induced by solute dissolution, as reported by Nakata et al. (*"camphor boat"*) [270]. Apart from this original realization, other similar scenarios have been experimentally analyzed, for instance the self-running of oil droplets on glass surfaces as observed by Sumino et al. [382].

Active droplets driven by Marangoni flows, as theoretically analyzed by Schmitt et al. [348, 349], are interesting realizations of a swimmer model known as **squirmer**. This well-known swimming archetype was introduced more than fifty years ago to study the cilia-based locomotion of microorganisms [229, 34]. In essence, this model amounts to prescribe a pure tangential velocity at the surface of the swimmer. For axisymmetric specimens, the tangential velocity is expanded in terms of polar derivatives of Legendre polynomials $P_n(\cos\theta)$,

$$v_\theta = \sum_{n=1} B_n \frac{2}{n(n+1)} \sin(\theta) \frac{dP_n(\cos(\theta))}{d\cos(\theta)} = B_1 \sin(\theta)(1 + \beta \cos(\theta)) + ..., \quad (3.28)$$

where $\beta = \frac{B_2}{B_1}$ is the squirmer parameter. Commonly the expansion is closed at the second-order ($B_n = 0$ for $n \geq 3$), B_1 fixes the magnitude of the swimming velocity, while the sign of B_2 distinguishes **pushers** ($B_2 < 0$) from **pullers** ($B_2 > 0$). Pushers correspond to a positive *force dipole*, and induce an outgoing flow field directed along their swimming direction, and an incoming flow field from their sides. Conversely, pullers correspond to a negative force dipole, inducing an incoming flow field along their swimming direction and an outgoing flow field along their sides.

Squirmer-based approaches have been mostly used in numerical simulations of collective swimming effects [175, 174]

3.1.3   Biological Microswimmers

3.1.3.1   Flagellated Bacteria

Bacteria use helical filaments for their locomotion, each species evidencing differences in the number and arrangements of these filaments, known as flagella. Flagella can be either localized, or appear distributed along the bacterium surface. Prominent examples, as reviewed by Berg, are *Escherichia coli* (see Fig. 3.7), *Salmonella typhimurium*, or *Proteus mirabilis* [26, 28, 25]. The single flagellum unit rotates as a motor protein complex.

Chemotaxis of *E. coli* and *Salmonella* is based on a *"run-and-tumble"* mechanism. In the "run" phase, rotating counter-clockwise flagella are bundled together following a helical left-handed winding, and bacteria display a forward motion along their long axis (see Fig. 3.7). When entering the "tumble" phase, the bundle disintegrates when one of the flagellum reverses motion to a clockwise rotation. This leads to a random reorientation of the bacterium motion. The tumbling interval ends by recomposing the bundle and reestablishing the counter-clockwise rotation. The run-and-tumble mechanism provides bacteria with efficient escaping or searching strategies under chemical or temperature gradients.

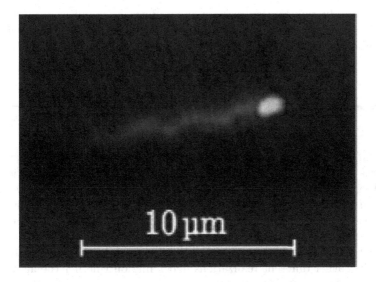

FIGURE 3.7
Body and flagella in Escherichia coli. *E. coli* is propelled at a velocity around 30 µm/s, by a bundle of 4–5 rotating helical flagella assembled by hydrodynamic forces. The green (GFP labeled) body is 2 µm length and the red flagella bundle (stained by Alexa Fluor 647) is around 8 µm (colored image in original version). Image provided by E. Clement, PMMH-ESPCI, Paris / ICMCS, Univ. Edinburgh.

Bacteria not only swim individually, but also show emergent patterns of collective motion upon cell shape modifications under contact with surfaces (bacterial **swarming**) [72], or they may equally display **gliding** behavior [306] to form biofilms.[17]

3.1.3.2 Other Biological Microswimmers

Eukaryotic ciliated cells, like *Paramecium*, propel by *cilia*, hair-like extensions of the cell that consist in bundled microtubules. Cilia, ten microns long and five times shorter than typical flagella, display characteristic non-synchronized beating patterns known as *"metachronal waves"*. Flagellated algae, like *Chlamydomonas reinhardtii*, swim by either a synchronous or an asynchronous breast-stroke-like motion of its two flagella [312, 90]. Sperm cells are propelled by a snake-like reptation of their flagellum, propagating as a bending wave [109, 110].

---

[17]Gliding motility is a type of microorganism translocation that is independent of propulsion appendages such as flagella.

An introduction to planktonic swimming can be found in a review paper from Lauga et al. [215]. A more recent account on biological microswimmers was published by Elgeti et al. [98].

3.2   Colloids Driven to Swim

Alternatively to autonomous swimming, realized through different self-phoresis processes that were briefly considered in the previous section, colloidal motion can be also achieved under the action of an externally applied external field. Also here one can distinguish a disparate variety of designs. I will classify them in two wide categories depending on whether forcing is of *magnetic* or *electric* nature. The two scenarios will be treated separately in the next sections, albeit electric forcing will be more extensively analyzed, as it admits rather different realizations depending on whether the actuation is based on a DC or an AC mode.

Although implicit when dealing with self-phoretic based swimmers, an important underlying concept that must be explicitly considered when addressing actuated microswimmers refers to the role of the *time-reversal symmetry*. This fundamental property is rooted in the Navier-Stokes equation under low Reynolds number conditions. This means that under these conditions, directed motion in Newtonian fluid environments requires non-reciprocal modes of deformation.[18] This is known as **"scallop theorem"**, a brilliant result announced by Purcell more than forty years ago [313]. Purcell himself proposed a minimal swimming design based on a three-arms body, periodically varying the spanned angles in a way to break time-reversal symmetry. Following the same principle that inspired Purcell's prototype, a swimmer with three collinear beads that follows a deterministic cycle of four deformations was proposed and theoretically solved by Najafi et al. [269].

3.2.1   Magnetic Forcing

As a paradigmatic example of swimmers forced by magnetic fields we should remember the inspiring design proposed by Dreyfus et al. [91] that was mentioned in the introductory remarks of this chapter. On what follows, I illustrate the characteristics of forced magnetic swimming at the microscale by referring to a simpler realization, a doublet roller, that admits a straightforward analysis both from the experimental and modeling points of view.

---

[18]See Appendix 2 9 to see how this constraint is relaxed when working with non-Newtonian fluids.

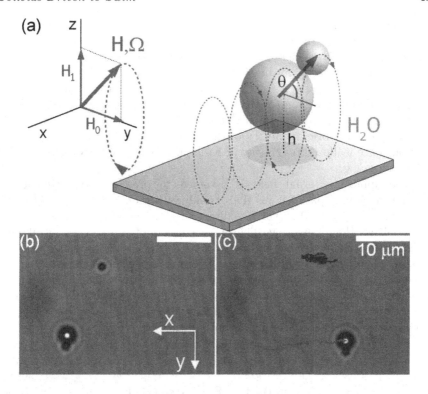

FIGURE 3.8
Externally rotated paramagnetic doublet moving above a glass plate. Schematic of the magnetically forced paramagnetic doublet, and microscopic images of a (transported) doublet and of an individual (practically at rest) particle. Snapshots (b) and (c) are separated 8.9 s. Image and text caption adapted from Tierno et al. [404].

3.2.1.1 A Doublet Roller

Based on dimer-assembled colloids, a microswimmer performing under a rotating magnetic field was proposed by Tierno et al. [404] (see Fig. 3.8). More in detail, DNA-linked doublets of paramagnetic particles of different size behave as rectified rotators, i.e. as rollers, when sedimented close to an enclosing plate. The small length of the DNA bridge ensures a strong and stiff linkage within the doublet, and makes the doublet rotate as a solid body. It is worth noticing from the very beginning that this design breaks time-reversal symmetry due to asymmetric dissipation by the solid boundary. In other words, if the doublet were uniformly surrounded by a homogeneous bulk fluid, rather than floating near a plate, its cyclic motion would never render a net displacement.

Achieved velocities are of the order of a few microns/second, and show always a maximum when plotted in terms of the forcing frequency for all precession ratios. Conversely, two distinct dependences of the doublet velocity on the precession ratio can be observed for low and high frequency regimes, corresponding, respectively, to a constant trend and to a non monotonic behavior. A detailed account of the results, and on the mechanism underlying doublet motion can be consulted in the original report. I prefer on what follows to present a brief summary of the analytical treatment of this rectified motion, since it illustrates the use of basic principles of the hydrodynamics of microswimming.

The applied external field is assumed to precess with frequency Ω around a y-axis parallel to the bounding bottom plate, i.e. $\mathbf{H}(t) = (H_1 \sin \Omega t, H_0, H_1 \cos \Omega t)$. Vertical direction is taken along z, and displacement follows the x direction. The dimer consists of two spheres of different radii denoted respectively a, b, with $\delta_{roller} = b/a < 1$ characterizing disparity in size. The doublet is parameterized in terms of the coordinates of the two spheres $(\mathbf{r}_a, \mathbf{r}_b)$, and of the (unit) director along the axis that links the doublet $\mathbf{n} = (\mathbf{r}_b - \mathbf{r}_a)/(a+b)$ (doublet length is assumed fixed). At low Reynolds number, the linearity of hydrodynamic equations permits to relate the velocity of each particle to the force acting on both of them,

$$\mathbf{V}_a = \mathbf{G}(\mathbf{r}_a, \mathbf{r}_a) \cdot \mathbf{F}_a + \mathbf{G}(\mathbf{r}_a, \mathbf{r}_b) \cdot \mathbf{F}_b, \qquad (3.29)$$

$$\mathbf{V}_b = \mathbf{G}(\mathbf{r}_b, \mathbf{r}_a) \cdot \mathbf{F}_a + \mathbf{G}(\mathbf{r}_b, \mathbf{r}_b) \cdot \mathbf{F}_b, \qquad (3.30)$$

where $\mathbf{G}(\mathbf{r}_a, \mathbf{r}_b) \equiv \mathbf{G}_{ab}$ denotes the *hydrodynamic mobility matrix*, i.e. the velocity of particle a generated by a (unit) force acting on particle b. Since one is interested in the motion of the doublet we neglect angular velocities and torques on individual particles.[19]

Motion is expressed in terms of the center-of-mass velocity $\mathbf{V} = d\mathbf{R}_{cm}/dt$, with $\mathbf{R}_{cm}$ simply given in terms of the mass weighted coordinates (equal density for both particles is assumed). Since the system is assumed globally force-free $\mathbf{F}_b = -\mathbf{F}_a = \mathbf{F}$, defining a single effective force $\mathbf{F}$ in terms of which we write formally the couple of dynamical equations for the doublet $\mathbf{V} = \mathbf{N} \cdot \mathbf{F}$ and $d\mathbf{n}/dt = \mathbf{M} \cdot \mathbf{F}/(a+b)$. Here, $\mathbf{M}$ and $\mathbf{N}$ denote matrices whose elements are linear combinations of the components of the hydrodynamic mobility matrices $\mathbf{G}_{ab}$.

Combining this pair of equations one obtains a simple relation between the doublet velocity and the precessing velocity of the director $\mathbf{V} = (a + b)\mathbf{N}\mathbf{M}^{-1}d\mathbf{n}/dt$. This equation is rendered explicit by neglecting at lowest order hydrodynamic interactions between particles (i.e. cross-mobility $\mathbf{G}_{ab}$ elements). The mobility matrices can be written in terms of two scalar functions separating in-plane and perpendicular contributions $\mathbf{G}(\mathbf{r}, \mathbf{r}) = C_{\parallel}(\mathbf{r})(\mathbf{I} - \hat{z}\hat{z}) +$

---

[19]As it is much used throughout the text, the $\cdot$ notation amounts to an index contraction (summation), in this case rendering a vectorial quantity.

$C_\perp(\mathbf{r})\hat{z}\hat{z}$. In order to arrive at analytic expressions, one introduces the far field expressions for the particle-surface mobility matrices which only depend on the distance z from each sphere to the plate,

$$C_\parallel(\mathbf{r}) = \frac{1}{6\pi\eta l}[1 - \frac{9}{16}(\frac{l}{z}) + O((\frac{l}{z})^3], \tag{3.31}$$

$$C_\perp(\mathbf{r}) = \frac{1}{6\pi\eta l}[1 - \frac{9}{8}(\frac{l}{z}) + O((\frac{l}{z})^3], \tag{3.32}$$

where l denotes the sphere radii (i.e. a or b). An explicit equation for the doublet velocity in terms of the precession dynamics of the doublet can be obtained after some algebra, the final result being,

$$\mathbf{V}(t) = \Upsilon\Big[\frac{1}{1 - \Lambda_-(h)n_z(t)} - \frac{\delta_{roller}}{1 + \Lambda_+(h)n_z(t)}\Big]\frac{d\mathbf{n}}{dt}. \tag{3.33}$$

In this last expression, Υ and $\Lambda_\pm(h)$ are geometric factors explicitly calculated in the original reference. In particular, two values of Υ distinguish between the two phases of doublet motion, respectively parallel and perpendicular to the bounding plate. Notice also that the second parameter contains an explicit dependence on the floating distance of the center-of-mass from the plate denoted h, assumed fixed, and considered as an averaged-value over the precession cycle.

The second part of the calculation permits to obtain the equation for the dynamics of the doublet director. The precessing velocity is imposed by the applied magnetic field. The latter exerts a torque on the doublet dipole $\mathbf{n}_d$ directed along $\mathbf{n}$ with the same precessing angle than $\mathbf{H}$. Thus $\mathbf{T}_m = \mu_0\,\mathbf{m}\times\mathbf{H} = (\mu_0 V_d\chi H)\mathbf{n}\times\mathbf{H}$, where μ_0 and χ account for the magnetic permeability of the doublet, and V_d denotes the doublet, and V_d its volume. This fixes the angular velocity of the director $\boldsymbol{\omega} = (1/\gamma_r)\mathbf{T}_m$, where γ_r stands for the doublet effective rotational friction coefficient. Finally, one employs the kinematic equation of motion,

$$d\mathbf{n}/dt = \boldsymbol{\omega}\times\mathbf{n}, \tag{3.34}$$

for $\mathbf{n}(t) = (\sin\theta\sin(\Omega t - \phi), \cos\theta, \sin\theta\sin(\Omega t - \phi))$, in terms of the precessing angle of $\mathbf{n}$ denoted θ. In this way, one gets an explicit expression for the angular variable θ in terms a characteristic frequency $\Omega_B = \mu_0 m H_0/\gamma_r$. The phase lag of the director with respect to the external fields is given as $\tan\phi = (\Omega/\Omega_B)\cos\theta$. Finally, averaging over a precession cycle, it turns out that the y and z components of the doublet velocity average to zero, while,

$$\frac{\langle V\rangle}{V_0} = \frac{\Omega}{\Omega_B}\Big(-\delta_{roller} - \delta_{roller}^3 + \frac{\delta_{roller}}{(1 - \Lambda_+^2\sin^2\theta)^{1/2}} + \frac{\delta_{roller}^{-3}}{(1 - \Lambda_-^2\sin^2\theta)^{1/2}}\Big), \tag{3.35}$$

in terms of a characteristic velocity scale $V_0(\Omega_B)$. Notice that this is a general expression valid also for equal-size spherical beads.

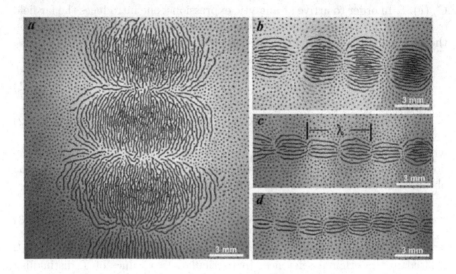

FIGURE 3.9
Snake structures of magnetic colloids assembled at a air/water interface under an alternating magnetic forcing. Panels from (a) to (d) correspond to increasing applied excitation frequencies. Image and caption text adapted from Snezhko et al. [362].

Comparisons with experiments for different values of the precession ratios of the magnetic field (i.e. H_1/H_0) permitted to estimate the average distance from the center-of-mass to the plate ($h \approx 2\mu$m), and to obtain the doublet rotational viscosity ($\gamma_r = 8.6 \times 10^{-19}$ N s m).

The previous discussion unveils the possibilities offered by simple designs towards the realization of controlled swimming at the microscale. It is honest to recognize, however, that one of the major drawbacks, singularly in relation to application strategies, is to gain directional control on swimming [288, 275]. In this respect, a striking development was reported by A. Ghosh et al. [127], based on the propulsion of magnetic nanostructured propellers (helical screw-like structures attached to spherical beads).

3.2.1.2   A Magnetically Driven Magnetic Snake

In relation with magnetic systems, a complementary scenario is provided by swimmers that are built from the self-assembling of individual magnetic colloids. This strategy was followed by Snezhko et al. [362, 363] (see Fig. 3.9). Notice that this is different from considering collective effects of individual microswimmers, discussed later on specifically in Sect. 5.1.

Ensembles of magnetic Ni spherical microparticles of around one hundred micrometers were assembled at the air-liquid interface forming "snake-like" structures. In a typical experiment, particles are energized by a vertically applied alternating magnetic field. In addition, an in-plane directed magnetic field is created with a pair of Helmholtz coils. Organized structures are induced by surface waves in the liquid triggered by the collective motion of the colloids in response to the magnetic forcing. Snakes are divided into segments, each segment consisting of chained and aligned ferromagnetic colloids. Segments themselves show long range antiferromagnetic response to in-plane fields, while their characteristic length is a decreasing function of the frequency. Surface waves needed for the assembly appear very much like *Faraday waves*. Colloids try to align with the external forcing while dragging the surrounding fluid. This induces local oscillations of the solution free surface and facilitates head-to-tail attractions of the corresponding magnetic dipoles. Above a critical field this attraction overcomes magnetic repulsion of the dipoles (vertically) aligned with the external field, if particles happen to be close enough. Incipient chaining enhances surface deformation and consolidates the chain formation [362].

These structures could be converted into swimmers by two different mechanisms: either by increasing the forcing frequency or by breaking the fore-aft symmetry after attaching a bead to one of the snake's ends [363]. In the first case, the symmetric quadrupolar structure of the flows elicited around the snake and located at their ends deform asymmetrically to propel the structure as a straight propagating interfacial swimmer. The swimming velocity is found to increase quadratically with the square of the applied magnetic field. On the other hand, when the magnetic suspension is prepared at the interface between two immiscible liquids, the self-assembly mechanism leads to localized asters and arrays of asters, which exhibit locomotion and shape change [361]. Other features of this system or variants of it will be commented later on when revising emerging collective swimming effects (i.e. the so-called active turbulence; see Sect. 5.3).

3.2.1.3 Magnetic Spinners

Still within the context of magnetic forcing, a new class of driven fluid was recently reported by Soni et al. [371] after preparing a millimeter-scale cohesive suspension of spinning colloidal magnets (see Fig. 3.10).

The colloidal micromagnets are micron-size hematite cubes coated with a thin silica shell dispersed in water and subjected to a rotating magnetic field of a few mT strength. The dispersion containing millions of spinning units is enclosed in a glass chamber. As a result, a millimeter-scale cohesive fluid monolayer is created from the interactions of individual rotors. Such fluid shows very interesting features, some of Newtonian nature, and others quite peculiar of chiral active fluids. Among the first, authors report droplets merging or spreading under gravity. The most distinctive features of chiral hydrodynamics is the existence of spontaneously developed unidirectional

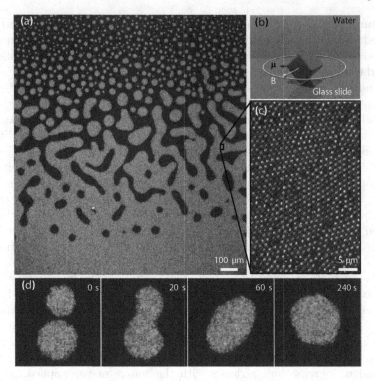

FIGURE 3.10
A two-dimensional monolayer of spinning magnetic colloids. (a) An
optical micrography of the colloidal magnets in bulk. (b) A schematic diagram of one colloidal particle. The 1.6 micron haematite colloidal cubes have
a permanent magnetic moment (μ, black arrow). They are suspended in water,
sedimented onto a glass slide and spun by a rotating magnetic field. (c) An optical micrography of the colloidal magnets in bulk at increased magnification.
(d) Particles attract and form a cohesive material with an apparent surface
tension that, over timescales from minutes to hours, behaves like a fluid, shown
here as clusters coalescence. Image and caption text adapted from Soni et al.
[371].

edge currents associated with the existence of a disipationless linear response
coefficient called *odd* (or equivalently, Hall) *viscosity* [16].[20] I choose to out-
lining the analytics of this system since it bears a loose resemblance with the
active fluids prepared form microtubules that will be much discussed in the

---

[20]Viscosity in its most generic formulation is expressed through a term η_{ijkl} that sets the
proportionality between the stress tensor σ_{ij} and the strain rate v_{kl}. Onsager's reciprocity
theorem ([78]) establishes that under time reversal, this coefficient must be symmetrical
under exchange of indices, i.e. $\eta_{ijkl} = \eta_{klij}$. In the case of a fluid of self-spinning particles,
it can be demonstrated [16] that breaking of time-reversal in two dimensions give rise to a
disipationless (antisymmetric) coefficient called odd, or Hall, viscosity $(\eta_0)_{ijkl} = -(\eta_0)_{klij}$.
In general grounds, it can be demonstrated that the odd viscosity in incompressible fluids
has no effect in the bulk, since it can be adsorbed in the pressure, but may elicit currents
at the boundaries.

next chapter. One must keep in mind, however, that there are two main differences: first, the latter are active by themselves, rather than being externally actuated, and, second, the former are distinctively (macroscopically) chiral as a consequence of this external driving, while chiral effects in active fluids are commonly disregarded, even if potentially present at the level of their elementary components.

I sketch on what follows the basic ingredients of the hydrodynamic description of a colloidal chiral fluid. I summarize the treatment proposed in the original reference [371]. The conventional set of two hydrodynamic variables, mass density of spinners $\rho(\mathbf{r};t)$ and momentum density $\rho(\mathbf{r};t)\mathbf{v}(\mathbf{r};t)$, is completed with the angular momentum density $I\Omega(\mathbf{r};t)$, where I_0 denotes the moment of inertia density of the spinners. Conservation of mass, linear, and angular momenta yields three basic two-dimensional hydrodynamic equations. The first one reduces to the usual incompressibility condition $\nabla \cdot \mathbf{v} = 0$. The conservation of linear momentum for a homogeneous and isotropic 2d fluid reads in this case,

$$\rho(\partial_t + \mathbf{v} \cdot \nabla)v_i = -\partial_i p + \eta\nabla^2 v_i + \eta_r \epsilon_{ij}\partial_j(2\Omega - w) + \eta_0\nabla^2\epsilon_{ij}v_j - \Gamma_{ij}v_j, \quad (3.36)$$

where ϵ_{ij} represents the components of the 2d antisymmetric Levi-Civitta tensor,[21] p denotes the pressure, η and η_r stand for the shear and rotational viscosities,[22] $w = \epsilon_{ij}\partial_i v_j$ denotes the (scalar) 2d vorticity of the fluid, η_0 is what is known as odd viscosity and, finally, $\Gamma_{ij}v_j = (\Gamma\delta_{ij}+\Gamma_\perp\epsilon_{ij})v_j$ introduces an anisotropic friction tensor. Apart from this last contribution, we recognize two important new contributions in the stress tensor: the first one comes from the odd viscosity, and in fact can be formally adsorbed into the pressure term for an incompressible fluid. The second contribution is introduced through η_r, and accounts for the deviation of the rotor spinning rate Ω with respect to half the local fluid vorticity. It originates in the coupling of the individual rotors in the cohesive fluid [16, 371]. This is a distinctive feature of such "active" fluid, since in torque-free systems, angular momentum conservation will preclude the existence of such a difference (Faxen's law).

Finally the third equation corresponds to the conservation of angular momentum,

$$I_0(\partial_t + \mathbf{v} \cdot \nabla)\Omega = -\Gamma^\Omega\Omega - 2\eta_r(2\Omega - w) + D_\Omega\nabla^2\Omega + \iota, \quad (3.37)$$

where Γ^Ω denotes a rotational friction, D_Ω represents an angular momentum diffusion constant, and ι stands for the torque density applied to the spinners. Neglecting material derivatives, retaining only the diagonal term in the friction

---

[21]The components of the Levi-Civitta tensor are defined as $\epsilon_{ij} = 1$ for $(ij) = (1,2)$, $\epsilon_{ij} = -1$ for $(ij) = (2,1)$ and $\epsilon_{ij} = 0$ for $i = j$.

[22]I preserve the original notation. In Chapter 6, I will use instead γ_1 to refer to the rotational viscosity of active fluids.

tensor, and invoking the incompressibility condition, one can transform the last couple of equations to,

$$\eta \nabla^2 v_i - \partial_i \tilde{p} + \eta_r \epsilon_{ij} \partial_j (2\Omega - \omega) - \Gamma v_i = 0, \tag{3.38}$$

where $\tilde{p}$ adsorbs the odd viscosity contribution, and,

$$(1/2)\Gamma^\Omega \Omega + \eta_r (2\Omega - \omega) = \iota/2. \tag{3.39}$$

A sort of adiabatic approximation is further invoked, i.e. one assumes that the torque exerted on each particle by the rotating magnetic field is instantly balanced to the frictional torques exerted by the neighboring particles and the solid substrate. Notice that this permits to decouple the equations for the linear momentum and the angular momentum. Working with the first one, using $-\epsilon_{ij} \partial_j \omega = \nabla^2 v_i$, and assuming Ω space-independent leads in a natural way to introduce an hydrodynamic screening length $\tilde{\delta} \equiv [(\eta + \eta_r)/\Gamma]^{1/2}$. The pressure term is further eliminated as usual by taking the curl operation, and invoking divergence-free condition one finally arrives at a particularly simple equation for the fluid vorticity in terms of the recently introduced length scale $\tilde{\delta}$,

$$(\tilde{\delta}^2 \nabla^2 - 1)\omega = 0. \tag{3.40}$$

In the presence of fluid domains, one would need to prescribe additional boundary conditions. The most obvious introduces the surface tension σ_s times the 2d curvature $\sigma \mathbf{n} = \sigma_s \kappa \mathbf{n}$ for the outward normal $\mathbf{n}$. The Helmholtz-like equation written above can be employed, for instance, to interpret the penetration length of edge currents, mentioned earlier, observed to occur in circular droplets [371]. Moreover, dedicated experiments looking at surface excitations permitted to extract the Hall viscosity, finally estimated to be comparable to the shear viscosity at typical frequencies of the experiment. I recommend the reader to go to the original reference for an extended account of the striking phenomenology observed and its assessment in theoretical grounds.

3.2.2   Electric Forcing: Quincke Rollers under DC Driving

The remaining subsections of this chapter are dedicated to analyze colloidal motion induced by electric forcing. Although this is a classical topic in colloidal science in relation to a variety of electrokinetic phenomena (see Sect. 2.1), I prefer to start this discussion by introducing a less known scenario, based on what is known as (colloidal) Quincke rotation. After this short detour, electrokinetic phenomena will be revisited later on in the two next subsections of this chapter. The reason for proceeding in this way is that it is precisely within this latter context that one can naturally refer to colloidal motion in structured fluids, bridging characteristics of motion in isotropic and anisotropic colloidal dispersions. In this particular respect, the use of Liquid Crystals, to which I want to dedicate the final words in this chapter, turns out to be specially attractive, as will be commented with some detail.

As just stated, I start by commenting on a phenomenon observed by G. Quincke at the end of the nineteenth century. It refers to the rotation of dielectric spheres or cylinders immersed in slightly conducting fluids when stressed by DC fields above a critical threshold. On what follows, I will refer to specimens propelling under this principle as **Quincke rollers**. This particular mode of colloidal motion has been recently revisited when referring to colloid ensembles, singularly as a prototypical mode of swarming/flocking, through contributions mainly from the group of D. Bartolo. This also justifies the interest in presenting Quincke's mechanism with some detail.

More precisely, early observations of the collective properties of Quincke rollers (PMMA spherical beads dispersed in hexadecane solution) were reported by Bricard et al. [44] (see Fig. 3.11). Different modes were analyzed for rollers confined to racetracks depending on their dispersed amount (area fraction). Rollers form an isotropic gaseous phase at very high dilution conditions. Increasing the area fraction, collective motion appears either in the form of **flocks** at low densities, or as a homogeneous **polar suspension**, at still higher densities.

The basic ingredient underlying Quincke's mechanism is the electrohydrodynamic instability created by the charge surface density accumulated at the colloid/fluid interface associated to the inherent dielectric contrast. This originates a dipole centered at the colloid. The collinear situation of this dipole and the externally applied electric field renders an unstable equilibrium condition. Any fluctuating displacement of the charge distribution carried in the direction of rotation by the interface results in a deflection of the dipole movement. The subsequently elicited torque tends further to increase the rotational velocity. A steady rotation regime is achieved when electric and viscous torques are balanced. Finally the rotational motion is rectified into directional displacement on a bounding surface. Paralleling what I did in relation to the fluid of magnetic spinners, and with similar motivations, I present on what follows a brief theoretical account of this fluid of electric rollers. Interested readers will find a detailed treatment in the original reference [44].

Similar to the analysis of a doublet roller sketched previously, the derivation here has two parts: first an equation for the dynamics of the polarization is obtained, that is further transferred to a propulsion equation for the colloid. I focus only on the first part of the calculation since this permits already to extracting the instability threshold. I illustrate it for a sphere of radius R under a DC electric field $\mathbf{E}_0$ applied along a direction perpendicular to the bounding plate. The sphere is supposed non-conducting but endowed with an electric permittivity ϵ_p. The liquid that disperses it has permittivity ϵ_l and conductivity σ_l. The thickness of the boundary layer is assumed thin compared to the colloid radius, and accumulates charge mediated by conductive

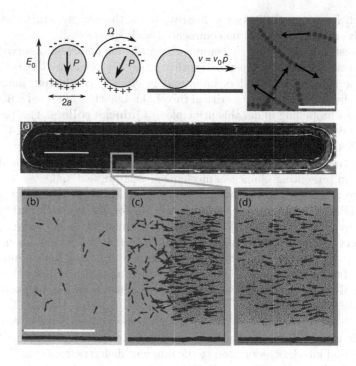

FIGURE 3.11

Colloidal flocking electrically driven by the Quincke principle. Top panels: Sketch of the Quincke rotation elementary mechanism for the self-propulsion mechanism of a colloidal roller characterized by its electric polarization, P, and superposition of ten successive snapshots of colloidal rollers. Time interval, 5.6 ms; scale bar, 50 µm. Bottom panels: (a) dark-field pictures of a roller population that spontaneously forms a macroscopic band propagating along the racetrack; scale bar, 5 µm. Panels (b) to (d): Close-up views for increasing roller fraction: isotropic gas (b), polar bands (c), polar liquid (d). The arrows correspond to the roller displacement between two subsequent video frames. Image and caption text adapted from Bricard et al. [44].

and advective processes. When both contributions are taken into account a dynamic equation for the sphere polarization **P** is obtained in the form,

$$\frac{d\mathbf{P}}{dt} + \frac{1}{\tau}\mathbf{P} = -\frac{1}{\tau}(2\pi\epsilon_0 R^3)\mathbf{E}_0 + \mathbf{\Omega} \times (\mathbf{P} - 4\pi R^3 \epsilon_0 \chi^\infty \mathbf{E}_0), \qquad (3.41)$$

where $\chi^\infty \equiv (\epsilon_p - \epsilon_l)/(\epsilon_p + 2\epsilon_l)$ denotes an effective (dimensionless) electric permittivity, and $\tau \equiv (\epsilon_p + 2\epsilon_l)/2\sigma_l$ a characteristic relaxation time. The interpretation of this equation is straightforward. In absence of rotation, a steady solution is given by $\mathbf{P}^{eq} = -2\pi\epsilon_0 R^3 \mathbf{E}_0$ oriented opposite to the applied external field, and attained after a relaxation time τ arising in the finite conductivity of the medium. Under rotation, advection competes with conduction and a new steady (rotating solution) applies. A solution for first order deviations $\mathbf{P} = \mathbf{P}^{eq} + \delta\mathbf{P}$ with $\Omega \sim \delta\mathbf{P}$ readily identifies the instability condition $\chi^\infty + 1/2 > 0$. The more precise treatment follows by obtaining explicit expressions for forces and torques, and further using them in the corresponding hydrodynamic equations

I skip all the details that can be found in the original reference to concentrate on the scaling forms for the rotation frequency and electric field threshold. The explicit expression for the elicited angular velocity is given by $\Omega = \frac{1}{\tau}\left[(E_0/E_{th})^2 - 1\right]^{1/2}$ for a threshold value $E_{th} = \left[4\pi\epsilon_l R^3(\chi^\infty + 1/2)\mu_r\tau\right]^{-1/2}$, where μ_r represents the inverse of the rotational friction coefficient ($\mu_r = (8\pi\eta R^3)^{-1}$) [44]. The translational velocity must be obtained by explicitly considering the existence of the bounding plate. In this case, the problem must be slightly reformulated since not only the problem gets hydrodynamically but also electrically modified. One would expect from pure dimensional analysis that the magnitude of translational velocity V_0 should be $V_0 \propto R\Omega$. A detailed analysis extracts the proportionality coefficient as a dimensionless ratio of mobility coefficients.

Notice that this Quincke rollers-based system is a particularly simple experimental realization of an **active fluid with polar symmetry**. Theoretical models applied to such fluids, both with and without momentum conservation, are going to be analyzed respectively later on in Chapters 6 and 7.

3.2.3 Electric Forcing: Classical Fixed-Charge Electroosmotic Flows

To better frame the discussion on electrokinetic flows that will basically occupy us for the remaining of this chapter, let's briefly introduce the phenomenon of **electroosmosis** in its classical DC version. I first review the classical Smoluchowski's treatment as can be found in standard textbooks on colloids [168].

A charged surface in contact with an electrolyte solution attracts a screening cloud of opposite charged counterions to form the electrochemical double layer. This was already commented in the introductory remarks on (self)-electrophoresis quoted in Sect. 3.1.1, and the potential distribution was obtained in the introductory chapter (see Sect. 2.1.2). As mentioned there, in the Debye-Hückel approximation corresponding to small surface charge density, the excess diffuse ionic charge follows a decreasing exponential distribution that screens the potential created by the surface charge up to a typical distance set by the Debye length. One rewrites the corresponding profile of the electrokinetic potential, taking z as the axis perpendicular to the charged surface,

$$\phi = \zeta \exp -(\kappa z), \qquad (3.42)$$

where ζ denotes the ζ-*potential* and $\kappa^{-1} = \lambda_D$ is the Debye length. An externally applied field exerts a body force on the electrically charged contact fluid and generates a slip velocity. On what follows, I illustrate a simple derivation that permits to obtain the velocity scale of this electrokinetic flow.

Balance of forces for a volume element within the fluid is readily established in terms of the charge density,

$$E_x Q = E_x \varrho dz = -\eta \frac{d^2 v_x}{dz^2} dz. \qquad (3.43)$$

This equation can be converted into a differential equation for the electrostatic potential using Poisson equation (i.e. $\frac{d^2 \phi}{dz^2} = -\frac{\varrho}{\epsilon}$),[23]

$$E_x \epsilon \frac{d^2 \phi}{dz^2} = \eta \frac{d^2 v_x}{dz^2}. \qquad (3.44)$$

Integrating twice with boundary conditions ($\phi = \zeta, v_x = 0; \phi = 0, v_x = v^s$) leads immediately to the classical result,

$$v_x = -\frac{\epsilon \zeta}{\eta} E_x, \qquad (3.45)$$

or more generically,

$$\mathbf{v}^s = -\frac{\epsilon \zeta}{\eta} \mathbf{E}_\parallel. \qquad (3.46)$$

Notice that when space charge is negative in the liquid, the ζ-potential is positive, and the slip velocity is established in the direction opposite to that of the applied electric field.

When referring to a moving dispersed colloid, the convenient picture is the reverse one. Thus, the electrophoretic velocity in the Smoluchowski limit (i.e. thin double layer relative to the colloid radius) reads,

$$\mathbf{V}_0 = \frac{\epsilon \zeta}{\eta} \mathbf{E}_\infty, \qquad (3.47)$$

in terms of an electrophoretic mobility $\mu_e = \frac{\epsilon \zeta}{\eta}$.

---

[23]To simplify the notation, I use here a simple symbol for the electric permittivity rather than the product of the vacuum and relative values as was proposed in Sect. 2.1.2.

3.2.4 Induced-Charge Electrophoresis under AC Driving

Electrophoresis in the DC version just outlined is one of the basic mechanisms of colloidal motion. On what follows, I will be more interested in non standard AC realizations that provide useful mechanisms of colloidal transport as well. As a matter of fact, we can prevent in this way the usual blocking electrode effects, or concurrent undesired electro-induced chemical reactions that may appear in normal electrophoresis. Moreover, this AC mode constitutes a transport mechanism for colloids dispersed in anisotropic fluids, as will be analyzed next. In this latter respect, I will pay special attention to *(nematic) Liquid Crystal-based colloidal suspensions* (in Chapter 2 referred to as nematic colloids).

In essence, the phenomena I am going to comment here are known under the generic name of **Induced-Charge Electrokinetics** (ICEK) (albeit a little bit more restrictive, one could similarly use the term **Induced-Charge Electroosmosis** (ICEO)), and, particularized to electrophoresis, of **Induced-Charge Electrophoresis** (ICEP). When the dispersing medium is a liquid crystal I adapt these terms to explicitly quote this particular state of the continuum phase and employ, respectively, the terms **Liquid Crystal-Enabled Electrokinetics** (LCEEK), and **Liquid Crystal-Enabled Electrophoresis** (LCEEP). A useful review in this latter context was published by O. D. Lavrentovich a few years ago [218]. I start by commenting on the essentials of ICEP and I turn later to the description of LCCEP.

3.2.4.1 Induced-Charge Electrophoresis

As commented in the previous section, electrophoresis of charged colloids is rooted on the separation of electric charges between the colloid and the diffuse ionic layer around it, and this is completely independent of the existence of any external field. The latter simply uses the preexisting charged boundary layer to propel the colloid. This is in marked contrast with the induced charge version. In ICEP, and in fact in any ICEK phenomenon, the boundary layer is created by the external electric field itself. One thus immediately realizes that the ICEK mechanism, compared to normal electrokinetic phenomena, will carry a higher order (quadratic) dependence on the intensity of electric forcing, whatever the quantitative observable we choose to describe it, either the elicited slip current, or, conversely, the speed of the phoretic particle.

To better frame the discussion let's consider first the case of a spherical metallic particle dispersed in a conducting electrolyte under a uniform electric field **E**. The electric field polarizes the sphere, field lines are expelled from its interior, and oppositely polarized charged double layers are created in both hemispheres perpendicular to **E**. While the steady-state electric field has no component normal to the sphere, its tangential component will create slip currents of quadrupolar symmetry with four vortices wrapping the sphere. It is important to recognize that these slip flows do not break the spatial symmetry, and are not able by themselves to propel the particle. However, they continuously bring ions that consolidate the boundary layers, and secure in this way a stable state of the colloid at rest [374]. Notice that is in contrast with what was commented earlier to be essential to the Quincke's instability.

However, a simple intervention may completely change this scenario and promote direct transport. It consists, for instance, in replacing the uniform metallic particle by a Janus, metallo-dielectric, colloid (more intense electric disturbance is expected in the metallic side). Intuitively, it is clear that ionic currents along the particle in this case no longer bear the perfect quadrupolar symmetry mentioned above, and, consequently, non-compensated slip flows can be easily generated. As the induced space-localized electric charge densities are proportional to the intensity of the electric field itself, the acting body force should go as the square of the electric strength. This is the crux of the induced-charge electrokinetic phenomena.

As a matter of fact, an experimental realization of ICEP in such a context was observed in the range of hundreds Hz to a few kHz by Gangwal et al. [121]. Strikingly enough colloid motion was observed to be perpendicular to **E**, as another distinctive feature of ICEP as compared to conventional electrophoresis. The possibility to rationally design the dynamics of active colloids powered by this mechanism was reported much more recently by Brooks et al. [45].

In the remaining of this subsection, I present a minimum of the theory of induced charge electrokinetics for isotropic electrolytes following the treatment by Squires et al. [374]. This will permit us to obtain the functionality of the elicited Induced-Charge Electroosmotic (ICEO) flows in terms of the parameters of the forcing, i.e. the intensity and frequency of the applied electric field. In the next subsection, I will particularize this analysis to ordered LC-fluids.

I consider flows around an uncharged (ideally polarizable) conducting sphere under an applied field $\mathbf{E}_0 = E_0\hat{\mathbf{z}}$, assumed time independent for the moment, under the assumption of very thin double layers. Dipolar charge clouds form as long as normal field lines inject ions (positive iones accumulate at the outer boundary layer at the south hemisphere with image negative charges inside the sphere; the arrangement is the opposite one in the north side). This continues until a steady state is reached when no field lines penetrate the double layers. For a sphere of radius R, the conducting surface is equipotential and the **r**-distorted electrostatic potential reads $\phi_0 = -E_0 z(1 - R^3/r^3)$. The steady state potential, corresponding to boundary condition $\mathbf{r}_s \cdot \mathbf{E} = 0$, reads $\phi_{ss} = -E_0 z(1 + R^3/2r^3)$. The potential drop at the interface, i.e. the ζ-potential is thus given as $\zeta = (3/2)E_0 R\cos\theta$ (θ is the normal polar angular variable measured from the equator).

In simple words, we are describing a standard electrokinetic flow but with a non-uniform ζ-potential that depends linearly on the applied field. While the steady-state electric field has no component normal to the sphere, its tangential component will create slip currents of quadrupolar symmetry. Since the slip velocity in electrokinetics is bilinear in the applied field and the ζ-potential, the natural velocity scale in ICEO is given as,

$$v_0 = \frac{\epsilon E_0^2 R}{\eta}. \tag{3.48}$$

When considering time-dependent effects as corresponds to AC forcing, one has to take into account two concurrent effects [374]. First, there is a time-dependent polarization of the sphere that comes with a factor $Re\left(\frac{\exp(i\omega t)}{1+i\omega\tau_c}\right)$, expressed in terms of the sphere charging time τ_c. Second, one must additionally consider the blockage time of the electrode that results in a time-dependent electric field $E_0(t) \propto \cos(\omega t) Re\left(\frac{i\omega\tau_e}{1+i\omega\tau_e}\exp(-i\omega t)\right)$. Both effects renormalize the velocity scale given above for ICEO flows according to the form,

$$v(\omega) = v_0 \frac{\omega^2 \tau_e^2}{(1 + \omega^2\tau_c^2)(1 + \omega^2\tau_e^2)}. \tag{3.49}$$

3.2.4.2 Liquid Crystal-Enabled Electrophoresis

Let's refer now to the realization of this principle in non-aqueous, anisotropic fluids, i.e. employing nematic liquid crystals as dispersing electrolyte media. There is a fundamental difference between ICEK and LCEEK that must be immediately noticed. In the latter, the spatial localization of electric charges does not need to occur at the expenses of inhomogeneities at the colloid surface, but can simply result from the spatial distortions of the nematic director around the colloid. More precisely, in LCEEK space-charge separation follows from the applied electric field, which drives positive and negative ions to different regions of the deformed liquid boundary around the particle, thanks to the anisotropic conductivity and dielectric permittivity of the dispersing medium. In this respect, it is worth noticing that the phase contrast at the interface (normally solid/liquid for normal sols) is not at all a requirement for LCEEP. This was demonstrated, for instance, by the driving of water droplets in stabilized LC-based emulsions, as reported originally by Hernàndez-Navarro et al. [162].

However, what turns out to be crucial to achieve the spatial symmetry breaking required to secure directed motion is the nature of the distorsions of the nematic director around the inclusion, i.e. the topological defects residing at the colloid/nematic interface (see Sect. 2.2.3). In this respect, defect arrangements of quadrupolar characteristics, as for instance Saturn ring-like dislocations for normal anchoring, or double-boojum for planar, do not break the symmetry around the colloid, and are not able to support motion. Conversely, dipolar arrangements, as correspond mostly to hedgehog for normal anchoring, may promote sustained colloid transport. More in particular, Lazo et al. [220] registered flow patterns on immobilized glass spheres dispersed in a liquid crystal mixture with tiny dielectric anisotropies (see Fig. 3.12), demonstrating that flows were puller-like with normal anchoring and pusher-like for planar (see Sect. 3.1.2).

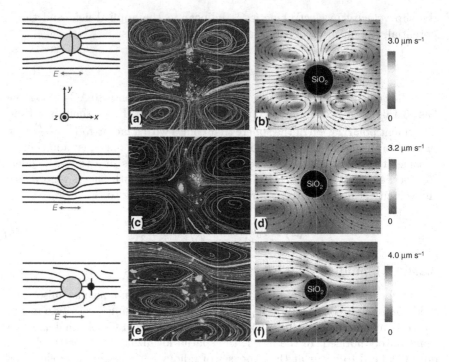

FIGURE 3.12
LCEEK flows around glass spheres. Panels (a) and (b): Saturn ring disclination creates quadrupolar flows puller-like. Panels (c) and (d): Double-boojum flows have opposite pusher-like nature. Panels (e) and (f): symmetry-broken situation of puller characteristics induced by a hedhehog point defect (colored image in original version). Image and caption text adapted from Lazo et al. [220].

The first experiment demonstrating LCEEP was reported by Lavrentovich et al. for metallic and dielectric particles as well[219].[24] Although other possibilities exist as well, see for instance [162], the director and the applied AC field were both aligned along the same direction in this case. On the other hand, a liquid crystal with (large) positive dielectric anisotropy was employed. This rules out any reorientation of the nematic layer that is enclosed in a cell with planar anchoring at the bounding plates. Two aluminum stripes serve as electrodes. Dielectric spheres of diameter between 5 and 50 μm, as well as

---

[24]More recently, Dhara and coworkers extended LCEEP to include spherical particles with metal-dielectric Janus characteristics. Regardless of the symmetry of the LC distortions around the spherical particle, net propulsion was observed under AC electric fields. Moreover, the classification of swimmers in either puller or pusher is not longer clear in this situation [334].

(neutral) gold spheres of diameter a few microns, were treated to guarantee a normal anchoring of the LC at the colloid surface. Applied electric fields were in the range $-20, +20$ mV μm^{-1}, and frequencies up to a few hundreds Hz were used. The colloids speed, moving along the director, was proved to fulfill a combination of linear and nonlinear contributions,

$$V_i = \mu_{ij} E_j + \beta_{ijk} E_j E_k. \tag{3.50}$$

In particular, this tensorial nature of the nonlinear electrokinetic phenomenon is nicely illustrated from complementary experiments reported in the mentioned original publication. More in detail, Lavrentovich et al. demonstrated colloidal motion normal to **E** along racetracks previously drawn by buffing in a circular fashion the coating layer that provides the planar alignment of the LC inside the cell. The values of the nonlinear coefficient β, in terms of characteristic size and electric parameters of the colloidal particle, were estimated as well in the original reference.

As predicted by the simple theoretical analysis in the preceding subsection, maximum observed velocities correspond to intermediate frequencies placed between the two basic time scales of the system, i.e., the particle charging time $\tau_c = \kappa^{-1} a/\mathcal{D}$ (metallic) or $\tau_c = (\kappa^{-1})^2 \epsilon/\mathcal{D}\epsilon_d$ (dielectric), and the electrode blocking time $\tau_e = \kappa^{-1} L/\mathcal{D}$, where $\mathcal{D}$ stands for the standard diffusion coefficient of the ionic species, and L for the distance separating the electrodes ([374]).[25]

Let us turn now to some simple analytics specific to an LCEEP scenario. I follow the treatment by Peng et al. [303]. As I did when introducing ICEK, I start by considering very low frequencies of the externally applied field. The key point in the analytical procedure is to assume an inhomogeneous director $\hat{n}(\mathbf{r})$ characterizing the local orientation within the nematic fluid. The applied electric field **E** drags the charges of opposite signs along the curved director lines, accumulating them in different regions of the liquid crystal bulk. This allows to build up a volume density of charge $\varrho(\mathbf{r})$ that depends on the conductivity and dielectric permittivity of the LC and its anisotropy. Further on, the electric field acts on this space charge and, in turn, creates electroosmotic flows.

To evaluate the created charge density I consider the simplest situation for a (nematic) director field minimally distorted in the x/y plane: $\hat{n} = (\cos\theta(x,y), \sin\theta(x,y))$, with θ accounting for the angle between $\hat{n}$ and the x axis. One starts from the *Maxwell's equation* for the magnetic field **H**,

$$\mathrm{curl}\mathbf{H} = \mathbf{\nabla} \times \mathbf{H} = \frac{\partial \mathbf{D}}{\partial t} + \mathbf{J}. \tag{3.51}$$

An applied low-frequency (harmonic) electric forcing $\mathbf{E}(t) = \mathbf{E}e^{i\omega t}$ creates a current density $\mathbf{J}(t) = \mathbf{J}e^{i\omega t} = \boldsymbol{\sigma} \cdot \mathbf{E}e^{i\omega t}$, and an electric displacement

---

[25]It should be remembered at this point that κ^{-1} is another way to denote the typical Debye length.

$\mathbf{D}(t) = \mathbf{D}e^{i\omega t} = \epsilon_0\epsilon \cdot \mathbf{E}e^{i\omega t}$.[26] The conductivity and dielectric tensors are represented as $\boldsymbol{\sigma} = \sigma_\perp\mathbf{I} + \Delta\sigma\hat{\mathbf{n}}\hat{\mathbf{n}}$ and $\boldsymbol{\epsilon} = \epsilon_\perp\mathbf{I} + \Delta\epsilon\hat{\mathbf{n}}\hat{\mathbf{n}}$, where, respectively, $\Delta\sigma = \sigma_\parallel - \sigma_\perp$, and $\Delta\epsilon = \epsilon_\parallel - \epsilon_\perp$, stand for the anisotropic difference in the electric parameters of the liquid crystal. Moreover, the vector product is understood in its tensorial form $[\hat{\mathbf{n}}\hat{\mathbf{n}}]_{ij} = n_i n_j$. One further assumes that the diagonal components of the conductivity tensor $\sigma_\parallel$ and $\sigma_\perp$, and those of the dielectric tensor $\epsilon_\parallel$ and $\epsilon_\perp$, are all frequency independent. We can rewrite the equation above as,

$$\text{div}\left(\frac{\partial\mathbf{D}}{\partial t} + \mathbf{J}\right) = \boldsymbol{\nabla} \cdot \left(\frac{\partial\mathbf{D}}{\partial t} + \mathbf{J}\right) = \boldsymbol{\nabla} \cdot (\tilde{\boldsymbol{\sigma}}(\omega) \cdot \mathbf{E}) = 0, \qquad (3.52)$$

where $\tilde{\boldsymbol{\sigma}} = \boldsymbol{\sigma} + i\omega\epsilon_0\boldsymbol{\epsilon}$ is an effective conductivity tensor. For low frequency $\omega \ll \sigma_\perp/\epsilon_0\epsilon_\perp$, $\tilde{\boldsymbol{\sigma}} \approx \boldsymbol{\sigma}$ and $\mathbf{E} = -\boldsymbol{\nabla}V$, whereas the electric potential satisfies $\boldsymbol{\nabla} \cdot (\boldsymbol{\sigma} \cdot \boldsymbol{\nabla}V) = 0$, or,

$$\sigma_\perp\nabla^2 V + \Delta\sigma\boldsymbol{\nabla} \cdot [(\hat{\mathbf{n}} \cdot \boldsymbol{\nabla}V)\hat{\mathbf{n}}] = 0, \qquad (3.53)$$

and the charge density $\varrho = \boldsymbol{\nabla} \cdot \mathbf{D}$ turns out to be expressed as,

$$\varrho = -\epsilon_0(\Delta\epsilon - \epsilon_\perp\Delta\sigma/\sigma_\perp)\boldsymbol{\nabla}[(\hat{\mathbf{n}} \cdot \boldsymbol{\nabla}V)\hat{\mathbf{n}}]. \qquad (3.54)$$

In the simplest situation, one may consider the external electric field applied parallel to the director field along say the x axis. Moreover, the sample is assumed to show weak anisotropies in both the conductivity and permittivity constants. The electric field in the bulk of the LC can be then represented as $\mathbf{E} = (E_0 + E'_x(x,y), E'_y(x,y))$, incorporating small corrections E'_x and E'_y arising in the director inhomogeneities that satisfy at the lowest order,

$$\sigma_\perp\left(\frac{\partial E'_x(x,y)}{\partial x} + \frac{\partial E'_y(x,y)}{\partial y}\right) + \Delta\sigma E_0\left(cos2\theta\frac{\partial\theta}{\partial y} - sin2\theta\frac{\partial\theta}{\partial x}\right) = 0. \quad (3.55)$$

According to the expression obtained above, the space charge density correspondingly reads,

$$\varrho = \epsilon_0(\Delta\epsilon - \epsilon_\perp\Delta\sigma/\sigma_\perp)E_0\left(cos2\theta\frac{\partial\theta}{\partial y} - sin2\theta\frac{\partial\theta}{\partial x}\right). \qquad (3.56)$$

This charge density is actuated by the electric field creating a bulk force of density $f = \varrho E_0 \propto E_0^2$ that drives the electro-osmotic flows. If R denotes the length scale of the spatial distortion of the director (typically the radius of the colloidal inclusion), the spatial derivatives of the electric field scale as $1/R$. The amplitude of the velocity of the electro-osmotic flows v_0 is then

---

[26]To respect the original notation, notice that in this section ϵ denotes the (relative) electric permittivity tensor, although no subindex r is employed. This permits later on to designate with explicit subindices its parallel and perpendicular components.

estimated to follow from the balance of the driving force density and the viscous resistance $\eta v_0/R^2$,

$$v_0 = \epsilon_0 \epsilon_\perp \eta^{-1} |(\Delta\epsilon/\epsilon_\perp - \Delta\sigma/\sigma_\perp)| RE^2. \qquad (3.57)$$

In the last expression, the LC viscosity refers typically to the component parallel to the director field which is the easy flow direction. Naturally, to get the final expression for the velocity, this result should come multiplied by the frequency-dependent factor evaluated previously, to properly account for the time scales associated to the charging and blocking times under AC forcing.

Notice that anisotropies both in the electric conductivity and permittivity enter into the last equation, and actually in a way that permits a change of polarity of the elicited flows. This specific question for the electrophoresis of dispersed colloids in LC, under changes of composition and temperature, was addressed in a paper by Paladugu et al. from Lavrentovich's group [293]. In general, $\Delta\epsilon$ is positive for most LCs. However, the sign of $\Delta\sigma$ can be easily tuned.

3.2.4.3 Anomalous Statistical Characteristics of Driven Nematic Colloids

Individual swimming under LCEEP offers other possibilities for study not centered in the mechanism of motion itself, but rather referring to specific characteristics of the elicited motion, either at individual or collective level. To be precise, I will be briefly commenting on some striking anomalies observed when looking at statistical characteristics of (individual) colloid displacement trajectories, as recently reported by Pagès et al. [289]. I will first comment on these anomalies, and later on I will provide some theoretical context to understand them.

To widen the discussion as much as possible I will consider colloids driven under the LCCEP mode, those we have been mostly concerned with in the preceding paragraphs, but also when motion does not come from an electric forcing but simply is gravity-forced (i.e. sedimenting nematic colloids).

The unexpected discovery, under specified circumstances described on what follows, is the observed *superdiffusion* of the colloid displacement when the latter is resolved in different components relative to the far field orientation of the nematic LC. The simplest scenario corresponds to an experiment with the driving direction parallel to $\hat{n}$. In this case, while motion is ballistic along $\hat{n}$, transversal fluctuations may become superdiffusive or keep normally diffusive, depending on the pattern of the topological defect around the inclusion.

Interestingly enough, this does not depend on the driving mode (LCEEP or gravity-induced), nor on the propulsion velocity of the colloid [289]. Spherical silica particles of different sizes (5 and 10 µm diameter) were used in the original paper. Specific chemical functionalization protocols guarantee dipolar hedgehogs for homeotropic anchoring, while bare non-functionalized

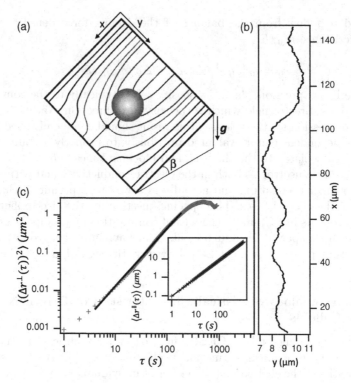

FIGURE 3.13
**Sedimentation of a particle dispersed in a nematic liquid crystal
with dipolar LC configuration.** (a) Sketch of the experimental geometry
and LC director surrounding the colloidal inclusion. (b) Particle trajectory
in the x/y plane. (c) Mean square displacement (MSD) in the transversal
direction vs. time window. The solid line is a power law fit that yields an
exponent $\nu = 1.41 \pm 0.04$. In the inset, the mean parallel displacement is
plotted vs. time, showing a ballistic trend. The solid line has a slope 1. Image
and caption text adapted from Pagès et al. [289].

particles, employed for comparison, generate quadrupolar distortions in the
form of dual surface point defects (boojums). The chosen LC has negative dielectric anisotropy and a considerably high nematic/isotropic T_{NI} transition.
Boundary conditions at the bounding plates are prescribed planar for samples
of thickness roughly twice the particle diameter.

Particularizing to sedimentation experiments, transversal displacements
were found superdiffusive (see Fig. 3.13) with exponent values ranging between
1.2 and 1,8, depending on particle size, and increasing with temperature, always below T_{NI}. LCEEP-driven particles were investigated in a similar way.
Despite more than two orders of magnitude larger speeds, results for lateral

displacement showed similar trends. To assess the influence of LC distortions, sedimenting experiments were repeated this time for bare particles. Notice that LCEEP drive is not a choice in this case, since the quadrupolar defect structure prevents fore-aft symmetry breaking. Compared with with the case of dipolar defects, exponents are now markedly smaller, and they barely differ from normal diffusion. This conclusion applies to the whole range of temperatures investigated. Although not fully understood yet, a qualitative interpretation of these results following [289] will be provided at the end of this section.[27]

Prior to that, I would like to refer first to some theoretical ideas that have been put forward more generically within the context of swimming in Liquid Crystals. In particular, I briefly mention a pair of studies by Lintuvuori et al. [231] and Daddi-Moussa-Ider et al. [71] reporting the behavior of model microswimmers (squirmers) in nematics.

By combining simulations and analytic calculations, it is shown that the hydrodynamic coupling between the flow field of a spherical squirmer ($V(\theta) = V_0 \sin\theta(1 + \beta\cos\theta)$) (see Sect. 3.1.2) and the LC director may lead to the reorientation of the swimmers [231]. More in particular, pushers are shown to swim parallel to the nematic director, while pullers swim perpendicular (see Fig. 3.14). These results apply irrespective of the anchoring conditions, providing evidence that the swimmer reorientation is rooted, in this case, in the large scale hydrodynamic coupling of the squirmer flow with the anisotropic viscosities of the host fluid, rather than reflecting short range elastic interactions within the nematic matrix.

An important physical parameter in this context is the *Ericksen number* (Er), which gives the ratio between viscous and elastic forces, $Er = \gamma_1 R V_0 / K$, where R is a typical length scale (colloid size), V_0 its velocity, and γ_1 and K denoting, respectively, the rotational viscosity and elastic constant of the LC. In the study of Lintuvuori et al., the Ericksen number is assumed of order unity.

The conceptual idea underlying the theoretical approach by Lintuvuori et al. is to use an isotropic approximation for the squirmer velocity in the limit of small Ericksen and Reynolds numbers, while the anisotropic viscosity of the LC is retained in order to evaluate the stress and further the torque exerted on the swimming particle. The obtained expression for this torque is written in terms of an effective viscosity $\tilde{\eta}$ as $\mathbf{T} = -8\pi\tilde{\beta}\tilde{\eta}V_0 R^2(\hat{\mathbf{n}} \cdot \hat{\mathbf{u}})\hat{\mathbf{n}} \times \hat{\mathbf{u}}$, where

---

[27] In a parallel development, not looking though specifically at dispersion modes, gravity-driven transport of spherical colloidal particles dispersed in a nematic LC (5CB) within microfluidic arrays of cylindrical obstacles arranged in a square lattice were considered by K. Chen et al. [62]. Homeotropic anchoring at the surfaces of the obstacles created periodic director-field patterns that strongly influenced the motion of the colloids, whose surfaces had planar anchoring. When the gravitational force was oriented parallel to a principal axis of the lattice, the particles moved along channels between columns of obstacles and displayed pronounced modulations in their velocity. As the angle between the gravitational force and principal axis of the lattice was varied, the velocity did not follow the force but instead locked into a discrete set of directions commensurate with the lattice.

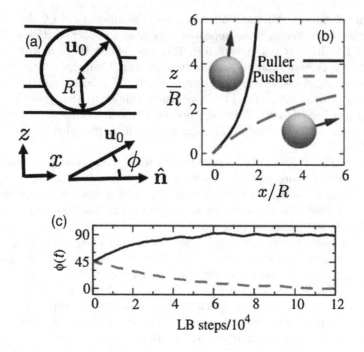

FIGURE 3.14
Motion of a squirmer in a nematic liquid crystal. Distinctive characteristics of pullers and pushers are depicted. Image and caption text adapted
from Lintuvuori et al. [231].

$\hat{\mathbf{u}}$ stands for the particle axis.[28] When $\tilde{\beta}\tilde{\eta} > 0$, the torque aligns the particle
axis with the far field. From viscosity values of common nematic LCs one has
$\tilde{\eta} < 0$, so the stable orientation of pullers ($\tilde{\beta} > 0$) is perpendicular to the
nematic orientation, whereas pushers ($\tilde{\beta} < 0$) will move parallel to the main
direction. This analytic finding is supplemented with 3d *lattice Boltzmann*-
based numerical simulations. Similar conclusions were reached in the parallel
study [71], although this latter case considered regimes of smaller Er numbers.

I come finally to the central question of the anomalous diffusion properties of motile colloids. The first theoretical paper I am aware of devoted to
this question was published by Toner et al. [411]. Authors considered active
point particles, i.e. defects were disregarded, moving in an anisotropic nematic
background assumed negligibly perturbed by the swimmer motion. Moreover,

---

[28]Notice that $\tilde{\eta}$ is a viscosity parameter that has not to be confused with a conventional
shear viscosity. It enters only into the anisotropic part of the stress tensor. It comes from a
combination of specific (direction dependent) viscosity parameters of the liquid crystal, the
so-called *Leslie coefficients*.

it is assumed that the swimmer carries no internal compass; any preference it exhibits for one direction of motion over any other must arise from the local nematic director at the instantaneous location of the swimmer. Under these postulates, Toner et al. reported ballistic behavior for the translational motion parallel to the director, while the long-time transverse motion was indeed demonstrated superdiffusive, although with an anomalous scaling proportional to $t \ln(t/t_0)$ rather than power-law, as in the experiments by Pagès et al. [289]. These analytic results were corroborated by numerical simulations using a simple lattice model for the nematic LC.

I briefly review Toner et al.'s approach, by starting with the equation of motion,

$$\dot{\mathbf{r}}(t) = V\hat{\mathbf{n}}(\mathbf{r}(t); t) + \boldsymbol{\xi}(t), \tag{3.58}$$

where $\xi(t)$ denotes a zero-mean fluctuating Gaussian force, and V represents the mean swimmer velocity. By averaging this Langevin-like equation one obtains,

$$\langle \mathbf{r}(t) \rangle = Vt\langle \hat{\mathbf{n}} \rangle = V_z t\hat{\mathbf{z}}, \tag{3.59}$$

with $\hat{\mathbf{z}}$ denoting the nematic director direction. Thus, the mean motion of the swimmer is purely ballistic, although the speed V_z can not be totally determined within the frame of the outlined theory. One then turns to consider fluctuations about this mean, and in particular we focus on the mean squared lateral displacements of the swimmers $\langle [\Delta \mathbf{r}_\perp(t)]^2 \rangle \equiv \langle |\mathbf{r}_\perp(t) - \mathbf{r}_\perp(0)|^2 \rangle$ (see Sect. 2.1.1). By projecting the equation of motion perpendicular to the mean direction one has,

$$\dot{\mathbf{r}}_\perp(t) = V\hat{\mathbf{n}}_\perp(\hat{\mathbf{r}}; t) + \boldsymbol{\xi}_\perp. \tag{3.60}$$

One would proceed further by integrating the latter equation for $\Delta\hat{\mathbf{r}}_\perp(t)$, squaring it and averaging. The terms that involve the stochastic contribution to the velocity (i.e. those involving correlations between $\hat{\mathbf{n}}_\perp$ and $\boldsymbol{\xi}_\perp$, or between $\boldsymbol{\xi}_\perp$ themselves), can be immediately evaluated, both giving normal diffusive terms, i.e. a linear behavior with t. The anomalous diffusion thus comes solely from the term,

$$I = V^2 \int_0^t dt' \int_0^t \langle \hat{\mathbf{n}}_\perp(\mathbf{r}(t'); t') \cdot \hat{\mathbf{n}}_\perp(\mathbf{r}(t''); t'') \rangle dt''. \tag{3.61}$$

On long time scales one can neglect the slow (diffusive) dynamics of the nematic, and neglect the explicit time dependence of the correlation function that appears in this last expression. In other words, for $|t' - t''| \geq t_0 = D/V_z^2$, one simply needs to calculate the correlation $\langle \hat{\mathbf{n}}_\perp(V_z\delta t\hat{\mathbf{z}}, 0) \cdot \hat{\mathbf{n}}_\perp(\mathbf{0}, 0) \rangle$, with $\delta t \equiv (t' - t'')$. This is easily handled by Fourier transforming,

$$\langle \hat{\mathbf{n}}_\perp(V_z\delta t\hat{\mathbf{z}}; 0) \cdot \hat{\mathbf{n}}_\perp(\mathbf{0}; 0) \rangle = \int \frac{d^2q_\perp dq_z}{(2\pi)^3} e^{iV_z q_z \delta t} \langle |\hat{\mathbf{n}}_\perp(\mathbf{q})|^2 \rangle. \tag{3.62}$$

The equal-time transversal fluctuations of the director are known to be readily evaluated in terms of the Frank-Oseen free energy constants (see Sect. 2.2.3) [77] (I omit writing down explicitly the volume of the sample that would appear as a common factor on the right-hand side),

$$\langle |\hat{\mathbf{n}}_\perp(\mathbf{q})|^2 \rangle = \frac{k_B T}{K_1 q_\perp^2 + K_3 q_z^2} + \frac{k_B T}{K_2 q_\perp^2 + K_3 q_z^2}. \tag{3.63}$$

One would perform the left integral in Eq. 3.62, and substitute in Eq. 3.61. Leaving apart technical tricks that can be found in the original reference, the final important result for the mean squared lateral displacement reads,

$$\langle [\Delta \mathbf{r}_\perp(t)]^2 \rangle = \frac{V k_B T}{2\pi} (K_1^{-1} + K_2^{-1}) t \ln(t/t_0) + D_0 t \tag{3.64}$$

rendering the logarithmic anomaly previously commented. An extension of this study to smectic LCs was published a couple of years later [104]. The logarithmic scaling is also extracted in this case for MSD perpendicular to the far-field orientation, but only for times smaller than a system-dependent crossover time. For larger times, the predicted behavior is instead of the form $t(\ln(t/t_0)^1)^2$.

This theoretical prediction clearly does not reproduce the experimental results for driven colloids [289], either under gravity or LCEEP, since no power-law scaling is predicted as was found there. Definitively a major difference is that experiments deal with finite-size particles, while the theory of Toner et al. [411] supposes point-like active particles. Notice in this respect that the particular defect structures around the colloids do not play any role whatsoever within this latter theoretical framework, while it is a key ingredient in experiments, i.e. superdiffusion anomalies are only observed for dipolar (hedgehog) distortions around motile colloids.

In the case analyzed by Pagès et al. [289], an ad hoc rationale permits, however, to interpret the observed superdiffusion exponents. The underlying idea is to use arguments borrowed from the study of dynamical equations for the transport of particles under random forces of disparate origins and dynamical correlations [47, 82]. In the experimental situation [289], these random forces correspond to thermal, but also principally to those that come from transient misalignment of the dipolar structure of the particle plus defect with respect to the far field orientation of the nematic director.

Formulated in simple terms as for a two-dimensional system, the instantaneous motion perpendicular to the far field can be written in terms of a Langevin equation of a form similar to that used to introduce the model of ABPs, but now expressed in terms of two separate stochastic forces, i.e.

$$\gamma \dot{r}_\perp(t) = f(t) + \xi(t), \tag{3.65}$$

where $f(t)$ stands for the non-thermal stochasticity inherent to the dynamics of the nematic colloid, while $\xi(t)$ represents the conventional thermal (Brownian) white noise. In particular, $f(t)$ is considered a (pseudo)torque written proportional to $\sin 2\theta(t)$, where $\theta(t)$ denotes the angle of misalignment. From accurate experimental measures, the time autocorrelation of the misalignment was evaluated, and demonstrated to follow a power-law,

$$\langle f(t)f(t+\tau)\rangle \equiv \Lambda(t) = \Lambda_0(\tau/\tau_0)^{-\alpha}. \tag{3.66}$$

Following [47, 82], the MSD associated to the above Langevin dynamics then reads,

$$MSD(\tau) = \frac{2k_BT}{\gamma}\left(\left(\frac{\tau}{\tau_0}\right) + \tilde{\epsilon}\kappa_{1,\alpha}\left(\frac{\tau}{\tau_0}\right)^{2-\alpha}\right), \tag{3.67}$$

where $\kappa_{1,\alpha}$ is a numerical factor, and $\tilde{\epsilon}$ introduces the ratio between the intensities of the two stochastic forces. For α smaller than unity, the first term, a purely diffusive contribution, is clearly subdominant in relation to the superdiffusive contribution with exponent $2-\alpha$. In spite of the crudeness of the approximations made, an striking agreement is found between the experimentally extracted $(2-\alpha)$ value, and the experimental superdiffusion scalings of Pagès et al., as can be seen in the original reference [289].

In any case, it is clear that we need more specific theoretical models, and precise numerical model simulations, to fully encompass the richness and complexity of motile colloids when dispersed in liquid crystals.

I will comment later on collective properties of colloids moved under LCEEP, singularly in relation to striking driven-assemblies characteristics (see Sect. 5.1.1).

Totally apart from scenarios of magnetic or electric forcing to which I have dedicated the entire second part of this chapter, and before closing it, I find worth mentioning a couple of additional scenarios of driven microswimming. Acoustic intervention has been demonstrated by W. Wang et al. to propel anisometric metallic microrods based on a **self-acoustophoretic** mechanism [321], while locomotion of photoresponsive soft microrobots using structured light has been reported by Palagi et al. [294]. Still very recently, driving of Janus particles from an X-ray source has been reported by Xu et al. [447]. In this latter case, propulsion follows from bubble growth that is enhanced by water radiolysis near the particle surface.

3.3 Brief Commented List of Selected Review Papers

The topic of microswimming in Newtonian fluids encompasses a large variety of aspects that have been summarized and revised at length in a considerable number of review papers during the last years. I believe it may be interesting for potential readers to have at hand a single list that compiles those publications that in my opinion have been more influential in the development of this striving field. On what follows, I have selected some of these references starting with three general reviews on colloidal-based microswimming that were published practically simultaneously around 2016.

A similar list more oriented towards active matter (continuous), mostly-fluid systems, is available at the end of Chapter 6.

Physics of microswimmers-Single particle motion and collective behavior: A review.

J. Elgeti, R. G, Winkler and G. Gompper, Reports on Progress in Physics 78, 056601 (2015).
The article reviews the physics of locomotion of biological and artificial microswimmers, from an individual and collective points of view as well. Singularly interesting is the physics-oriented way to present the various propulsion mechanisms at the beginning of the review.

Emergent behavior in active colloids.

A. Zöttl and H. Stark, Journal of Physics Condensed Matter 28:253001 (2016).
Generic features of microwimming at the level of individual particles are first presented. Special attention is dispensed to Active Brownian Particles, including confinement effects, or their motion under external flows and external fields. The second part summarizes different aspects of the emergent collective behavior of active colloidal suspensions.

Active particles in complex and crowded environments.

C. Bechinger, R. di Leonardo, H. Löwen, C. Reichhardt, G. Volpe and G. Volpe, Review of Modern Physics 88, 045006 (2016).
Comprehensive review on the classical aspects of artificial microswimming with special emphasis on their collective aspects. Singularly relevant is the interest to assess new features when addressing swimming in realistic environments incorporating complex and crowded ambients (with similar motivation, see the here specially dedicated Appendix 1 8).

The hydrodynamics of swimming microorganisms.

E. Lauga, Reports on Progress in Physics 72, 096601 (2009).
A brief overview is given first relative to the mechanisms of swimming motility, and of the basic properties of flows at low Reynolds number. This is followed by a review on classical theoretical work on microorganisms motility, in particular revisiting early calculations of swimming kinematics and the theory of flagellar locomotion.

Colloidal transport by interfacial forces.

J. L. Anderson, Annual Review Fluid Mechanics 21, 61 (1989).
Classical paper that reviews theoretical aspects of motion triggered by interfacial effects. The basic message conveyed by the paper is that the interface is a

region of small but finite thickness that support dynamical processes that lead not only to interfacial stresses but also to an apparent "slip velocity". This idea is elaborated in depth in the review covering a variety of scenarios typical of colloid science.

In pursuit of propulsion at the nanoscale.
S. J. Ebbens and J. R. Howse, Soft Matter 6, 726 (2010).
The paper reviews developments in self-propelling nano- and microscale swimming devices. Particular emphasis is placed on describing autonomously powered devices driven by asymmetrical chemical reactions (self-diffusiophoretic swimmers).

Chemical locomotion.
W. F. Paxton, S. Sundararajan, T. E. Mallouk and A. Sen, Angewandte Chemie Int. Ed. 45, 5420 (2006).
One of the earlier reviews on chemically powered microswimmers. The review highlights developments and discusses principles of the transduction of chemical into mechanical energy at the microscale. A more recent update by some of the same authors is W. Wang et al, **Small power: Autonomous nano and micromotors propelled by self-generated gradients,** *Nano Today 8, 531, (2013).*

Chemicall powered micro-and nanomotors.
S. Sánchez, LL. Soler and J. Katuri, Angewandte Chemie Int. Ed. 54, 1414 (2015).
The paper reviews the major advances in the growing field of catalytic micromotors. Specially interesting is the discussion on the fabrication of the corresponding devices (nanowires, Janus spheres and microtubes), and on strategies pursuing controlled motion of microswimmers. A quite large section is devoted to applications (actuators, biomedical applications, pumps, sensors and environmental remediation.

Fabrication of micro/nanoscale motors.
H. Wang and M. Pumera, Chemical Reviews 115, 8704 (2015).
Emphasis is devoted to fabrication and functionalization techniques of microswimmers. Among the principal techniques authors distinguish electrochemical procedures, physical vapor deposition, strain engineering, 3d direct laser writing, assembly of materials and biohybrid techniques.

Nano/Micromotors in (Bio)chemical science applications.
M. Guix, C. C. Mayorga-Martinez and A. Merkoçi, Chemical Reviews 114, 6285 (2014).
The differential trait of this review is a first section devoted to biological motors encompassing both linear-motion motors (for instance cytoskeletal molecular motors), and rotary motors (in particular the bacteria rotary flagellar motors).

The environmental impact of micro/nanomachines: A Review.
W. Gao and J. Wang, ACS Nano 8, 3170 (2014).
The paper highlights the opportunities and challenges in the applications of nanomotors-based technology in the context of environmental research. It emphasizes nanomachine-enabled degradation and removal of major contaminants or nanomotor-based water quality monitoring.

Microrobots for minimally invasive medicine.
B. J. Nelson, I. K. Kaliakatsos and J. J. Abbott, Annual Review Biomedical Engineering 12, 55 (2010).
Early review aimed at providing a comprehensive survey on the technological state of the art, potential impact and challenges in the application of microrobots in medical contexts. More recent reviews on the same thematics have been published by Z. Wu, Y. Chen, D. Musaka, O. S. Pak and W. Gao, Medical micro/nanorobots in complex media, Chem. Soc. Rev. 49, 8088 (2020), and by Q. Wang and L. Zhang, External power-driven microrobotic swarm: From fundamental understanding to imaging-guided delivery, ACS Nano 15, 149 (2021).

Active Brownian particles: From individual to collective stochastic dynamics.
P. Romanczuk, M. Bär, W. Ebeling, B. Lindner and L. Schimansky-Geier, European
Physical Journal Special Topics 202, 1 (2012).
Several theoretical models of individual motility, as well as collective properties and pattern formation of active particles are reviewed. The chosen methodology is that of nonlinear and stochastic dynamics. The main underlying assumption is that the additional inflow of energy with respect to the canonical Brownian particles can be effectively described by a negative dissipation in the direction of motion (negative friction).

Finally, from the topical collection "Chemical systems out of equilibrium" appeared in Chem. Soc. Rev. (2017), I mention a couple of recent review papers published by P. Illien et al., **Fueled motion: phoretic motility and collective behavior of active colloids**, Chem. Soc. Rev. 46, 5508 (2017), and J. Zhang et al., **Active colloids with collective mobility status and research opportunities**, Chem. Soc. Rev. 46, 5551 (2017).

4

Protein-based Active Fluids

If Chapter 3 was entirely dedicated to particle-based dispersions, with distinctive features of swimming motion at individual or collective level, in this chapter I abandon this discrete picture to set foot in the ambit of continuous active systems. In parallel, we move from a context that from a physicochemical point of view could be considered as typical of colloidal (particulate) systems, to enter a, loosely speaking, bioinspired complex fluids perspective, that of **active fluids**. To be more precise, the systems that will be examined are based on minimal *in-vitro reconstitutions of the cell cytoskeleton*. In addition, these active formulations suppose an extremely rewarding arena where to look at characteristics of forced self-assembly under the non-equilibrium conditions secured by (adenosin triphosphate) ATP feeding. These declared potentialities make these systems hold big promises in relation to fields as diverse as cell biophysics or that of functional materials [274].

This chapter is entirely devoted to active fluids that are prepared from dispersions of filamentary proteins polymerized from **actin** and, mainly, **tubulin**, with their corresponding motor constructs, respectively, **myosin** and **kinesin**. Mimicking the cell milieu, these preparations are driven far from equilibrium, and thus kept active, when fueled with ATP. A central attention will be devoted to their performing as interfacial dense suspensions to highlight their *nematic symmetry*, that qualify them as **active nematic fluids**, or in short **Active Nematics**.

Still, before starting with **(two-dimensional) active nematics** and to better frame the contents of this chapter, I introduce in a first section the characteristics of filamentary and motor protein-based active fluids when assembled as three-dimensional gels in the form of **active gels**. This part itself is separated into two subsections respectively dedicated to active gels based on actin filaments and on tubulin polymers, i.e. **microtubules** (MTs).

An historical precedent of the protein-based systems here discussed refers to the so-called **motility assays**. Motility assays are quite enlightening to reveal self-organizing properties of filamentary proteins under the non-equilibrium conditions secured by ATP or analogue feeding. An overview of the most significant results in this context from the perspective of active fluids is compiled in Appendix 3 (see 10). The last appendix, Appendix 4 (see 11), is devoted to tackle from the same perspective another parallel ambit to active fluids, i.e. the particularly fascinating terrain of cell tissues.

DOI: 10.1201/9781003302292-4

4.1　Active Gels Based on Filamentary Proteins

4.1.1　Active Gels Based on Actin Filaments

The literature based on actin-based gels is abundant, singularly in relation to the actin-cortex and its crucial implication in the cell structure and motility.[1] I completely refrain here from entering into this much investigated topic. On what follows, I limit myself to review a few references appeared in the last years devoted to actin-networks, and that were aimed at evidencing striking patterns of self-organization. I avoid to review as well the specific literature that is devoted to actin gels when analyzed from a mechanical or rheological point of view (see [260, 199, 364] for useful references).

The first report I am aware of featuring self-assembling patterns based on actin was published by Backouche et al. [15]. In vitro studies reported in this paper showed that myosin actively reorganizes actin into mesoscopic structures, but only in the presence of bundling *fascin*, a family of cross-linking proteins. The nature of the reorganization process is complex, exhibiting diverse patterns from active networks, asters and even rings depending on motor and bundling protein concentrations.

A later work was published from Bausch's group by Köhler et al. [200]. It reports the formation of active networks from a minimally reconstituted system, with a formulation similar to that of Backouche et al.. The chosen control parameter is the ratio of molecular motors (myosin), and (passive) cross-linker components (fascin). Actin filaments in the presence of fascin at a molar ratio 1:1, in absence of motors, assembled a network of stiff and rigid bundles, with a well-defined bundle thickness of a few tens of filaments per bundle, spanning lengths up to several hundreds of micrometers.

After switching from a passive to an active state by adding the molecular motors and ATP, the structure and dynamics of the network changes considerably. A prevalence of passive cross-linkers leads to a quasi-static network with minor reorganizations. Contrarily, in the regime where passive and active (myosin) cross-linkers are balanced, major changes appear. Instead of well-defined bundles, condensed and interconnected actin-fascin moving clusters with a broad range of size distributions and variable shapes emerge. Exploring a broad regime of cross-linker concentration, a complementary study reported critical behavior, manifested by a power-law distribution of cluster sizes, at the transition between the regime of dynamic clusters and that of an elastic connected network [6].

---

[1]For a comprehensive account see the review by Blanchoin et al. [38].

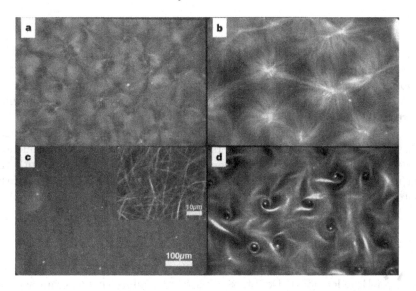

FIGURE 4.1
Different large-scale patterns formed through self-organization of microtubules and motors. (a) A lattice of asters and vortices. (b) An irregular lattice of asters. (c) Microtubules forming bundles. (d) A lattice of vortices. Image and caption text adapted from Nédélec et al. [273].

4.1.2 Active Gels Based on Microtubules

4.1.2.1 Historic Antecedents

Let's start with a couple of paragraphs reviewing some pioneering studies on the self-organization of microtubules in the form of active gels. Self-assembly of microtubules mediated by (kinesin) motors was reported in a seminal paper by Nédélec et al. [273] (see Fig. 4.1), following earlier observations by Urrutia et al. [417], and Stearns et al. [375].

Starting from an isotropic gel prepared in a quasi-two-dimensional geometry, asters were reported to form within a few minutes as kinesins accumulate in their centers. Asters locate the plus ends of the MTs at their center, contrarily to what would be observed by employing minus-end directed *dyneins*, when the polarity of the aster would be reversed.[2] Kinesin constructs were secured with streptavidin and can be seen as mobile cross-links between neighboring MTs. The length of the microtubules was stabilized with Taxol, although authors stated that this is not strictly necessary for aster formation. As a matter of fact, aster formation was proved to be robust in front of assembly and

---

[2]According to what is mentioned in the paper, the latter is the generic scenario in cells and cell-free extracts.

disassembly caused by the dynamic instability of MTs. In experiments conducted in absence of Taxol and performed in quasi-two-dimensional chambers right- and left-hand vortices were reported following from asters instabilities.

In unconfined geometries, a surprising variety of large-scale assemblies was reported that are built from asters and vortices, both considered as elementary motifs for the spatial organization of the microtubules (see Fig. 4.1). The final patterns depend on the initial concentration of the protein components. At low motor concentration a lattice of vortices was observed, but at slightly larger kinesin concentration lattices of asters formed. This study was complemented a little bit later with new experiments and computer simulations with the purpose of analyzing how the concentrations and dynamic parameters of the motors contribute to the collective behavior of the microtubule assembling [384].

4.1.2.2   The Brandeis Approach

Wide attention to protein-based active fluids spurred from the work of Dogic's group in Brandeis [337]. Doubtless, this report published ten years ago has turned to be central to the rapid development of the field of *biophysically inspired active matter*.

A typical preparation popularized by this group consists in formulating an aqueous solution of the microtubule/kinesin system with an added non-adsorbed depleting polymer. The robustness and versatility of this recipe has permitted to explore in the last years many different experimental aspects of this preparation, and, at the same time, has motivated numerous theoretical analysis. As a matter of fact, the observations reported by Sanchez et al. refer both to the active gel preparation, when assembled in bulk (three-dimensional) samples, as well as to its, more often portrayed, denser reconstitution in the form of a (two-dimensional) active nematic. This second design results from interfacing the original gel-preparation with a passive oil, and forms at the interface of the two immiscible phases. I will comment in this and next sections, respectively, the main characteristics of either preparation.

Stabilized MTs[3] with an average length of c.a. 1.5 μm get bundled under the action of a depleting agent based on polyethylene glycol (PEG). The employed molecular motors are biotin-labeled double-headed processive kinesins (K401) assembled into multimotor clusters by tetrameric streptavidin. These motors simultaneously bind and move along multiple MTs. Molecular motors take discrete steps (c.a. 8 nanometers length), a single kinesin motor moving in average about a micrometer along a filament before detaching. Kinesin clusters generate sliding forces between microtubules of opposite polarity, whereas no sliding force is induced between microtubules of the same polarity.[4]

---

[3]Stabilization of microtubules is achieved using GMPCPP. This non-hydrolysing analogue of GTP reduces the microtubule nucleation barrier, resulting in very short filaments [337].

[4]Let's comment on passing that the same protein mixture had been assayed a little bit earlier by the same group trying to mimic cilia-like beating of microtubules [338].

As a whole, this process triggers local extension of MT bundles, and lead to a permanent source of internal shear stresses fed into the system through ATP consumption. Average bundle length is 9 µm long, and their average cross-section contains between 10 and 20 MTs. Tubulin is copolymerized with fluorescent labeled monomers, producing MTs with 3% of labeled monomers for fluorescence observation. To stabilize the system an ATP regenerator is added, including enzyme mixture pyruvate kinase/lactate dehydrogenase (PK/LDH) and phosphoenol pyruvate (PEP).

Dilute suspensions were confined in quasi-two-dimensional chambers. No sliding at the level of isolated bundles indicates those that are polarity sorted. However, bundle domains of opposite polarity likely merge leading to stretching and eventual bundle disassembly (see Fig. 4.2). This is a clear signature of **extensile** behavior, in contrast with the more predominant **contractile** nature of actin/myosin gels, as observed in previously commented references [6].Thus, bundle dynamics undergo cyclically repeated phases of extension, buckling, fracture, and recombination, in permanent search for an ideal fully polarity-sorted state. Increasing the concentration of MTs leads to the formation of active networks, permeated with large-scale chaotic-like

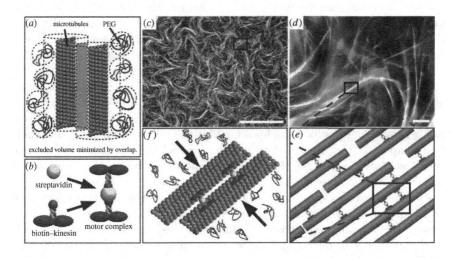

FIGURE 4.2
Active PEG-assembled microtubule network. (a) and (b): Schematic illustration of the basic components of the system. (c) Large field of view of the active gel under fluorescence, scale bar 250 µm. (d) Zoom-in region reveal classical polidispersivity in bundle structure, scale bar 10 µm. (e) Zoom-in showing microtubules and binding motors. (f) Scheme of the bundled MTs cross-linked with kinesin constructs. Image and caption text adapted from Henkin et al. [159].

fluid flows. These active networks turn to be very robust and easily tuned by varying the concentrations of their minimal components (microtubules, kinesin/streptavidin complexes, PEG and ATP).

Characterization by tracer dispersion in absence of ATP probed a passive viscoelastic network, with corresponding subdiffusive mean squared displacement (MSD). At intermediate ATP concentrations a crossover was observed from sub- to superdiffusive behavior at longer time scales, as a claimed signature of advective flows. Finally, ballistic behavior is characteristic of saturating ATP conditions.

From the same group and published later by Henkin et al. [159], an exhaustive study of the spatio-temporal characteristics of the active flows was performed, and dependences on the material parameters were analyzed in depth. More precisely, the recorded observables were the MSD, velocity autocorrelation and probability distribution functions, all of them based on trajectories of tracer particles. Valuable trends were extracted from these systematic observations, with, however, two main limitations.

First, in a top-down view, it is straightforwardly recognized that basic control parameters in the experiments might have very complex roles, influencing simultaneously several of the coarse grained attributes of the active system. For instance, changing the ATP concentration modifies the kinesin speed and, in turn, the intrinsic activity of the system, but might also expectedly alter the processivity of the motors and, as a result, the cohesiveness of the elastic material. In contrast from a bottom-up perspective, the emerging complexity of the assembled system makes it very difficult to assess an accurate quantitative evaluation of its constitutive properties from existing parameter values obtained at the level of single (motor and filamentary) proteins.

The alternative that consists in employing theoretical modeling has certainly proved highly successful for a qualitative understanding of active fluids (see Chapter 6), but, to our understanding, modeling is at this time not able yet to estimate consistently the rheological/mechanic constants of these active gels (viscosities or elastic constants to mention a few). Some of these considerations motivated a specific experimental analysis of the multiscale dynamics of microtubules in active nematics as recently reported by Lemma et al. [224]. From a theoretical perspective, a useful review dedicating a generic overview to the modeling of microtubule/motor protein assemblies was published by Shelley [356].

I finish this subsection dedicated to MT-based gels with a final remark. Similarly to our position in the previous section, I am not going to consider here microtubule-based active gels from the point of their rheological properties. However, interest readers may find worth consulting a very recently published paper by Gagnon et al. [117]. Using a combination of microscopy and rheology, authors quantify the relationship between the microscopic dynamics and the bulk mechanical properties of these non-equilibrium networks, and discover a principle of shear-induced gelation. More precisely, the main result is that the network viscosity first increases with the imposed shear rate before transitioning back to a low-viscosity state. Moreover, the speed of molecular motors controls the non-monotonic shear-dependent viscosity.

4.2 Two-dimensional Active Nematics

As commented in the previous paragraphs, active gels provide striking realizations of filamentary protein-based active fluids. However, deeper insights, singularly from a quantitative point of view, can be obtained by working with two-dimensional denser realizations that have become an archetype in theory and experiments. Their textures display remarkable orientational order, albeit commonly punctured by a considerable number of topological defects. This qualifies these preparations as active nematics (ANs). Just for the sake of continuity in the discussion, I start by examining ANs based on microtubules, which is the most conventional scenario we find in the literature. At the end of this section, I will comment on experiments with active nematics prepared from actin.

4.2.1 Active Nematics Based on Microtubules

The classical active nematic phase of the microtubules/kinesin system can be obtained in a flow cell containing a flat two-dimensional oil-water interface, stabilized with a PEG-based surfactant, in contact with the active gel, as demonstrated by Sanchez et al. [337]. With the course of time, the active material is progressively depleted from the bulk towards the interface, eventually forming a dense liquid crystalline-like phase, similar to a monolayer of locally aligned bundles.

However, two striking characteristics distinguish active nematics from their counterparts in passive liquid crystals: i) the existence of intrinsic large-scale flows that permanently stream the bundled MTs, and ii) the presence of topological defects (see Sect. 2.2.3), that appear as regions depleted of MTs, that are continuously created and destroyed following bundle reorganization. Lifetime of defects, depending obviously on activity conditions, corresponds typically to tens of seconds, as an order of magnitude. Notice also that these episodes of pair defect creation and annihilation, taking place at equilibrated steady rates are always consistent with the preserved balance of the total topological charge.

Defects in ANs bear semi-integer topological charges, plus or minus 1/2, according to their nematic symmetry (polar systems would display, conversely, integer-like disclinations in the form of asters or vortices commented in the previous section). As a whole, the system of bundled MTs appears to be permanently trapped in a sort of turbulent-like state. This is a striking example of what has been known as **active turbulence**, commented earlier in relation to forced colloids (see Sect. 3.2.1.2), and one of the most celebrated concepts emerging from the study of active fluids (see Sect. 5.3). In spite of its apparent unpredictability, this chaotic regime is endowed with a characteristic length scale for defect separation and flow organization. One refers to it as **active length scale**, denoted l_a, a concept that will be much discussed on what follows.

Positive defects have a parabolic-like shape, while negative ones adopt a triangular-like configuration. This confers an intrinsic motility to the former (at speeds typically of a few micrometers per second), while the negative disclinations are simply advected by the active flows present in the system. This singular defect dynamics is again noticeably different from what is known in standard liquid crystals. In a classical context, defects mostly appear as a consequence of internal frustration originated, either from extended or local boundary conditions, or from external sample-scale interventions, like temperature, or flow-induced, texture changes. Here, contrarily, defects are created in pairs from the extension and buckling of uniformly aligned nematic domains. Lines connecting defect pairs with compensated charge appear sometimes very prominent and seem to fracturing randomly the system. These *crack lines* first extend (extensile behavior) and later self-heal above a critical length. Isolated defects meanwhile remain unbound and, after seemingly erratic motions, eventually annihilate by pairs, restoring in this way (local) uniform alignment (see Fig. 4.3).

Ordering of defects has become itself a question of debate in the recent literature. DeCamp at al. published a paper with some contradictory results [79]. The reported observation pointed out to an apparent (nematic) orientational order of the active defects. Thousands of defects over centimeter-scale distances were tracked to extract both their position and orientation. This was possible by using standard fluorescence microscopy in connection with a technique (LC-PolScope) that measures both the orientation of the nematic director at a pixel resolution and the effective thickness (in terms of the birefringence amount or equivalent optical retardance) of the nematic layer.

The complementary computational model employed Brownian dynamics of rigid spherocylinders, omitting long-range hydrodynamic interactions. To mimic the extensile behavior of the microtubule/kinesin system, the length of the constituent rods was supposed to increase at a constant rate, up to a maximum length when they split, with simultaneous merging of two other rods in order to keep total particle number fixed. In simulations, $+1/2$ defects also display system-wide orientational order. However, in marked contrast with experimental observations, defects appear to align with polar, rather nematic symmetry.

This question has been reassessed very recently in a experimental and modeling joint work published by Pearce et al. [302]. These authors report that $+1/2$ disclinations have short-range antiferromagnetic alignment, as a consequence of the elastic torques originating from their polar structure. The presence of intermediate $-1/2$ disclinations, however, turns this interaction

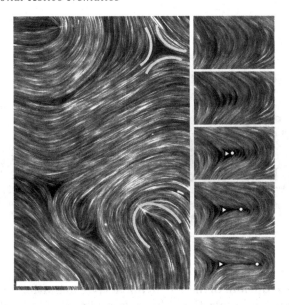

FIGURE 4.3
Defect dynamics in a two-dimensional active nematic. Left panel: Positive and negative defects as viewed in the two-dimensional nematic assembly of microtubules. Right panels: Fracturing of a locally ordered nematic domain following the dynamics of a pair of (marked) defects of opposite charge and topology. Image and caption text adapted from Guillamat et al. [150].

from antialigning to aligning at scales that are smaller than the active length scale. More importantly, no long-range orientational order is observed.

Up to now I have been essentially concerned with characteristics of the active nematic that have mostly to due with the orientation textures of the material. However, it is worth remembering that the active nematic preparation is equally distinctive in its flowing features. An exhaustive theoretical study by Giomi [132] dedicated to analyze the geometrical aspects of the turbulent regime of the AN predicted that this flowing state is characterized by an exponential distribution of vortex sizes (see also Sect. 6.2.3). In turn, this exponential distribution is built on a length scale that is identified precisely as the active length scale mentioned earlier. I next examine this question from a couple of experimental results that were inspired, and in fact confirmed, Giomi's predictions.

The simplest approach to quantitatively describe the regime of active turbulence consists in locating and measuring the area of each vortex in the AN, as reported by Guillamat et al. [149]. To this end, the local instantaneous velocity of the active flow is evaluated from a sequence of micrographs. Velocimetry data are used to obtain values of the Okubo-Weiss parameter,

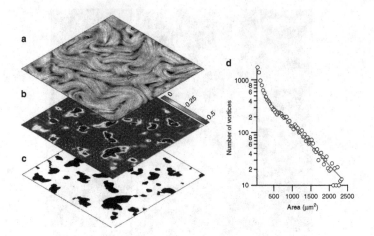

FIGURE 4.4
Vortical distribution of flows in an active nematic: Experimental determination and statistical characteristics. (a) Confocal fluorescence micrograph (375×375 µm^2) of the AN in the turbulent regime when in contact with an isotropic oil. An example of the proliferating $+1/2$ (parabolic) and $-1/2$ (triangular) defects is sketched on top of the image. (b) Instantaneous flow field (vector plot) and computed Okubo–Weiss parameter field (density plot, arbitrary units) (see text). (c) Binary image corresponding to the Okubo–Weiss field. (d) Statistical analysis of the distribution of vortex sizes. Image and caption text adapted from Guillamat et al. [149].

$OW = (\partial_x v_x)^2 + \partial_y v_x \cdot \partial_x v_y$. The latter provides with a standard criterion for vortex location, by considering the spatial extension of each vortex to be bound by the condition $OW < 0$ ([132] and references therein). A statistical analysis reveals indeed an exponential distribution of vortex sizes, as predicted by Giomi [132] (see Fig. 4.4). From this distribution one can extract a characteristic vortex area, A^*, and, subsequently, the active length scale is identified as $l_a \propto (A^*)^{1/2}$.

A second more exhaustive analysis along this direction was published more recently by Lemma et al. [223]. Again, experimental results confirmed the prediction by Giomi [132], and the earlier just commented results reported in [149]. Moreover an interesting added value of Lemma et al.'s work was that the dependence of l_a on the ATP concentration was probed, suggesting a possible mapping between the ATP concentration and the active stress. Such relationship is argued to be based on the Michaelis-Menten kinetics that governs the motion of individual kinesin motors. Equal-time velocity and vorticity autocorrelation functions were evaluated as well. Vortex size distributions, and velocity and vorticity correlators provide three independent methods of evaluating l_a. It was found that values extracted from these analyses scale similarly with system control parameters. In particular, at low ATP concentrations the measured active length scale decreased with increasing ATP, and plateaued above a critical ATP concentration.

An important issue towards the understanding of this distinctive turbulent-like dynamical regime in ANs is to unveiling its mechanism, and in turn to decipher the origin of its characteristic length scale l_a. I examine this question in the following paragraphs following results reported by Martínez-Prat et al. [253]. As a matter of fact, quite a number of theoretical studies had predicted that the extensile nature of active nematics makes their aligned configuration prone to undergo **bend-like instabilities**. More precisely, one might find interesting to consult in this respect the paper by Aditi Simha et al. [1] on a theoretical study of active suspensions, or the contribution by Voituriez et al. [423] in relation to models of active gels to be commented later on (see Sect. 6.2). Moreover, this instability of aligned MT bundles is also apparent from direct observations of the interfaced active nematic, as mentioned above in relation to Sanchez et al. original experiments (see Fig. 4.3).

The ultimate goal of Martínez-Prat et al. was to fetch the instability precisely at its onset, and to confirm whether it brings about a selection mechanism that could explain the build-up of the active length scale l_a.[5] The experiment by Martínez-Prat et al. [253] (see Fig. 4.5) consisted in choosing as specific initial condition a well-aligned configuration of the AN. Although this may appear a trivial task, it is quite a subtle issue since one must remember that the active nematic is conventionally prepared from the interfacial depletion of a completely unstructured three-dimensional and non-aligned gel. In the experiments by Martínez-Prat et al., initial alignment was achieved by replacing the original closed-cell design in [337] by a more versatile open setup. A custom polydimethylsiloxane (PDMS) block containing a cylindrical well and bound to a support plate was used. After filling the well with silicone oil, the aqueous gel was injected between the bottom plate and the oil. This results in an aqueous layer between 100 and 200 µm depth underneath an oil phase of $1 - 2$ mm depth. The AN is progressively formed at the PEG-stabilized interface as expected, and displays the usual regime of seemingly chaotic streaming currents and permanent defect dynamics typical of an unbounded active nematic.

AN alignment is further imposed following the radial flows that are originated at a contact between a capillary tube and the active layer. After removing the tube, the initially radial structure starts to be disrupted by the spontaneous buckling of the aligned extensile material, leading to the local development of bend-type distortions. The instability organizes into a pattern of periodically spaced concentric crimps that evolve into circular walls.

---

[5]The bend instability in 3d-active liquid crystals is fundamentally different from what is analyzed here. In absence of confinement, the fastest-growing deformation has an infinite wavelength. However, confinement effects have been demonstrated to introduce a boundary-dominated wavelength selection mechanism, as recently reported from experiments and modeling by Chandrakar et al. [56].

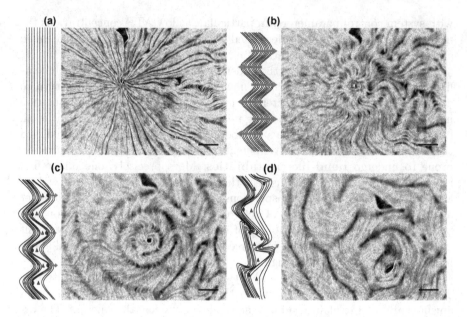

FIGURE 4.5
Route to the two-dimensional turbulent state of the active micro-tubule/kinesin system . Fluorescence micrographs showing the spontaneous evolution of a radially aligned active nematic towards the turbulent regime. (a) Radially aligned active nematic. (b) Onset of bend-like instability. (c) Nucleation of pairs of defects within the crimps. (d) Final dismantlement of the radial structure towards the turbulent regime. Image and caption text adapted from Martínez-Prat et al. [253].

Elastic stress accumulated in the walls is released through the nucleation of pairs of defects that align along dark circular lanes. Subsequent defect unbinding dismantles completely the circular pattern and reverts the system into a turbulent regime. At low activity it may also happen that the instability is repeated several times in cascade along orthogonal directions before yielding the fully turbulent state.

This instability is assessed quantitatively in terms of the most unstable wave number q^* and its corresponding growth rate Ω^*. Varying the complete set of control parameters (protein, ATP and depleting agent concentrations) over the widest possible ranges, yet allowing AN formation, it is found that those determinations collapse into a single universal curve $\Omega^* \propto (q^*)^2$. Moreover, qualitative experimental arguments and theoretical modeling permit to conclude that q^* scales similarly to l_a^{-1} ([253]).

The turbulent state of the AN somehow echoes an scenario of *chaotic mixing* [11]. In this respect, the idea of applying canonical concepts typical of the theory of chaos to active systems was raised very recently from the paper published by A. J. Tan et al. [392]. Local observables such as Lyapounov exponents, as well as more global quantities like the so-called *topological entropy* admit to be evaluated. Lyapunov exponents measure the rate at which nearby fluid parcels separate from one another.[6] On the other hand, the topological entropy measures the asymptotic (in time) exponential growth rate in the length of a material curve as it is stretched within the fluid. The latter is further associated with *braiding modes* of the (positive) defects that are conceptually thought as virtual stirring rods. In this particular respect, the singularly interesting idea that is invoked in the paper is that the self-driven active fluid spontaneously creates a set of defects that move in a particular way to produce exactly the topological entropy needed to accommodate local stretching at the level of bundled MTs.

According to the quoted results, the topological entropy is slightly larger than the Lyapounov exponent, demonstrating a residual departure from statistically homogeneity of the pattern of active flows. As expected, such measures increase with activity (ATP-contents). However, strikingly enough, adimensionalizing these observables in terms of a characteristic time scale (prior to adimensionalization they bear inverse time dimensions), turn out to be independent of ATP concentration. As claimed by the authors, this suggests that these dimensionless quantities may be universal signatures of the fully developed turbulent state of the active nematic.

4.2.2 Active Nematics Based on Actin Filaments

The preparation of actin-based nematics, assembled in their two-dimensional form totally similar to the most conventional microtubule system, was reported by R. Zhang et al. [453]. More specifically, these authors used filamentous actin (F-actin), with a width of c.a. 7 nm and persistence length between 10 and 17 µm (two orders of magnitude smaller than for microtubules). Methylcellulose was used as crowding agent, instead of PEG. F-actin is further depleted onto an oil/water interface thereby forming a quasi-two-dimensional nematic system. F-actin's average contour length can be tuned from less than 1 µm to more than 10 µm by the addition of variable concentrations of a capping protein that limits polymer growth.

As a matter of fact, the system in Zhang et al. preparation can not be considered strictly as active, since myosin motors are absent, and ATP is used solely in the polymerization protocol. Nevertheless, this initial study allowed to extract elastic parameters from defect morphology. The main result in

---

[6]Lyapunov exponents were obtained from the velocity-gradient tensor field, computed from particle image velocimetry (PIV) using fluorescence microscopy images. The topological entropy was measured in up to three different ways: directly using beads attached to the microtubules, by computing the separation of neighboring topological defects, and finally, as a completely independent way, by tracking the braiding motion of the topological defects about one another.

[453] is that the defect morphology transitions from a U shape to a V shape as the filament length increases, indicating the relative increase of the material's bend over the splay modulus (see Sect. 2.2.3). Furthermore, through the sparse addition of rigid microtubule filaments, authors demonstrated a linear increase of the bend elastic constant as a function of microtubule filament density. Observations were rationalized in terms of a classical theoretical approach for conventional liquid crystals, beyond the common one-elastic constant approximation, and incorporating back-flow effects. More precisely, a hybrid lattice Boltzmann method is used to simultaneously solve a Beris-Edwards–like scheme for the nematic order parameter and the corresponding momentum equations.[7]

In an accompanying next paper [211], the system was reexamined, this time under myosin activity. Myosin assembles into bipolar filaments of several hundred motor heads that exert (extensile) stresses on antiparallel actin filament pairs. Length of actin filaments is short in this case, of the order of 1 μm, this likely explaining the relative dominance of extensile vs. contractile forces in this preparation. An apparent difference with the conventional microtubules/kinesin system is the lower defect density in the latter. Moreover, it is shown that the apparent elastic properties of the system (for instance considering the ratio of bend-to-splay elastic moduli) are altered considerably by increased activity, leading to an effectively lower bend elasticity. Not only changes at the level of defect morphology can be attributed to activity fine tuning, but also the defect dynamics can be similarly altered. As a matter of fact, unbinding of defects, i.e. the apparent repulsion of oppositely charged disclinations is observed above an activity threshold. Notice that this defect unbinding is a non-classical distinctive feature stressed as well when describing the dynamics of the microtubule-based active nematics. Again, experimental observations are complemented with numerical simulations of a theoretical scheme similar to that provided in [453].

4.3 The Effect of the Interface on Two-Dimensional Active Nematics

4.3.1 Aqueous Active Nematics Interfaced with Isotropic Oils

Up to now active fluids with built-in nematic symmetry have been described as totally independent of the characteristics of the contacting fluids forming the interface where ANs reside. It is worth remembering that in the conventional preparations commented so far, the active layer is in close contact with two

---

[7]The active version of this theoretical approach is discussed at length in Sect. 6.2.

distinct fluids, an aqueous phase that contains the active material (filamentary and motor proteins and their corresponding buffer solutions), and a totally passive oily phase. It is thus reasonable to envisage some important effects on the textures and flow patterns of the active nematic that may arise from the characteristics of the pair of ambient fluids. Particularly, we expect differences singularly originated in the viscosity contrast between the aqueous and the oil phases. On what follows, I report on some experimental efforts trying to unveil this important, and, unfortunately, many times forgotten aspect of the behavior of the two-dimensional active material. The most obvious intervention is to change the oily phase, to tune in a controlled way such a viscosity contrast. I postpone for later consideration a deeper, and more suggestive, modification of the rheological properties of the contacting non-aqueous phase after replacing the isotropic oil with a passive liquid crystal material.

An original exhaustive experimental study within this announced perspective was published by Guillamat et al. [150]. The aim of this research was twofold. First to investigate qualitatively, and in quantitative terms, the role of the hydrodynamic coupling of the active layer with the oil phase. The chosen observables were those that are more characteristic of the active material, i.e. the average speed of positive defects and the defect density itself. The second and more ambitious idea was to use these experiments, in connection with an hydrodynamic model, to probe the shear viscosity of the AN.

The employed experimental setup was again based on an open-cell design, similar to that reported above in relation to the work by Martínez-Prat et al. [253]. From a practical point of view, this specific protocol does not demand the use of a low viscosity oil, as in the conventional flow cell, thus allowing to explore nearly five orders of magnitude of viscosity contrast between the interfacing oil and the aqueous bulk phase. Activity (ATP contents) and the rest of components of the active solution were kept at unaltered concentrations for the whole series of experiments. Easily tracked positive defects were used as tracers of the active flows.

Remarkable effects were detected on the textures and streaming currents as oil viscosity changes. More specifically, the number of defects increase, while their velocity decrease, as viscous damping of the active film raises for oils of higher viscosities. Textures appear largely fractured with defect core sizes slightly larger under higher viscosity contrast (see Fig. 4.6) Speculatively, one could argue that both features might be compatible with a smaller amount of condensed microtubules and corresponding motors at the monolayer as oil viscosity increases. Yet the fluorescence signal, albeit integrating the response from the accumulated thickness of the active boundary layer, is apparently enhanced when increasing oil viscosity.

Inverse trends in the density (n) and velocity (v) of the defects could be combined to extract an invariant under oil changes expressed as $nv^{1/2}$. A simple scaling argument justifies this particular system invariance. One starts from the already commented balance between defect creation and annihilation

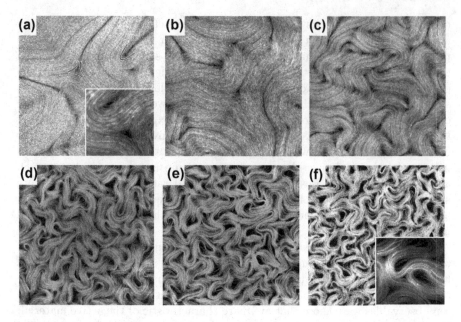

FIGURE 4.6
Active nematic preparation in contact with silicone oils of different viscosities. Starting with a contacting oil of 5×10^{-3} Pa s, viscosity is increased from panel (a) to panel (f) by nearly one order of magnitude in each panel. Insets corresponding to extreme conditions zoom in topological defects of either charge. Image and caption text adapted from Guillamat et al. [150].

rates. The former, denoted per unit area $\mathcal{R}_c$, is expressed as $\mathcal{R}_c \propto (l_a^2 \tau_a)^{-1}$, where l_a stands for the active length scale l_a mentioned above, and τ_a denotes a characteristic active time scale. Physical arguments, apart from existing theories (see Chapter 6), suggest a simple dependence for the active length scale in the form $l_a \propto (K/|\alpha|)^{1/2}$, with K the elastic constant of the active nematic material, and α the activity-dependent coefficient of the stress. The active time scale τ_a is assumed to go as $|\alpha|^{-1}$ times a viscosity parameter. Conversely, the annihilation rate $\mathcal{R}_a$ can be estimated in terms of an effective cross-section σ as $\mathcal{R}_a \propto \sigma v n^2$. Balancing both contributions leads to the formal scaling (neglecting viscosity and cross-section terms) $n^2 v \propto (\alpha^2/K)$, a ratio independent of the oil viscosity that it is confirmed constant in the experiments.

A comment on the piece of modeling that permitted to estimate the value of the viscosity of the active film is postponed to the discussion of a couple of theoretical approaches, specific to interfaced active nematics, that will be presented in Sect. 6.3.

If active turbulence is one of the hallmarks of the dynamics of an active nematics, as commented in the previous section, it seems natural to investigate this issue in relation with the properties of the interface where it resides. This was the goal of experimental studies reported very recently by Martínez-Prat. [252]. The idea of this letter work was to extract experimentally the spectrum of kinetic energy, and to analyze its eventual dependence on the viscosity contrast established at the oil/water interface. Similarly to the experiments reported earlier by Guillamat et al. [150], the designed cell permitted to change the viscosity of the oil by more than three orders of magnitude. A series of representative results are shown in Fig. 4.7.

Upon increase of the oil viscosity, the entire kinetic energy spectrum decreases, which is consistent with the previously observed decrease of flow speed [150]. At low oil viscosities, the spectrum features at least three regimes: a large-scale (small-q) regime that is followed by a peak, an intermediate regime, and a crossover to a small-scale (large-q) regime [panel (a)]. As oil viscosity increases, the peak shifts to smaller scales, expanding the range of the large-scale regime and shrinking the intermediate regime until it can no longer be observed for high oil viscosities [panels (b) and (c)]. In parallel to these changes in the flow properties, we also observe a higher density of defects for higher oil viscosities (top insets), a feature commented earlier in relation with the observations by Guillamat et al. [150].

Other quantities related to the flow structures were extracted as well from these experiments [252] (see Fig. 4.7). In particular, the vortex area distributions display exponential tails [panel (d)], in agreement with both the theoretical assumption [132] and previous experiments by Guillamat et al. [149]. Here, we find that this feature does not change with the oil viscosity. We also measure the correlation functions for the velocity and vorticity fields [panels (f) and (g)], and obtain the corresponding correlation lengths (see insets in the corresponding panels), which exhibit dependencies on the oil viscosity that are very similar to that of the vortex size. This observation suggests that these lengths are all proportional to one another. Later in Sect. 6.3, all these results will be discussed in relation to an appropriate model of interfaced active fluids.

4.3.2 Aqueous Active Nematics Interfaced with Anisotropic Oils

Having in mind what I just commented relative to the effects that the interfacing oil has on the textures and flows of an active nematic, it seems natural to take a step further and consider replacing the isotropic fluid by an anisotropic oily component, i.e. a (thermotropic) liquid crystal (LC). This is precisely the context of a couple of works published by Guillamat et al. [149] and [147].[8] In both cases, the employed thermotropic LC displayed a smectic (SmA), rather

---

[8]Additional details can be found in the pair of complementary publications [148] and [146]).

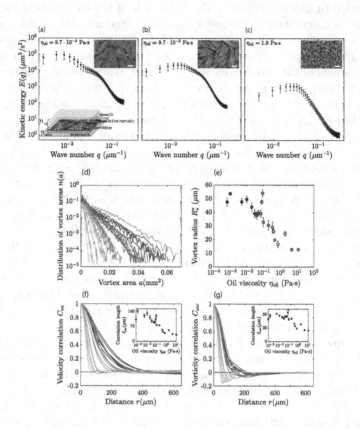

FIGURE 4.7

Kinetic energy spectra, vortex size distribution and velocity correlation functions of an active nematic interfaced with oils of different viscosities. Kinetic energy spectra of turbulent flows in an active nematic film in contact with an oil layer of low (a), intermediate (b), and high (c) viscosity. In each panel, the top inset shows a representative microtubule fluorescence micrograph. Scale bar is 100 μm. The bottom inset in (a) shows a schematic of the experimental system. Panel (d): Vortex area distributions in the active turbulence regime for 20 different oil viscosities. Panel (e): Mean vortex radius obtained from the exponential tails of the vortex area distributions in (d). Panels (f) and (g): Spatial autocorrelation functions of the velocity (f) and vorticity (g) fields, for different oil viscosities. The insets show the corresponding correlation lengths, defined at maximum half value (colored image in original version). In panels (d), (f) and (g), curves from right to left correspond to increasing oil viscosity. Image and caption text adapted from Martínez-Prat et al. [252].

than the most conventional nematic, phase, the difference being that in the second reference the effects of an externally imposed magnetic field were analyzed. Smectic, compared with nematic configurations, are lower temperature liquid crystalline phases depicting a certain degree of layered positional order on top of the characteristic orientational order of the nematic phases (see Sect. 2.2.2). The rationale behind employing smectic phases in this context is the marked anisotropic viscosities along and perpendicular to the layers of the molecular orientation.

Let's consider first the active nematic prepared in the usual way and interfaced with octyl-cyanobiphenyl (8CB) (see Fig. 4.8). This mesogen features mesophases at temperatures compatible with protein activity, in particular it adopts a smectic-A phase between 21.4°C and 33.4°C (the latter corresponds to the smectic/nematic transition temperature). At room temperature, free energy minimization results in the SmA phase spontaneously self-assembling into polydisperse domains, known as *toroidal focal conic domains* (TFCDs) [287], that organize into a fractal tiling known as *Apollonian gasket* [36] (see Fig 4.8 (a)). At the water/LC interface, TFCDs have a circular footprint, and are formed by concentric SmA planes perpendicular to the interface. Thus, 8CB molecules, which are both parallel to the water layer due to the interaction with the used surfactant, and perpendicular to the SmA planes, orient radially in concentric rings.

Crucially, molecules in the SmA phase can diffuse freely within a given smectic plane but their transport is severely hindered in the direction perpendicular to the planes. Although 8CB is, from a macroscopic point of view, a liquid, the local interfacial shear stress probed by the active material when in direct contact is markedly anisotropic, forcing active stretching of the MT bundles to occur preferentially along circular trajectories, centered in TFCDs. Swirling laminar currents now evolve within the interfacial domain limits, segregated from the rest of the large-scale flows in the system (see Fig. 4.8 (b, c)). This results in the size distribution of swirls to become commensurate with the Apollonian gasket (see Fig. 4.8 (d)), at odds with the exponential distribution reported earlier for the conventional active nematic in contact with an isotropic oil (see Sect. 4.2.1).

Looking at the figure 4.8 it is clear that trapping of the streaming currents is only possible for TFCD above a minimal size. This can be understood as a pure topological effect. Circular entrainment of the active material creates a total defect topological charge +1, most easily realized in terms of a pair of +1/2 defects. Thus, the minimal size is nothing but a measure of the distance between defects, of the order of the active length scale . Other more complex configurations of trapped defects were also observed and reported in the original paper. Finally, a striking observation for ANs in such soft constraining conditions is illustrated in the second series of panels of Fig. 4.8. What is observed there is the previously commented bend-like instability affecting aligned, albeit this time curved, bundled microtubules, time-periodically

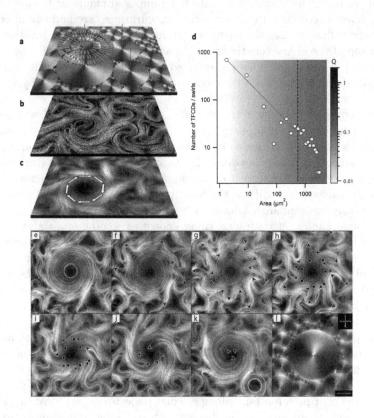

FIGURE 4.8
An active nematic in contact with a passive smectic-A liquid crystal.
(a) Confocal reflection micrographs of the water/liquid crystal interface. The diagram illustrates the arrangement of the 8CB molecules at the interface, (b) Confocal fluorescence micrograph of the AN, (c) Time averaged micrograph. Arrows indicate the direction of the circular flow. Line segments showing swirls in registry with underlying TFCDs. Field of view is 240 µm wide, (d) Size distribution of TFCDs in a 1 mm² window. The solid line is a power law fit to the data. The background grayscale (color code in original version) corresponds to measures of the winding number Q for bundled microtubules, as a function of domain area. The dashed vertical line marks the characteristic area of the unconstrained turbulent regime. The second series of panels shows micrographs of the AN swirls constrained by a TFCD, displaying the instability of aligned MTs. Scale bar 50 µm. Image and caption text adapted from Guillamat et al. [149].

destabilizing the outer corona assembled within the domains imprinted by the passive SmA liquid crystal interface.

The scenario reported in the second paper [147] is even more remarkable. The active nematic is again interfaced with 8CB but this time the passive LC is allowed to thermally transit from the nematic to the SmA phase under the orienting effect of a magnetic field acting through the positive diamagnetic anisotropy of the LC. There is no observed difference with respect to the conventional isotropic oil when the passive LC is preliminarily configured in the nematic phase (see Fig. 4.9 panels (a) and (b)), but a very abrupt change happens when it adopts the SmA configuration by lowering the temperature. The contacting active nematic rapidly rearranges due to the new boundary conditions, so that the chaotic filament orientation is now regularized into parallel stripes of uniform width aligned perpendicularly to the magnetic field (see Fig. 4.9 panels (c) and (d)). Fluorescence microscopy indicates that the bright stripes consist of densely packed MT bundles, whereas intercalated dark lanes incorporate the cores of moving defects. Strings of alternated $+1/2$ and $-1/2$ defects align and move in antiparallel directions in adjacent lanes (see Fig. 4.9 (d)).

The alignment process is reversible and versatile. By cycling the temperature above and below $33.4°$C, the active nematic returns to the disordered state when it is free from the interfacial alignment, and a new steering direction can be arbitrarily chosen by rotating the magnetic field (see Fig. 4.9 panels (e) to (h)).

Interpretation of these results follows the same ideas announced previously. The SmA planes are perpendicular both to the 8CB/water interface and to the field (see Fig. 4.9 (i)), conforming a *bookshelf* arrangement [287] that defines an easy-flow direction when sheared along the planes, but responds as a solid to stresses exerted in the orthogonal direction. As a consequence, the active nematic encounters an interfacial viscosity that is much higher for flow along the magnetic field than perpendicular to it, resulting in the observed alignment (see Fig. 4.9 (panels (j) and (k)). Scaling relations for the lane periodicity and flow velocities can also be extracted in terms of the ATP contents. Periodically, this well-aligned structure is interrupted by bursts of new born defects originated in the aligned regions. These episodes feed the circulating lanes with new pairs of defects in a continuous manifestation of the underlying and always present bend-like instability of the aligned material.

The just commented steering scenario lead us to raise two important remarks. First, I claim that the reported observations may open new promises for the eventual control of biological fluids by interfacing them with liquid crystals. As a matter of fact, a very recent realization of this principle, in the context of a proposed strategy of control of human fibroblast cells monolayers by a liquid crystal elastomer, was published by Turiv et al. [416].

The second remark is dedicated to a more subtle, and yet totally unexplored question. Beyond the usual consideration of aqueous/oil interfaces to assemble biofilament-based active nematics, one could wonder whether there

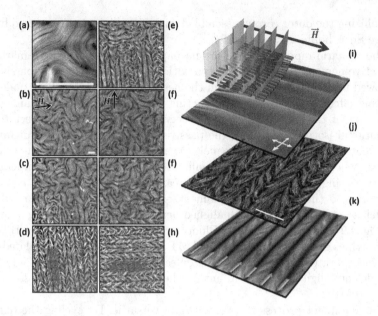

FIGURE 4.9

Alignment of active flows with a magnetic field. (a) Fluorescence micrograph of the active nematic with a pair of +1/2 and −1/2 defects highlighted. Panels (b) to (h): Fluorescence micrographs with different configurations of the active nematic in the presence of a 4 kG uniform magnetic field. (b) The active fluid is initially in contact with nematic 8CB, which is transited into the lamellar smectic-A phase (c) under a horizontal magnetic field. (d) The active nematic aligns perpendicularly to the field. By temperature cycling above (panels (e) and (f)) and below (panels (g) and (h)) T_0 under a vertical magnetic field, the active nematic is now realigned in the orthogonal direction (h). Pairs of aligned defect lanes are highlighted in panels (d) and (h). (I) Polarizing optical micrograph and configuration of the underlying molecular planes in the SmA phase of the passive liquid crystal. (j) Fluorescence confocal micrograph revealing the correlation between the aligned active nematic and the anisotropic SmA phase. (k) Time average of the dynamic pattern. The arrows depict the antiparallel flow directions along the lanes of defect cores. Scale bars 100 μm. Image and caption text adapted from Guillamat et al. [147].

is a similar effect in case of frictional damping coming from a solid substrate in contact with the microtubule preparation. Although to my understanding this issue is pending of a deeper scrutiny, our own preliminary observations seem to indicate a more dramatic effect in this situation than the just reported when replacing isotropic with anisotropic oils. Strikingly enough, totally reproducible, though non-systematically designed, experiments seem to indicate that the AN itself does not form in contact with a solid substrate, but rather the active preparation remains in the bulk gel phase without being depleted to the interface. In a sense, this seems to suggest that the textures and flows largely referred to in this chapter for the microtubule/kinesin system need to be realized in contact with a stress-absorbing fluidized phase.

I close this section by referring to a more severe intervention on interfaced microtubule-based active fluids. It consists in considering the influence of externally imposed stresses created by a magnetically actuated rotation of disk-shaped colloids proximal to the active layer, as considered recently by Rivas et al. [327]. The impact on the local motion of $+1/2$ defects is demonstrated through the merging of two such defects leading to the formation of a $+1$ topological vortex.

4.4 Effects of Spatial Confinement

In this section, I will examine the role of spatial confinement on protein-based active fluids. I distinguish two kinds of confinement that I loosely qualify respectively as *soft* and *hard*, treated separately in the following subsections. The first scenario refers essentially to active fluids when encapsulated in the form of active droplets. The primary attention will be devoted to situations when depletion, totally analogous to what occurs on planar interfaces, accumulates the active nematic at a curved interface. Topological defects present in these cortical active flows will necessarily have to accommodate specific topological constraints that will be examined with some detail.

In the second context, I will refer to effects of hard (geometric) constrains applied by solid bounding surfaces on either active nematics or active gels. The perspective there will be to analyze eventual taming conditions of the, otherwise, chaotic active flows typical of unbounded situations.

4.4.1 Encapsulated Active Nematics

The consideration of encapsulated AN flows has a basic interest arising, as announced previously, in the interplay between topology and activity . Moreover, this analysis may open new venues to design robust experimental protocols to prepare **active emulsions** [171]. On what follows, I examine this scenario by separating the cases of spherical and toroidal topologies.

Already Sanchez et al. [337] in their original report minimally addressed the possibility to encapsulate the tubulin/kinesin active gel as (conventional) oily emulsions.[9] The main reported observation was that for big enough droplets (typically with diameters of a hundred microns) the AN can be effectively depleted towards the interface, while the droplet interior is essentially devoid of microtubules. This results in droplet-organized cortical flows, that even might eventually lead to persistent autonomous droplet motion when the latter are in contact with contacting surfaces. The scenario of encapsulated AN, this time using vesicles, was analyzed in depth in a later specific report by Keber et al. [193]. I will comment more extensively this situation on what follows.

It is well-known in a classical Liquid Crystals context that nematic arrangements on curved surfaces are locally frustrated and, thus, they unavoidably generate defects. The *Poincaré-Hopf theorem* implies that the accumulated topological charge must be equal to the *surface Euler characteristic* $\tilde{\chi}$ [189]. On the other hand, according to the Gauss-Bonnet formula, the surface Euler characteristic can be calculated as $\tilde{\chi} = (1/2\pi) \int \mathcal{K} dA$, where $\mathcal{K} = (R_1 R_2)^{-1}$ denotes the *Gaussian curvature*, written in terms of the principal radii of curvature of a curved surface. For a sphere of radius R, $\tilde{\chi} = 2$. In the case of a two-dimensional nematics, the most trivial way to satisfy this constraint supposes to split the total $+2$ charge on four topological $+1/2$ defects.

This naturally applies also to an active nematic, with the obvious difference that in this case defects are intrinsically propelling on the curved surface. Still, in equilibrium conditions, it is known that defect arrangements are largely degenerated in liquid crystal samples. In particular, the final adopted configuration will minimize the free energy of the system that in any case will depend on the relation of elastic constants. For a two-dimensional (cortical) classic nematic, under the usual assumption of equal elastic moduli for bend and splay distortions, the chosen configuration corresponds to the four $+1/2$ defects located at the vertices of a tetrahedron inscribed within the sphere. This was theoretically predicted in a couple of papers by T. Lubensky and D. R. Nelson [241, 276], and further experimentally confirmed for spherical nematic shells by Lopez-Leon et al. [239]. This result admits a straightforward interpretation since defects located at the corners of the tetrahedron maximize their mutual separation and minimize elastic distortions of the liquid crystal matrix and, in turn, the associated pair-repulsion forces.

The question is how this latter result translates into an AN preparation. Active stresses propel the positive defects and thus hinder them to satisfying simultaneously minimal interactions, while preserving fixed vector velocities, the final result being the appearance of complex defect trajectories. The latter can be tracked with confocal microscopy revealing in some cases an oscillatory dynamics that switches the defect configuration between tetrahedral and planar, arrangements, with a frequency that depends on the ATP concentration and the droplet size (see Fig. 4.10).[10]

---

[9]Droplet confinement of active microtubule networks inducing large-scale rotational cytoplasmic flows has been examined by K. Suzuki et al. [385].

[10]Naturally this applies to small enough vesicles. For large specimens, curvature effects

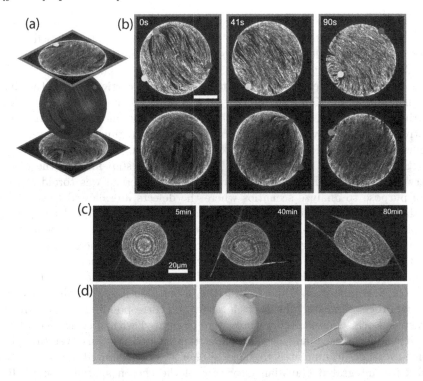

FIGURE 4.10
Dynamics of an active nematic at the inner surface of a spherical droplet. (a) Hemisphere projection of a 3d confocal stack of a nematic vesicle. The positions of four $+1/2$ disclination defects are identified. (b) Time series of hemisphere projections over a single period of oscillation in which the four defects switch from tetrahedral (t = 0 s) through planar (t = 41 s) and back to tetrahedral (t = 90 s) configurations. Scale bar, 20 µm. Bottom panels show vesicle shape changes driven by the dynamics of cortical defects. (c) Confocal images showing the z-projection of the vesicle shape, with corresponding 3d schematics shown in (d) (colored image in original version). Image and caption text adapted from Keber et al. [193].

In addition to this periodic dynamical mode, Keber et al. were also able to detect pronounced shape deformations under particular stability conditions of the active vesicles. This was achieved by applying a hipertonic stress that caused a water efflux and subsequent vesicle deflation. The shape of slightly deflated vesicles continuously fluctuated around a mean spherical shape and is characterized by the continuous growth and shrinkage of the major and minor

---

become progressively irrelevant and the population of defects of both topological charges start to resemble that of a planar AN. Threshold presumably should depend on activity.

axis of an ellipse, with a periodicity set by the defect speed. In addition, these vesicles exhibited four motile protrusions that are tightly coupled to the dynamics of the underlying defects (see bottom panels in Fig. 4.10).

An exhaustive numerical work by R. Zhang et al. [455] (see also Sect. 6.4.1) was entirely dedicated to reproduce the dynamics of cortical defects that had been reported by Keber et al. Low activity conditions reproduced the experimentally observed behavior. Authors also analyzed intermediate and high activity conditions, with indication of some chaotic regimes. Contractile systems were also considered for the sake of comparison.

Still considering encapsulated ANs but going a step further, the geometry considered by Ellis et al. [99] rather than spherical was toroidal, with the purpose to analyze scenarios where the defects are allowed to move on surfaces with non-uniform curvature, and, more in particular, letting them explore regions with positive and negative curvatures. The experimental system consisted on water-based toroidal droplets containing the active material and stabilized with an oil-based *yield-stress fluid* as the dispersing phase (see Fig. 4.11). The lower half of the toroidal droplet was imaged over time using confocal microscopy, and from it the orientation map, together with the tensor order parameter, the defect location and corresponding charge were extracted. Local Gaussian curvatures were computed with specific algorithms. By tracking positive and negative defects in selected regions over time, the time-averaged accumulated topological charge was computed and correlated with the integrated Gaussian curvature of the chosen spatial domain. Results confirmed positive slope, i.e. defects are attracted by regions of like-sign Gaussian curvature. This is consistent with curvature-induced defect unbinding. Experimental observations were supplemented with numerical simulations, where defects are modeled as massless particles moving on the torus. As in the case of spherical droplets, this scenario will be also commented from the point of view of modeling in Sect. 6.4.1).

A different and original perspective of encapsulated ANs is to look at the coupling of active and passive topological defects. This means that unavoidably we need to design a system composed of two liquid crystalline phases, one being the household of the active defects, while the second corresponding to the passive material. This scenario is simply realized when emulsifying the active material with a passive (nematic) liquid crystal, as reported by Guillamat et al. [151]. I recall from Sect. 2.3 that colloids dispersed in LCs (not necessarily suspensions of solid inclusions, i.e. sols) induce topological disclinations in the liquid crystal matrix, either point or line defects. We may anticipate that active cortical flows inside droplets will cause a distorting effect on the passive outer disclinations, while it could be equally imagined, in return, that the exterior defects would constrain as well the dynamics of the interior active defects. I examine this issue on what follows.

In the quoted experiments, the dispersing medium of the aqueous active preparation is the hydrophobic thermotropic LC 4-pentyl-4-cyanobiphenyl (5CB) enclosed in a closed cell of 140 μm gap between two parallel glass

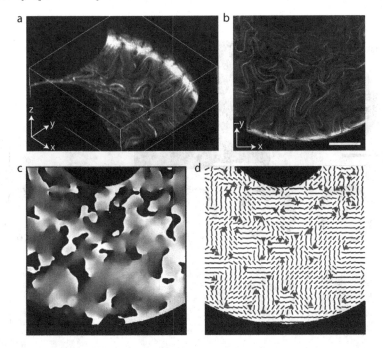

FIGURE 4.11
An active nematic confined to the surface of a toroidal droplet. Panel
(a): Confocal stack of a portion of a nematic toroid at a given time. Panel (b):
Intensity projection along $-\hat{z}$ of the data in (a). Scale bar is 200 μm. Panels
(c) and (d): Pattern of orientational coherence and defects, respectively. In
(c), the orientation is measured clockwise from the horizontal, with black
representing 0 and white representing π (colored image in original version).
Image and caption text adapted from Ellis et al. [99].

plates treated to impose a homogeneous in-plane alignment of the passive LC.
Anchoring of the passive LC depends on the surfactant used to stabilize the interface. More precisely, when using a PEGylated phospholipid, 5CB molecules
organize perpendicularly to the droplet surface (homeotropic anchoring). This
results in the formation of a localized (Saturn) ring disclination of quadrupolar character. Conversely, using the standard Pluronic surfactant for planar
interfaces one obtains planar alignment on the exterior of the droplets and
double-boojum defect structures. On what follows, I focus on the more interesting case that corresponds to the coupling of a passive outer Saturn ring
(SR) with active inner defects.

Bright-field and fluorescence imaging (see Fig. 4.12 a (left) and b (middle))
permit to track at any time the behavior of the SR disclination and to monitor
the dynamics of defects in the active nematic (AN) shell, although only two of

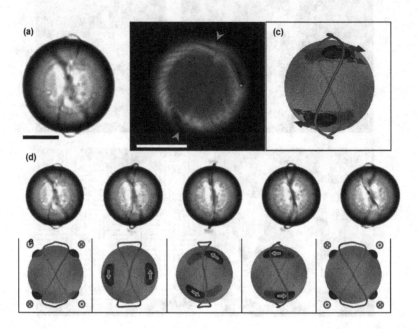

FIGURE 4.12
Periodic oscillation of a Saturn ring dragged by active cortical flows.
(a) Bright-field micrograph of an oscillating SR at the maximum amplitude
of the oscillations. (b) Fluorescence micrographs allow simultaneous visual-
ization of the active shell and the SR. The SR is subtended between the two
arrowheads, and it is dragged by two active defects, whose tips are marked by
a dot. (c) Snapshot of a simulated SR in the oscillating state. Velocity field
(arrows) in the passive nematic and isosurface of its magnitude, indicating
the position of the active defects. An image sequence showing a half-period of
the oscillation of an SR is shown for experiments (d) and for simulations (e).
Image and caption text adapted from Guillamat et al. [151].

the four $+1/2$ active defects can, at most, be observed simultaneously. Panel
c (right) in Fig. 4.12 corresponds to numerical simulations that accompanied
the experimental observations. A complete cycle of deformation is shown for
experiments and simulations in the remaining panels of the figure.

Strikingly enough, some SRs, singularly under low activity, feature simple and periodic in time oscillations, while in other situations more complex
scenarios are observed, with SR being distorted in multi-mode and/or non-
periodic distortions, or even getting multi wrapped around the droplet, and
eventually, collapsed at one of the poles. Regular oscillations are a signature
of the tight coupling between the AN dynamics and the SR. This regime is

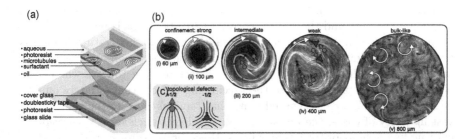

FIGURE 4.13
Active nematics under lateral confinement. (a) Schematic of sample cell: A micropattern of cylindrical holes was imprinted into a photo-resist layer bound to a microscope slide. (b) Fluorescence images of confined active nematics with increasing diameters as noted. White arrows indicate direction of circulation. Line overlaying the 200 µm disk highlights the double spiral configuration of the nematic director observed for intermediate confinements. (c) Structure of the nematic directoir field around topological defects. Image and caption text adapted from Opathalage et al. [285].

demonstrated to be stable only through a feedback mechanism that orients the AN defect manifold with respect to the SR. Beyond this simple scenario, the large variety of more complex observed behaviors is likely due to both a quite wide range of droplet size distribution, added to the difficulty to guarantee a uniform partition of the components of the active sample and of the chemical surfactant. More details can be found in a later paper published by Hardoüin et al. [155].

4.4.2 Geometric Confinement of Active Nematics

The first systematic report on a geometrically confined active nematic I am aware of was published by Opathalage et al. [285] (see Fig. 4.13). The chosen geometry consisted in circular disks of varying diameter adapted to the original close cell design [337]. Planar alignment and, apparently, no-slip flow conditions were assumed, although the latter are difficult to assess unambiguously given the preparation protocol. Circular geometry enforces a total topological charge of +1, that was realized with diverse defect configurations. For the strongest confinements, up to 100 µm, the disk diameter is below the minimal separation distance between defects for the chosen activity conditions. This favors microtubule alignment and accumulation at the boundaries. For intermediate sizes up to 200 µm, a pair of +1/2 defects form arranged into an asymmetric double spiral configuration. For weak confinement, from 300 to 600 µm, proliferation of additional defects is observed. In all the three cases, persistent circular flows were observed of equally realized handedness.

Increasing further the disk diameter leads to complete disorganization of the streaming currents and the typical turbulent-like behavior is recovered.

Focusing specifically on defect dynamics, the most interesting regime is that of intermediate confinements. Nucleation of defect pairs at the boundary occurs periodically, with the positive defect moving towards the center of the disk and replacing one of the preexisting defects that is lately annihilated with the negatively charged mate that remains close to the wall. In the vast majority of cases, the boundary-induced defect nucleation preserved the existing handedness of the circular flows. However, on rare occasions, flow reversal was observed. Active stresses were extracted from the evaluation of the spatial gradients nematic order tensor. Once projected along the azimuthal direction, the most important conclusion of this analysis is that nucleation of defects at the boundary is correlated with the time where the active force in the azimuthal direction reaches a minimum. In other words, sufficiently strong flow alignment generated by the circular flows effectively suppress defect nucleation.

A parallel development was published a little bit later by Hardoüin et al. [156], in a combined experimental and simulation work for a different geometry, i.e. a channeled active nematic (see top panels in Fig. 4.14). Apart from the chosen geometry and the preparation protocol of the channeled AN, a major difference with respect to the work of Opathalage et al. [285] is the different material characteristics of the enclosing polymeric matrix that favors in the case of Hardoüin et al. slip conditions of the bundled microtubules at the walls.

In addition to the expected turbulent-like flows typical of the smallest aspect ratios[11], a couple of interesting and originally observed regimes occur at intermediate channel widths w. A shear flow regime is found for $w < 80$ μm, which is transiently defect-free. Moreover, a *dancing defects regime* (a particular realization of the defect braiding dynamics mentioned earlier [392]) appears for $w > 90$ μm. The transition between these two flow modes is not sharp. As a matter of fact, for values of w in the range between 80 and 90 μm, the direction of the shear is not uniform along the channel, but rather the flow pattern is composed of patchy domains, where shear flow spontaneously arises with a random direction, and is then quickly disrupted by instabilities intrinsic to the aligned extensile material. To characterize the distribution of defects, a mean separation distance between defects λ was evaluated and demonstrated that it is largely conditioned by the channel width, and practically independent of the ATP contents at the range of considered experimental conditions. More precisely, and except for the widest channels, this characteristic separation distance is markedly different from the active length scale. It is thus interpreted as largely coming from the boundaries-induced flow screening. Organized shear-like flows when decreasing w occur for values $\lambda/w \approx 1/2$, while

---

[11]The aspect ratio is defined as the ratio of the large to the small dimensions of the channel.

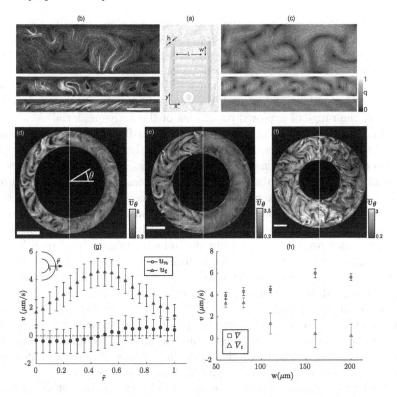

FIGURE 4.14
Flow states of an active nematic in rectangular channels and annuli. (a) Top view of the experimental setup in channel configuration. (b) Confocal fluorescence micrographs of regimes for increasing channel width; from bottom to top: unstable shear, defect braiding, and turbulent. Scale bar: 100 μm. (c) Corresponding simulations of the experimental system. Gray scale corresponds to the computed nematic order parameter. Lower panels display annular confinement. Panels (d) to (f), corresponding respectively to widths of 60, 160, and 200 μm, are composed of a fluorescence micrograph in the left half, and the intensity map of the time-averaged azimuthal component of the velocity in the right half. Scale bar 100 μm. (g) For the annulus of width 60 μm, time-averaged profile of the velocity tangential to the walls, as a function of the normalized radial coordinate. (h) Mean global speed and azimuthal velocity as a function of annulus width. Image and caption text adapted from Hardoüin et al. [156].

the crossover between the dancing-defect state and the shear-flow regime takes place around λ/w practically unity.

Dynamically, the system in the defect-dancing regime behaves as if two distinct populations of positive defects were traveling along the channel in opposite directions, passing around each other in a sinusoidal-like motion. At the same time, the flow pattern adopts in these situations a configuration similar to a vortex array with alternating vorticities. This state had been qualitatively predicted a couple of years earlier by Shendruk et al. [358], but never reported previously in experiments. The extended numerical simulations in [156] are in good agreement with the experimental observations and permit to extend the range of explored conditions of the active material.

A final remark is in order concerning the shear-flow regime. As mentioned, it is actually non-permanent and it gets transiently destabilized by defect nucleation events that occur most likely at the walls. This is again an indication of the intrinsic unstable nature of the aligned extensile nematic, a signature of active fluids that have been mentioned several times earlier in this chapter. On the other hand, here, at odds with the scenario considered in Opathalage et al. mentioned earlier, defects are nucleated massively. To explain this difference, one may conjecture that, apart from the different preparation condition, the planar aligned active nematic at the walls is likely more labile in the channel configuration as compared to the curvature-enforcing case of the disks.

A theoretical description of the channeled active nematic using a Leslie-Ericksen modeling approach (see Sect. 6.1) has been published recently by L. Zhao et al. [456]. This work correctly predicts the shear-flow regime, with shear profile independent of activity, and analyze its stability limit. More specific comments on the theoretical modeling of hard geometric confinement conditions on active fluids will be provided later on in Sect. 6.4.2.

Closer to the circular geometry considered by Opathalage et al., although employing again open-cell designs, I briefly mention some more recent results on geometric confinement applied to ANs reported by Hardoüin et al. in [157] (middle and bottom part of Fig. 4.14). The analyzed situation this time is that of a network of connected annular microfluidic channels that demonstrates the possibility to engineer controlled directional flows and autonomous transport.

In single annular channels and for narrow widths, the typically chaotic streams transform into well-defined circulating flows, whose direction or handedness can be controlled by introducing asymmetric corrugations on the channel walls. The dynamics is altered when two or three annular channels are interconnected. These more complex topologies lead to scenarios of synchronization, anti-correlation, and frustration of the active flows that I briefly comment on what follows.

The geometries/ topologies tested by Hardoüin et al. belong to the class of 2d handle-bodies composed of connected elementary annuli. They are simply classified using their genus number, g, which indicates the number of holes in the device: $g = 1$ indicates a single annulus, $g = 2$, two connected annuli, and $g = 3$, the maximum considered case, a triangular arrangement of overlapping annuli. In the cases of $g = 2$ and $g = 3$, the stabilization of topological defects is demonstrated, paralleling what had been reported earlier for passive nematic liquid crystals flowing through microfluidic devices [137]. In short, this result simply announced means that the geometry of the channels may lead to spatial distributions of topological defects that are at odds with those configurations that are energetically favored for the liquid crystal at rest. However, a fundamental and obvious difference with the scenario in [137] is that the coupling between topological defects and the orientation field is here spontaneously induced within the active fluid while interacting with the confining boundaries. More intricate flow patterns and dynamical states are reported in larger microfluidic networks of connected annuli, as exemplified for the case of $g = 3$ handle-bodies, where anti-synchronization and frustration of dynamic states come into play. The results presented uncover the richness of active topological microfluidics, paving the way for the design of fully autonomous fluid circuits capable of performing complex tasks without external driving. This idea was also explored in theoretical approaches by Woodhouse et al. [442].

4.4.3 A New Concept: Active Boundary Layers

In the study of active nematics under geometric confinement, a striking new concept emerges that reveals novel symmetries and dynamic modes of boundary-resident topological defects. This concept has been termed **Active Boundary Layer** (ABL), and appeared first in the literature of microtubule-based active nematics following observations by Hardoüin et al. In Fig. 4.15 results are shown for a active nematics confined in an annulus, a scenario reported in the precedent subsection, but now looking more precisely at what happens to the active two-dimensional fluid close to the confining circular walls. Fluorescence micrographs display the typical active turbulence regime away from the walls (Fig. 4.15 (a)), whereas walls are preferential sites for defect nucleation and unbinding of defect pairs, as commented earlier. The interesting observation here is that $+1/2$ defects are ejected from the walls into the bulk as expected, while negative counterparts remain at the boundary (Fig. 4.15 (b)), and display interesting new dynamics as commented in what follows.[12]

To better analyze the process, one may build kymographs by measuring the fluorescence intensity along circumferences parallel to the boundary, and stack vertically the resulting pixel lines obtained at increasing times (Fig. 4.15 (c)). Close to the inner wall, kymographs appear anisotropically patterned, with tree-like lines, separated by smoother regions of relatively uniform intensity. Branches originate from dark "blossoms" that pop up from uniform regions, and correspond to spontaneous events of boundary defect nucleation, while branching points correspond to defect merging events (Fig. 4.15 (d)). Branches are often short, corresponding to ephemeral defects, although the existence of long-lived branches are the signature of resident wall defects that experience long-distance attraction (Fig. 4.15 (e)). This hierarchical dynamics gradually

---

[12] A somewhat related effect that consists in the accumulation of negative defects in the vicinity of virtual boundaries, created using submersed micropatterned structures, was reported recently by Thijssen et al. [401].

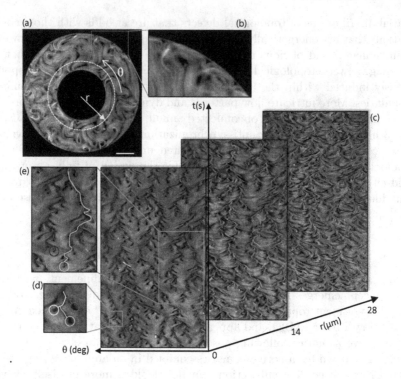

FIGURE 4.15
Active nematic dynamics near a wall. (a) Fluorescence micrograph of
the active nematic in an annular channel. Scale bar: 100 μm. (b) Magnified
image showing three of the −1/2 defects that form the outer ABL. (c) Kymographs corresponding to the fluorescence intensity profile along circumferences
at increasing distances, r, from the inner channel wall ($r = 0$). For each plot,
circular profiles at different times are stacked vertically, for a total duration of
300 s. (d) Three "blossoms" that merge into a single "branch". (e) Long-range
attraction between branches.

vanishes away from the wall. Indeed, we find that kymograph patterns are
highly correlated close to the wall but they become structureless around 25 μm
away from the wall. The situation is similar at the inner and at the outer
annulus walls, and also near flat walls in rectangular channels, indicating
that the ABLs form regardless of the channel geometry and, in particular,
independently of the sign of the curvature of the confining walls.

Isolated wall defects are best tracked in disk-shaped pools, where the
AN can display rather uniform textures and system-wide quasi-laminar flows
(Fig. 4.16 (a)). Corresponding kymograph of the boundary texture display
scarce blooms and branches (Fig. 4.16 (b)). The single wall defect either fluctu-
ates around a random fixed position for an undetermined period of time (first

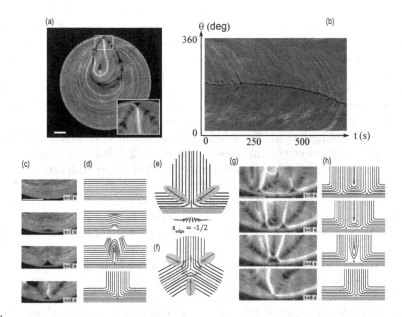

FIGURE 4.16
Birth and death of wall defects. (a) Fluorescence micrograph of the active nematic in a circular pool. In the inset, a magnified view of the region around the single wall defect highlighting the lateral crimps (see text). (b) Kymograph built along the contour circumference in (a). (c) Time-lapse of a wall defect nucleation in the disk (1 s between frames) and corresponding sketches. Active forces near the boundary defect are marked by arrows. (d) Sketches of boundary (edge or surface charge $-1/2$) and bulk ($-1/2$) defects. Active forces are marked by arrows. (e) Time-lapse of the merging between two wall-defects in the same system (2 s between frames) and corresponding sketches. Scale bars, 100 µm. In the sketches, symbols denote the core of $+1/2$ defects (circles), $-1/2$ defect (triangles), and -1 defect (square).

250 s in Fig. 4.16 (b)), or it drifts autonomously along the circular boundary, clockwise or counter-clockwise handedness having similar likelihood. Occasionally, this dynamics is interrupted by brief episodes of wall defect nucleation and annihilation.

The birth of a wall defect is observed as a time lapse image sequence in Fig. 4.16 (c) and is depicted in each accompanying sketch. When a pair of semi-integer defects unbind (see second panel in Fig. 4.16 (c), the defect that remains at the wall accumulates a $-1/2$ (edge or surface) charge and the conventional three-fold symmetry is transformed into a bilateral one about a plane perpendicular to the wall (Fig. 4.16 (c) bottom panel). Because of this, stress accumulates along the symmetry axis of wall defects, hindering their escape from the ABL. The defect core is prolonged by a plume of high-density fluorescent filaments (Fig. 4.16 (a)), and flanked by bent filaments where the orientation field turns by $\pi/2$. This amount is larger than the standard $\pi/3$ bending in bulk $-1/2$ defects, resulting in higher elastic stresses (Fig. 4.16 (d)). Because of their rigidity, filaments cannot accommodate the higher curvature, and additional void regions appear as crimps that steadily flank the boundary defect (Fig. 4.16 (a)). A decrease in the number of defects in the ABL typically proceeds through the merging of equal sign wall defects, facilitated by a bulk $+1/2$ defect that forms as the two $-1/2$ defects get closer. This is illustrated as a time-lapse image sequence in Fig. 4.16 (e) with accompanying sketches. Eventually, the two $-1/2$ merge to form a single -1 boundary defect, which quickly recombines with the extruded $+1/2$ defect. The net topological charge is conserved in the process. On the other hand, direct recombination of a single wall defect with a bulk $+1/2$ defect, although rare, is also possible. This type of event manifests itself in the kymographs as branches that suddenly end, without connecting to another branch. In summary, not only defects of different topological charge appear segregated, a feature already mentioned in relation to different curvature landscapes, but the symmetries of the boundary-residing defects is largely different as compared with their bulk counterparts.

In the study reported by Hardoüin et al., collective effects of defects residing at the walls are also analyzed. More precisely, the just reported dynamics of boundary-residing topological defects echoes a description of spatio-temporal chaos as modeled by *Kuramoto-Sivahinski–like equations*. Whether confirmed with more experimental and numerical work, this analogy would permit to establish an interesting connection between the AN dynamics and classical scenarios of pattern formation in classical Condensed Matter. I invite the interested reader to consult the original reference for details.

4.4.4   Geometric Confinement of Active Gels

Large-scale flows can result as well from particular confinement conditions of active gels. As a matter of fact, this question was addressed in three-dimensional preparations earlier than in two-dimensional active nematics I have been referring to in the previous section. More in particular, this issue was experimentally analyzed in a systematic study by K. T. Wu et al. [444]. These authors demonstrated that three-dimensional confinements and boundaries robustly transform turbulent-like dynamics of bulk active fluids into self-organized coherent macroscopic flows that persist on length scales ranging from micrometers to meters and time scales of hours.

The transition from turbulent to a coherently circulating state is not determined by an inherent length scale of the active fluid, but is rather controlled by a universal criterion that is related to the aspect ratio of the confining channel. Coherent flows robustly form in channels with square-like profiles and

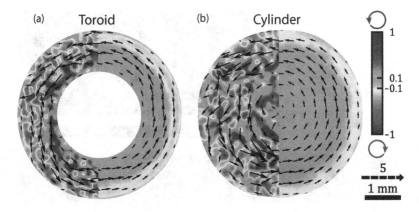

FIGURE 4.17
Coherent flows in confined active gels. Flows of active fluids in toroidal (a) and cylindrical (b) confinements showing persistent circular patterns. The greyscale map (color code in original version) represents the normalized vorticity distribution, with extreme (blue and red in original) tones representing CCW and CW vorticities, respectively. Left and right halves are instant and time-averaged plots, respectively. Image and caption text adapted from K. T. Wu et al. [444].

disappear as the confining channels become too thin and wide or too tall and narrow (see Fig. 4.17). Analysis of the microtubule network structure reveals that the transition to coherent flows is accompanied by the increase in the thickness of the nematic layer that wets the confining surfaces. Ratchet-like chiral geometries establish geometrical control over the flow direction. Overlapping toroids were also considered, as well as closed racetrack geometries. Very recently a modeling study of the situation considered by K. T. Wu et al. was published by Chandragiri et al. [55]. Authors find that an scenario of net flow is only possible if the active nematic is flow-aligning, and that, in agreement with experiments, the appearance of the net flow depends on the aspect ratio of the channel cross-section.[13]

A different striking manifestation of confinement effects, a little bit in between three- and two-dimensional realizations of active fluids, was considered by Senoussi et al. [352] (see Fig. 4.18). Senoussi et al. reported how a three-dimensional solution of kinesin motors and microtubule filaments spontaneously forms a 2d free-standing nematic active sheet that actively buckles out of plane into a centimeter-sized periodic corrugated sheet that is stable for several days at low activity. Importantly, the nematic orientation field does not display topological defects in the corrugated conformation. At higher

---

[13]Still an even more recent paper modeling this situation was published by Varghese et al. [420].

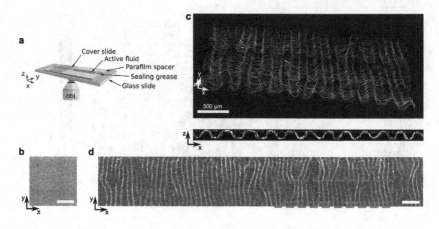

FIGURE 4.18
Corrugated active nematic sheet under confinement. (a) Schematics of
the experimental setup. (b) Epifluorescence image of the fluid at initial time.
(c) Confocal images in 3d (top) and cross-section in the xz plane (bottom)
of the fluid after 300 min. (d) Epifluorescence image of the sample after 24
h. The red dashed rectangle and the red dotted line respectively indicate the
region where top and bottom images in (c) were recorded. Image and caption
text adapted from Senoussi et al. [352].

activities these patterns are transient and chaotic flows are observed at longer
times. In comparison with the normal experiments with microtubule-bundled
active nematic that I have reviewed at depth, two variants are introduced in
Senoussi et al.'s preparation: the length of the microtubules is sensibly larger
(86 µm instead of 1 µm in the standard preparations referred previously in
most of the chapter), and the motor is a different variant of the kinesin.

Initially, the density of microtubule bundles is homogeneous in 3d while
aligned along the long axis of the channel. This nematic order arises spontaneously during the filling process of the channel by capillarity. With the course
of time in the range of a few hours, the fluid contracts anisotropically along
its two shortest dimensions to form a thin sheet of gel that freely floats in
the aqueous solution, mainly due to passive depletion forces. Simultaneously,
the extensile active stress generated by the motors buckles the sheet along the
direction perpendicular to its plane, forming a corrugated sheet of filaments
with a well-controlled wavelength of the order of a few hundreds of micrometers. Strikingly enough, a simple theoretical argument explains the selected
wavelength in terms, once more, of the active length scale $q^* \propto (\alpha(c)/K)^{1/2}$,
with $\alpha(c)$ the (positively defined) activity-dependent parameter that incorporates this time the motor concentration. A similar scenario was considered in
a recently published paper by Strübing et al. [381], where the theory is more
elaborated as it is based on standard models to describe active fluids.

Since this section summarizes some strategies aiming at controlling active fluids one way or another, I believe it is worth to dedicate the last paragraph to comment on a theoretical paper by Norton et al. [284], where this idea is given a new twist. Authors aim at applying principles of *optimal control theory* to an hydrodynamic model of an AN. This is demonstrated by the capability to switch the system between dynamical attractors in an efficient and rapid manner. More precisely, the considered scenario corresponds to a disk confinement, as analyzed by Opathalage et al. [285], when the system displays two stable limit cycle attractors characterized by two motile $+1/2$ disclinations that perpetually orbit the domain at fixed radius in either the clockwise or counterclockwise direction. As an exemplar application of optimal control theory, authors identify the spatiotemporal actuation of either applied vorticity or active stress that rearranges the nematic director field and moves the system from one attractor to the other.

4.5 Recent Advances in the Preparation of Active Gels and Active Nematics

I finish this chapter with some very recent experimental realizations of active fluids prepared from microtubules that open new and promising possibilities for the future development in the field.

The first advancement I would like to mention was reported in a paper published by Ross et al. [329]. Advancing into the perspective of bioinspired engineering of materials, these authors replace the conventionally employed streptavidin-mediated irreversible linking of (biotinylated) kinesin pairs ([337]) by a reversible variant after fusing kinesin I with optically dimerizable iLID (kinesin) proteins (see Fig. 4.19 (a, b)). Light patterns are projected into the sample throughout its depth and determine when and where motors link. Outside the light-excitation volume, the microtubules (mean length around 7 μm) remain disordered, whereas inside the light volume they bundle and organize. In particular, for a cylindrical pattern of light excitation, the microtubules contract into a three-dimensional aster (see Fig. 4.19 (c)).

The dynamics of aster formation and decay was analyzed through time-lapse imaging. During aster formation, microtubule distributions contract. The contraction speed was found to grow with the diameter of the excitation cylinder. During decay, microtubule distributions spread in a manner consistent with diffusion. The effective diffusion coefficient is independent of the characteristic aster size and is consistent with what is expected for free microtubules. Asters can also be moved by repositioning the light patterns relative to the sample slide. Another important observation is that asters formed near each other interact by spontaneously merging. The speed at which asters merge was observed to increase as a function of linking distance. Aster

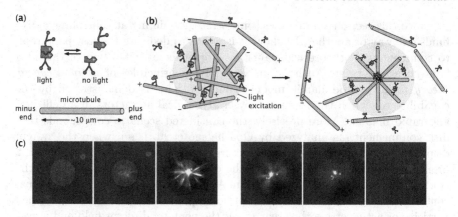

FIGURE 4.19

A light-switchable active matter system. (a) Schematic of light-dimerizable motors that walk towards the plus ends of stabilized aster size. (b) Schematic of light-controlled reorganization of microtubules into an aster, where locally dimerized motors pull microtubule plus ends towards each other. (c) Images of labeled microtubules during aster assembly and decay. Image and caption text adapted from Ross et al. [329].

merging, moving and trajectory experiments, demonstrate subsequent fluid flow of the buffer, as inferred by the advection of microtubules. Flow organization can thus be selected by prescribing appropriate illumination patterns as a new strategy in the design and control of active preparations based on microtubules and kinesin motors.

Another variant of optical control of the microtubule self-assembly that holds big promises in the context of designing active gels is based on azobenzene-based microtubule stabilizers, as very recently published by Müller-Deku et al. [265].

Still within a perspective of an illumination-based control of active fluids, although moving from the system tubulin/kinesin to the analog actin/myosin, I mention the recent paper published by R. Zhang et al. [454]. These authors introduce the concept of a spatially-structured activity as a means of manipulating an interfaced active nematic made of actin filaments and light-sensitive myosin motors. Different from the Ross et al. approach mentioned above [329], who used reversible cross-linking of motors to induce stress, here authors employ motor proteins with light-dependent gliding velocity. First, by illuminating selectively large portions of the whole sample, Zhang et al. show how high activity regions are self-contained revealing a sort of spatial confinement. In particular, tracked trajectories of $+1/2$ defects appear largely contained within the illuminated region, only rarely crossing illumination boundaries. Moreover, imposing structured stress following this illumination protocol, defect nucleation, as well as the capability of steering defect trajectories, is demonstrated. These observations are supported by numerical simulations.

Another extremely remarkable development, although totally different in spirit to those just mentioned, was published by Duclos et al. [92]. In fact, there is a very obvious way to motivate the interest for this paper. One needs only to consider a question that for a long time had been flowing around the field of active systems prepared from the microtubule/kinesin mixture. In spite of referring to their most usual preparations as active nematics, and many times stressing their similarities with nematic equilibrium counterparts, we should always keep in mind that Liquid Crystals in nematic phases are, for the most part, three-dimensional systems, while their active analogues are not. The latter, as many times emphasized in this chapter, are assembled as quasi-two-dimensional layers (interfaced with two fluids), whose thickness, albeit never precisely determined, is probably not larger than a few microns. Thus the obvious, and for many years pending, challenge was to prepare a three-dimensional protein-based active fluid still displaying nematic symmetries.

The original idea developed by Duclos et al. was to prepare a composite system by mixing microtubules and filamentous viruses to achieve such a purpose. To guarantee the necessary three-dimensional nematic scaffold, the latter should be kept properly disassembled. For this reason, authors first replaced the non generic and broadly acting depletant PEG, used in conventional preparations, by a specific microtubule cross-linker, PRC1-NS. This component assured the stability of a mixture of low-density extensible microtubule bundles (0.1% volume fraction) dispersed within a passive colloidal nematic phase made of filamentous viruses (see Fig. 4.20 (a)). Birefringence of the composite material indicates local nematic order, in contrast to three-dimensional active preparations that lack the passive liquid crystal component. Using advanced microscopic techniques, the spatiotemporal evolution of the nematic director field $\hat{n}(x, y, z; t)$ was extracted from a stack of fluorescent images (see Fig. 4.20 (b)). Spatial gradients of the director field identified regions with large elastic distortions. Three-dimensional reconstruction of such maps revealed that large elastic distortions mainly formed curvilinear structures, which could either be isolated loops or belong to a complex network of system-spanning defect lines (Fig. 4.20 (c)).

Similar structures were observed in simulations of three-dimensional active nematic dynamics that accompany experimental observations in Duclos et al.'s paper. The numerical study aims at characterizing closed loop disclinations. As a matter of fact, isolated loops were observed to nucleate and grow from undistorted, uniformly aligned regions. Likewise, loops also contracted and self-annihilated, leaving behind a uniform region. Other more complex dynamics supposing splitting and merging episodes are also reported. The structure of these disclinations was investigated from the premise that in 3d active nematics, an isolated disclination loop as a whole has two topological possibilities: It can either carry a monopole charge or be topologically neutral, depending on its director winding structure. Because charged topological

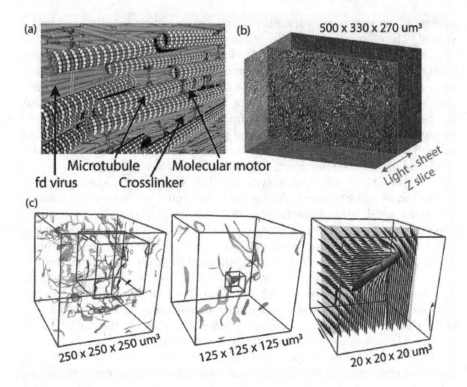

FIGURE 4.20
A three-dimensional active nematic. (a) Schematic of the 3d active ne-
maticsystem: extensile microtubule bundles are dispersed in a passive colloidal
liquid crystal. (b) Multi-view light sheet microscopy allows for 3d imaging of
millimeter-sized samples with single-bundle resolution. (c) Three-dimensional
elastic distortion map revealing the presence of curvilinear singularities. Image
and caption text adapted from Duclos et al. [92].

loops can only appear in pairs, nucleation of isolated loops as observed in this
system implies their topological neutrality. The topological nature of disclinations was throughly analyzed both from experimental determinations and
numerical simulations.

Finally, another remarkable development that remains so far largely un-
exploited in the context of engineering active gels is the possibility to use
DNA-based programming techniques for the building up of transport systems
using kinesin motors and microtubules. Preliminary results on this line were
reported quite a time ago by Wollman et al. [440], and more recently this
concept was put forward by Tayar et al. [394]. These authors consider the
actual active nematic preparation but replacing the streptavidin irreversible

clustering of kinesin motors by a reversible cluster of these motors (both processive, i.e. double-headed, and non-processive, i.e. single headed, versions were considered) based on a double-stranded DNA linker. The hybridization length (from 3 to 200 base pairs) controls the cluster binding strength, while the DNA is internally labeled with a fluorophore. In this way, a reversible DNA-based force-sensing probe is assembled that, by an optical readout, reveals the molecular arrangements and the force loads experienced by kinesin motors.

No better way to close this section than to refer to a concept that permeates nowadays many different basic and applied scientific disciplines. *Machine learning* is nowadays a fashionable technology that holds promise to revolutionize quantitative modeling in physical sciences and engineering [51]. It thus seems natural that this concept has been recently invoked looking for a better understanding, predictability or control capabilities of complex systems such as the usual realizations of active fluids we are considering in this chapter.

The best account of this recent perspective I am aware of is the report by Colen et al. [68]. In this work, authors demonstrate that neural networks can be used in relation to active nematics prepared from biofilaments to extract multiple hydrodynamic parameters of the system (i.e. the elastic constant or the activity parameter), as well as to forecast its chaotic dynamics solely from image sequences of its past. Interestingly, the only necessary input is the orientation of the biofilaments, but not the associated velocity field which is much more difficult to obtain in experiments.

More precisely, and following what is quoted in the paper, I briefly sketch the procedure to extract hydrodynamic parameters. One first generates a library of director fields, both in two and three dimensions, using lattice Boltzmann simulations of the continuum set of equations usually employed to describe active nematics (see Chapter 6). With this simulated library one trains the neural system, and once trained this expert system can be used to obtain hydrodynamic parameters in simulations as well as in experiments. Single and combined parameters (for instance the active length scale in terms of the ratio of elastic and activity coefficients) can be reliably obtained in this way. On what respects to forecasting, authors use another algorithmic basis different from the previous one, and thus without need of any modeling, to predict the spatiotemporal evolution of the director field including singular events such as defect annihilation and nucleation. A detailed account of the algorithms employed can be found in the main text and accompanying material of the mentioned paper.[14]

---

[14]Another perspective on the same issue can be found in the paper published by Zhou et al. [457].

5

Emerging Concepts in Active Matter

The previous two chapters, emphasizing experimental realizations, have set the map for our journey through active matter. Chapter 3 introduces disparate examples of microswimming, including the minimal account of modeling necessary to each reported scenario. Chapter 4, exclusively oriented towards protein-based active fluids, and their most prototypical instances of dynamic self-organization, was planned along a similar "case-based" perspective, i.e. renouncing to look for any sort of universality in the described system-dependent behaviors. In the present central chapter of this manuscript, I rather take a different transversal, and say more "phenomenology-dedicated" viewpoint, with the declared aim to establish generic trends between different classes of active systems. This is to be continued later on with the two final chapters of the book, both of them presenting different (generic) levels of theoretical modeling dedicated to active matter in their most distinctive realizations.

I can easily justify the sort of unifying perspective adopted throughout this chapter. No doubt all the systems studied so far, albeit active at their own way, share some properties that confer them with unifying characteristics, the principal being their collective motion nature performing under non-equilibrium conditions. In any case, this common basis has permitted to build over time, not only general theories commented in the following chapters, but also a list of important concepts underlying the study of active matter, whatever the view we take, whether as swimmer suspensions or as active fluids. The choice of these generic concepts is certainly a matter of taste, but a few of them have received higher attention within the community, and will be those discussed at some length in this chapter. I recognize that some of these emblematic concepts, say for instance the celebrated motility-induced phase separation, were brought about directly from studies on active colloidal dispersions, with no counterpart I am aware of in the context of protein-based fluids. Others, contrarily, have found an ample audience in both subfields, probably the most characteristic being the idea of active turbulence mentioned in different contexts of the two previous chapters, and much discussed in the following one.

The two first sections of the present chapter are devoted to two complementary manifestations of the intrinsic non-uniformity of active systems. The first section features (dynamical) clustering effects, while the second one refers to the just mentioned scenario of motility-induced phase separation. Still,

DOI: 10.1201/9781003302292-5

another facet of the anomalous spatial (density) heterogeneities that is typical of some active (colloidal) systems, and known under the name of giant number fluctuactions, could be equally added to the contents of this section. However, I prefer to confer it a special attention, and I postpone to comment on it until Sect. 7.1.1.1 in the context of the discussion on models of dry active matter, since it was in this altter context where this concept was initially formulated.[1] On the other hand, the non-equilibrium nature of active systems is so obvious that is often overlooked. This seems to preclude any reference to thermodynamics. However, this latter framework is so appealing, and so much rooted in Physics, that this chapter finishes with a brief section reporting some attempts to bring thermodynamic concepts into the realm of active systems.

5.1   Dynamic Clustering and Swarming Behavior

Active colloids display striking individual characteristics, as previously commented, but still more remarkable are their collective properties. Even dilute suspensions can show a plethora of effects that add to the complexity of the mechanisms of individual self-propulsion, and that arise from the different types of possible interactions among specimens. These latter include, to mention those more apparent, alignment or volume excluded interactions, hydrodynamic forces mediated by the solvent where they are dispersed, or long-range effects that are subsumed in the corresponding (self-)phoretic field in the case of phoretic swimmers.

Dynamic clustering of active units, coexisting with sparsely populated parcels within the same system, is probably the simplest among these complex collective effects. The term dynamic brings the idea that these clusters are not permanent, neither in their composition nor in their location. Instead, (active) clusters continuously recruit or expel individual particles, rearrange in space, move, collide or disintegrate. What is crucial and should be recognized from the very beginning is that, unlike classical systems in thermodynamic equilibrium, clustering in active systems appears independently of the existence of any sort of attractive interaction. It is solely mediated, one way or another, by the self-propulsion nature of the individual components of the system. There is a very simple argument that justifies the appearance and universality of this phenomenon.

Giving their intrinsic motion, active particles are destined to collide much more frequently than their passive counterparts. Keeping rotational diffusion constant, collision frequency increases linearly with the Péclet number defined

---

[1]Dry refers to friction-dominated active matter, while a wet description emphasizes the fluidic characteristics of the studied systems. A discussion on the difference between "dry" and "wet" approaches, the latter underlying mostly the contents of this monograph, is postponed to the introductory paragraphs of Chapter 7.

in Sect. 3.1.1.3 [324]. In simple words, we are describing a kind of *active crowding* of active particles that get blocked under contact and scape only on the time scale of their reorientation diffusivity. I will comment first experimental observations of this effect on different scenarios. Later on I will briefly consider modeling approaches to active clustering.

5.1.1 Experimental Observations of Dynamic Clustering

Collective behavior of self-propelled catalytic micromotors has been reviewed extensively from an experimental point of view [428]. In the same context, the complementary possibility of considering dynamic self-assembly was explored by W. Gao et al. [124], even looking for some purposed functionality, as suggested by J. Zhang et al. and Aubret et al. [452, 13].

More classically, dynamical clustering in sedimenting active suspensions with chemical signaling was reported by Theurkauff et al. [400]. A novel cluster phase, in addition to solid-like and gas-like states, was observed with a highly dynamic nature, i.e. showing permanent merging and separation of spherical Janus colloids, while lacking any eventual alignment interaction. As an important feature, the cluster average size grows with activity, as a linear function of the self-propelling velocity. Furthermore, the behavior was mathematically modeled by the same authors through a chemotactic aggregation mechanism.

Still using Janus colloids whose phoresis is induced by light-induced demixing of a binary mixture near critical conditions (see Sect. 3.1.1.2), but under experimental conditions at which the influence of phoretic motion or Van der Vaals attractions is largely reduced, clustering was observed by Buttinoni et al. [48] (see Fig. 5.1). These authors claimed that aggregation is in this case

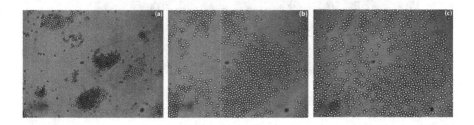

FIGURE 5.1
Dynamical clustering of self-propelled colloidal particles. The particles are carbon-coated Janus particles, which are propelled due to diffusiophoresis in a near-critical water-lutidine mixture. At thigh enough area fractions, phase separation into a few big clusters and a dilute phase occurs. Aggregation is completely reversible: Snapshots show how the clusters dissolve after illumination is turned off. Image and caption text adapted from Buttinoni et al. [48].

caused solely by dynamical self-trapping of the self-propelled particles. From another perspective, I already mentioned the concept of living crystals as introduced by Palacci et al., [292] referring to colloidal surfers activated by light (see Fig. 5.2 and Sect. 3.1.1.2). In this case, clustering is suggested to follow from a balance of long range osmotic and phoretic effects. A more quantitative analysis, looking for instance at the distribution of cluster sizes in Janus-like particles was published by Ginot et al. [130]. In particular, the size dependence of fragmentation and aggregation was obtained, and the motiion of individual clusters characterized.

In any case, experiments with self-propelled colloids reporting dynamic clustering seem to predict a limiting size for the self-assembled clusters. In other words, this is a situation of micro- (or arrested-) phase separation rather

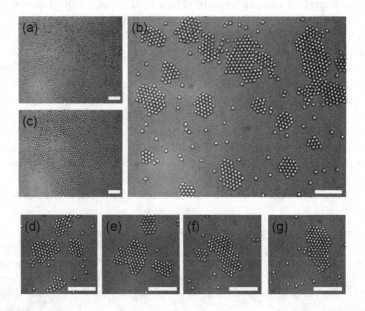

FIGURE 5.2
Living crystals of light-activated bimaterial colloids. (a) Initial state before illumination. (b) Aggregation following illumination. (c) Disassembling after turning off the illumination. The clusters are not static but rearrange, exchange particles, merge (panels (d) to (f)), break apart (panels (e) to (f)), or become unstable and explode (small cluster, panels (f) to (g)). Scale bars indicate 10 μm. Image and caption text adapted from Palacci et al. [292].

different from the scenario of motility-induced phase separation that will be commented in the following section.

Completing these earlier approaches, distinction between pure clustering and cohesion has been reported in a very recent experimental account on active particles with visual perception-dependent motility. This strategy is implemented with an external feedback on ABPs (active Brownian particles) such that individually they vary their velocity depending on the visual perception of their peers. Lavergne et al. demonstrated that this strategy is sufficient to obtain single cohesive nonpolarized groups from any number of particles, without the need for pair attraction or active reorientation [216] (a related modeling approach had been published a few years earlier by Barberis et al. [17]). Quorum sensing rules were realized based on the same system as reported by Bäuerle et al. [22]. One step further has been recently taken by Bäuerle et al. [23] after controlling not only the magnitude but also the direction of the particle propulsion allowing to add alignment interactions between ABPs.

Leaving self-propelled specimens, we find striking different forms of clustering for driven particles as well. Kaiser et al. [187] reported on the emergence of flocking and global rotation in a system of rolling ferromagnetic microparticles energized by a vertical alternating magnetic field. Likely, the most paradigmatic example of swarming in a system of colloids forced to move is based on the Quincke mechanism that was commented in Sect. 3.2.2 [44]. In this system, dilute populations of millions of colloidal rolling particles self-organize to achieve coherent motion in a unique direction, giving rise to colloidal flocks traveling through an isotropic phase, with markedly asymmetric density profiles (see Fig. 3.11).

In this latter context, a coarse-grained (hydrodynamic) theoretical modeling [54], anticipates the existence of periodic waves, soliton bands or polar-liquid droplets cruising in isotropic phases, similar to those observed in experiments [44]. Confined realizations originating vortices were reported and simulated a little bit later for the same system [43], as well as the response of these polar flocks to external fields (flows), as published recently by Morin et al. [261]. Simultaneously, coupled density (sound waves) and velocity fluctuations, propagating along all directions in such system were analyzed by Geyer et al. [126].

Needless to say, a higher potentiality of the scenarios just commented would be gained, provided control of the location of the formed clusters would be proved, or, even more, by achieving some degree of steering on them. This possibility was nicely realized with the help of liquid crystal-based dispersions, profiting from the alignment conditions rendered by the anisotropic fluid. I briefly comment on this LCEEP-based scenario on what follows (see Sect. 3.2.4).

Features of directed self-assembly of steered solid inclusions dispersed within a nematic liquid crystal were for the first time reported by Hernàndez-Navarro et al. [161, 163] (see Fig. 5.3). One of the plates was functionalized

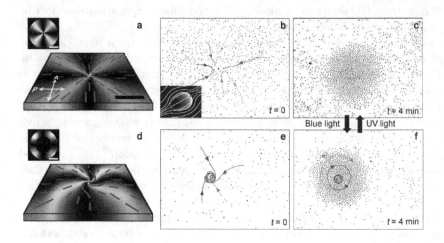

FIGURE 5.3
**Photo-patterning cluster assemblies of nematic colloids driven by
the mechanism of LCEEP.** Panels (a) to (c) correspond to the assembly of
an aster-like assembly following a local radial pattern of the nematic director.
Panels (d) to (f) feature the assembly of a rotating-mill assembly after the
enforcing of a spiral-like distortion of the nematic director. Image and caption
text adapted from Hernàndez-Navarro et al. [161].

with a photosensitive self-assembled azosilane monolayer to allow reversible
transitions between hometotropic and planar anchoring of the LC. The counter
plate was coated with a polyimide compound to secure hometropic anchoring
at all time. By irradiation with UV light from an incoherent source, the azosilane monolayer is forced to adopt a *cis* configuration (planar anchoring), easily
reverted to a *trans* form (homeotropic anchoring) when illuminating the cell
with blue light. By using a point-source light, this procedure allows the easy
and on-time addressability of the LC director locally at predesigned spots.
Local irradiation triggers a sub-millimeter spot of splay (radially spread) distortion of the nematic director that propagates for several millimeters into the
bulk of the cell following the application of the electric field. The region with
radial alignment acts as a basin of attraction of the dispersed particles, that
assemble a jammed aster-like aggregate before coming to rest.

The reversibility and quick response of the photoaligned layer enables
straightforward cluster addressability. More in particular a preformed cluster of arbitrary size can be relocated to a preselected place with minimum
dismantlement by changing the location of the UV spot. Swarming motion is
thus easily realized following predesigned arbitrary paths, or in general circuits
of arbitrary complexity and topology.

A quantitative account of this swarming behavior with the aid of numerical simulations can be found in a more recent paper by Straube et al. [380]. In particular, this study points out to a distinctive effect of the different ingredients participating in LCEEP clustering in comparison with the equilibrium assembly of nematic colloids. Elastic distortion of the nematic matrix that are the dominant specific forces in the latter situations are, however, sub-dominant here in front of pure hydrodynamics effects.

As a matter of fact, the most important clue justifying this study comes from the discovery of the intrinsic spatial inhomogeneity of the clusters assembled in experiments. This spatial inhomogeneity shows up in the coexistence of different aggregation states within a single cluster, including a central jammed core where short-range elastic forces dominate, surrounded by a liquid-like corona where particles come to rest by effectively reaching a (mechanical) equilibrium that results from the balance between the centripetal phoretic drive and pairwise repulsions. Finally, at the outer part of the cluster, particles obey a very diluted, gas-like, distribution. The compressible intermediate liquid-like region features linearly decaying density profiles whose slope can be tuned with the field frequency (as a matter of fact the slope scales with the phoretic velocity). At the same time, a bond-orientational order is extracted that reaches a maximum at intermediate packing densities, where elastic effects are minimized [380, 290].

5.1.2 Modeling Approaches to Clustering of Microswimmers

This issue has been extensively studied in the literature from multiple methodological perspectives and different aims. Given the generic presentation adopted throughout this monograph, I limit myself to propose a few simple considerations in the paragraphs that follow, while addressing the interested reader to specific papers that review this subject at depth. In this respect, I find particularly comprehensive the contribution by Zöttl et al. in [462], and similar approaches that are quoted in the list of selected review papers in Sects. 3.3 and 6.5.

As mentioned earlier when introducing the Active Brownian Particle model (see Sect. 3.1.1.4), nonequilibrium clustering of self-propelled rods was considered among others by Peruani et al. [305] and Baskaran et al. [18]. In particular, Peruani et al. study collective motion in a two-dimensional model of active units interacting through volume exclusion. Particle clustering is facilitated by a sufficiently large packing fraction or length-to-width ratio. The transition to clustering in simulations is well captured by a mean-field model for the cluster size distribution. Nevertheless, the simplest system for studying collective motion of active Brownian particles consists of active Brownian disks interacting only via hard-core repulsion. I refrain to follow this line here, and postpone it until next section devoted to motility-induced phase separation, since this is most properly the context where these studies were proposed.

I mention a few references as illustrative of the analysis of clustering effects for chemotactic active colloids (for a recent review interested readers can consult [227]). Pohl et al. [311] used Brownian dynamics simulations, motivated by the experiments by Thuerkauff et al. [400] on sedimenting (diffusio-phoretic) Janus colloids mentioned at the beginning of this section. The diffusiophoretic interaction has translational and orientational distinct contributions for these chemically asymmetric particles. From this basic idea, authors demonstrate that dynamic clustering occurs only when these two contributions give rise to competing attractive and repulsive interactions.

From a different point of view, this time using continuum-based models of the sort discussed in the following two chapters, and including a polarization variable added to the colloid number density, Liebchen et al. [228] analyzed a situation of chemorepulsive active colloids, predicting a generic route to self-limiting clustering. A continuum model was also proposed by Saha et al. [333], concluding that both the positional and the orientational degrees of freedom of the active colloids can exhibit condensation, together with the formation of clusters and asters.

Finally, the role of hydrodynamic effects in determining the collective behavior of active colloids modeled as squirmers (see Sect. 3.1.2) in quasi-two-dimensional confinement were analyzed at length by Zöttl et al. [461]. Importantly, the distinctive features between pullers, pushers or neutral largely determine the characteristics of the cluster behavior. More precisely, neutral squirmers phase separate in a gaslike and cluster phase. In contrast, strong pushers and pullers gradually develop hexagonal clusters.

5.2   Motility-Induced Phase Separation

Motility-induced phase separation, mostly recognized by its acronym MIPS, is a concept that has become popular in the study of discrete active systems. In some sense, it parallels the principle of active crowding or self-trapping announced in the previous section when referring to dynamic (active) clustering. Stated shortly this principle could be announced by saying that, in absence of *detailed balance*, active particles accumulate where they move more slowly [388, 52]. This is quite obvious if one looks at a steady solution of the conservation equation for the probability density of an ensemble of active particles (think for instance of an equation like Eq. 3.25 considered in absence of gravitational effects and orientational degrees of freedom). If the reverse is also true, i.e. considering for instance particles that show a decreasing velocity (at high enough rate) as density increases, nothing excludes to have a coexistence between a dilute active gas phase and a dense liquid assembly that displays a much smaller motility. Let's make these simple ideas analytic following the simple argument put forward in the review by Cates et al. [53] dedicated to this topic.

Let's consider the stationary and uniform reference situation $\rho_0 = C/V(\rho_0)$ with C a constant, and the particle's velocity assumed to be density-dependent as implicitly indicated. We look at the effect of a small perturbation in the density $\delta\rho(\mathbf{r})$. This results in a first order perturbation on the velocity $V = V(\rho_0) + V'(\rho_0)\delta\rho(\mathbf{r})$ (the prime superindex denotes here derivatives with respect to the functional variable density) that translates into a first order perturbed steady profile,

$$\rho_0 + \delta\rho'(\mathbf{r}) = \frac{C}{V(\rho_0) + V'(\rho_0)\delta\rho(\mathbf{r})} \simeq \rho_0 \left(1 - \frac{V'(\rho_0)}{V(\rho_0)}\delta\rho(\mathbf{r})\right). \qquad (5.1)$$

The condition for the onset of the instability of the uniform profile is then simply $\delta\rho' > \delta\rho$, or, equivalently,

$$\frac{V'(\rho_0)}{V(\rho_0)} < -\frac{1}{\rho_0}. \qquad (5.2)$$

The issue of motility induced phase transitions has been discussed by several authors. I choose first to summarize this topic following a sort of chronological perspective. A more detailed, albeit necessarily brief, theoretical analysis will be provided at the end of this section.

To the best of my knowledge, MIPS was first studied as a density-dependent mobility effect for run-and-tumble bacteria by Tailleur et al. [388]. Later on the analysis was extended to active Brownian particles as well by the same authors [52]. More or less simultaneously to this last contribution, Fily et al. [106] proposed both a numerical and an analytic study of a model of self-propelled polar disks on a substrate (no momentum conservation). Interactions were assumed soft (elastic) and isotropic (no alignment), and as a result particles do not order in a moving state. However, the isotropic fluid phase separates well below close packing into solid-like and gas phases. Furthermore, it was claimed that the phase separation was athermal, i.e., it can not be be simply described by an effective temperature, in agreement with conclusions from the simultaneous work by Bialké et al. [33] on crystallization of active particles, and in contrast with what been argued in sedimentation experiments of Janus colloids ([291]) that I referred to in Sect. 3.1.1.3.

Specifically, the main result in [33] was to predict the freezing of particles, interacting via a Yukawa potential, into crystalline lattices displaying long range orientational order, despite that energy is incessantly injected into the system. Compared to equilibrium freezing of passive particles, it was strikingly emphasized that the freezing density is significantly shifted to higher values (in terms of a strength parameter coupling speed and density), and, more important, the freezing location depends on the criterion used, either structural or dynamical.

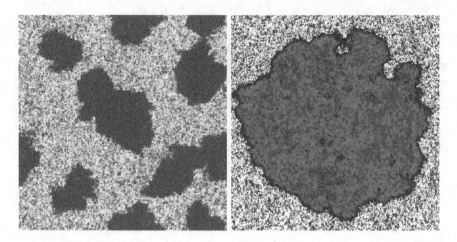

FIGURE 5.4
Phase-separating active colloidal fluid. Self-propelled Brownian spheres
that interact purely through excluded volume with no aligning interaction.
Left image depicts the phase separated state into dense and dilute phases
beyond critical density and activity levels. Right image displays a heat map
in greyscale of the pressure in the active solid material. Image and caption
text adapted from Redner et al. [324].

A complete phase diagram (packing fraction/Péclet number), and a study
of the separation kinetics of self-propelled Brownian spheres interacting purely
through excluded volume and no aligning interactions, was obtained by Redner
et al. [324] (see Fig. 5.4). It was shown that the system undergoes an analogue
of an equilibrium continuous phase transition with a *binodal curve* beneath
which the system separates into dense and dilute phases whose concentrations
depend on activity. The main conclusion is that the traditional role played by
an inverse temperature is played here by the orientational Péclet number.

Beyond simulations, continuum models were also employed in the discussion of MIPS from some of the authors quoted above, principally, by M. Cates's
group. Based on previous results from Tailleur et al. [388, 52], Stenhammar et
al. [376] proposed to account for the structural dynamics of a motility-induced
phase separation by coarse-graining the microscopic dynamics of a system of
ABPs to obtain the evolution for their density field. This equation contains an
effective diffusivity and a chemical potential with a Laplacian term that stems
from a functional non-local dependence $V(\rho)$. Solving this scheme numerically
one finds domain growth dynamics and morphologies in accordance with simulations. Interestingly, the authors point out that, in spite of the fact that
the effective chemical potential postulated in this way does not come from

an effective free energy, and thus violates detailed balance, the coarsening dynamics closely resembles equilibrium phase separation.

This mapping with classical theories on phase separation was an idea much pursued by other authors. Based on previous microscopic theories for the phase separation of self-propelled repulsive disks [31], a linearization of the corresponding long-wavelength equations for the number density and orientation fields lead to an effective *Cahn-Hilliard equation* for a system of ABPs [373], as analyzed by Speck et al. The claimed conclusion is that the phase separation close to the dynamic instability cannot be distinguished from the standard one for passive particles under attractive interactions.

A first-principles theory for systems of (interacting) active Brownian spheres was reported by Farage et al. [103]. These authors demonstrated explicitly how an effective many-body interaction potential is induced by activity. The required input quantities are the passive ("bare") interaction potential, the rotational diffusion coefficient, and the particle propulsion speed. The theory generates as output the static correlation functions and the phase behavior of the active system. For a repulsive bare interaction, activity generates an attractive effective pair potential, thus providing an intuitive explanation for MIPS. For an attractive bare potential, one finds that increasing activity first reduces the effective attraction, consistent with experiments of Schwarz-Linek et al. [350] for bacterial baths with a depleting agent, before leading at higher activity to the development of a repulsive potential barrier [103].

The just mentioned approaches were expanded later on opening new perspectives in the field. Wittkowski et al. [437] formulated a scalar ϕ^4 theory (or phase field model) for the order parameter field to study phase separation and coarsening dynamics, generalizing the classical model B of phase separation, into an active B-model, which minimally violates detailed balance. Extensions to include hydrodynamic effects gave rise to an active model H [405]. By extending the ϕ^4 field theory of passive phase separation to allow for all local currents that break detailed balance at leading order in the gradient expansion, Tjhung et al. demonstrated that, beyond standard scenarios in MIPS, more complex steady states can be also predicted, comprising a dynamic population of dense clusters in a sea of vapor, or dilute bubbles in a liquid [407].

In our view, the most relevant aspect of MIPS is the possibility to map this scenario, typical of active suspensions, to what is well-known from classical phase separation theories. As just mentioned, several papers have been published along this direction. I choose in my view what is likely the simplest approach to MIPS, following the work by Speck et al. [373] mentioned earlier. These authors propose to study the dynamics of a collection of repulsive self-propelled discs in two-dimensions. The following brief sketch, and indeed the justification for it, is that it also illustrates the use of hydrodynamic-like equations, i.e. a coarse-grained description, applied to a system of interacting active Brownian particles under those conditions that promote MIPS. In a sense, it parallels from a different motivation the interest to model particle-based dry models we will be concerned with in Sect. 7.2 of the last chapter. Another reason to present this piece of theory here is that, in this way, I extend to collectivities the use of the model of active Brownian particles (ABPs) that was introduced in Sect. 3.1.1.4.

The starting minimal model is based in the set of equations,

$$\dot{\mathbf{r}}_i = -\boldsymbol{\nabla} U + V_0 \mathbf{p_i} + \boldsymbol{\xi}_i, \tag{5.3}$$

in addition to consider free rotational diffusion parameterized with D_r. Effective hydrodynamic equations can be obtained starting from the full N-body Smoluchowski equation for the evolution of the joint probability distribution of all particle positions and orientations [31]. In this way, one obtains a pair of equations for the density and average orientation fields,

$$\partial_t \rho = -\boldsymbol{\nabla} \cdot [V(\rho)\mathbf{p} - D\boldsymbol{\nabla}\rho], \tag{5.4}$$

$$\partial_t \mathbf{p} = -\frac{1}{2}\boldsymbol{\nabla}[V(\rho)\rho] + D\nabla^2\mathbf{p} - D_r\mathbf{p}, \tag{5.5}$$

where D denotes the long-time diffusion coefficient of the passive system. Notice that the first term in the r.h.s. of the second equation introduces an effective (active) pressure $P(\rho) = (1/2)V(\rho)\rho$, whose derivation is justified in the original reference. As a matter of fact, I will refer later on in this chapter to the concept of active pressure with a similar meaning. This set of equations is obtained under the assumption of a slowly varying number density $\rho(\mathbf{r}; t)$ and orientation field $\mathbf{p}(\mathbf{r}; t)$. It is also worth remarking that the effect of the interaction potential is subsumed in the expression of the crowding term $V(\rho)$. Close to the instability of the uniform distribution it was demonstrated [31] that a simple linear relation $V(\rho) = V_0 - \rho\tilde{\zeta}$ captures in an effective way the effect of self-trapping.

One proceeds further by choosing convenient time $1/D_r$ and length units $l = (D/D_r)^{1/2}$, and normalize both fields by an averaged density $\bar{\rho}$, i.e. with the identifications $\rho \rightarrow \bar{\rho}(1 + \delta\rho)$ and $\mathbf{p} \rightarrow \bar{\rho}\mathbf{p}$, one transforms the pair of dynamical equations into the more convenient form,

$$\partial_t \delta\rho = -\alpha\boldsymbol{\nabla} \cdot \mathbf{p} + \nabla^2\delta\rho + 4\kappa\boldsymbol{\nabla} \cdot (\mathbf{p}\delta\rho), \tag{5.6}$$

$$\partial_t \mathbf{p} = -\beta\boldsymbol{\nabla}\delta\rho + \nabla^2\mathbf{p} - \mathbf{p} + 4\kappa\delta\rho\boldsymbol{\nabla}\delta\rho. \tag{5.7}$$

The set of dimensionless parameters entering in these last equations are respectively defined as $\kappa \equiv \frac{\bar{\rho}\tilde{\zeta}}{V_0}$, $\alpha \equiv 4(\frac{V_0}{V_*} - \kappa)$, and $\beta \equiv 2(\frac{V_0}{V_*} - 2\kappa)$, in terms of a characteristic speed $V_* = 4(DD_r)^{1/2}$.

The next step is to perform a linear stability analysis of the reference solution $\delta\rho = 0$ and $\mathbf{p} = 0$. The resulting dispersion relation, $\sigma(q) = -\frac{1}{2} - q^2 +$

$\frac{1}{2}\sqrt{1 - 4\alpha\beta q^2} \approx -(1 + \alpha\beta)q^2$, identifies instability conditions for the smallest wave numbers whenever $1 + \alpha\beta < 0$. The instability onset at $1 + \alpha\beta = 0$ defines a critical value of the self-trapping dimensionless coefficient,

$$\kappa_c = \frac{3}{4}\left(\frac{V_c}{V_*}\right) - \frac{1}{4}\sqrt{\left(\frac{V_c}{V_*}\right)^2 - 1}. \tag{5.8}$$

The analysis just past the instability proceeds by considering a velocity $V_0 = V_c(1 + \epsilon)$. Expanding correspondingly α and β, one finds that the fastest growing mode is of order $\epsilon^{1/2}$. This suggest to rescale length with $\epsilon^{-1/2}$ and time by ϵ^{-2}. Within the spirit of a *multiple scale expansion* typical in the study of pattern forming processes, one proposes the expansion,

$$\delta\rho = \epsilon c + \epsilon^2 c^{(2)} + ..., \tag{5.9}$$

$$\mathbf{p} = \epsilon^{1/2}[\epsilon\mathbf{p}^{(1)} + \epsilon^2\mathbf{p}^{(2)} + ...]. \tag{5.10}$$

To lowest order $\mathbf{p}^{(1)} = -\beta_0\boldsymbol{\nabla}c$. Matching powers in ϵ, one finds at the next non-trivial order the pair of equations,

$$\partial_t c = -\alpha_0\boldsymbol{\nabla}\cdot\mathbf{p}^{(2)} - \alpha_1\boldsymbol{\nabla}\cdot\mathbf{p}^{(1)} + \nabla^2 c^{(2)} + 4\kappa_c\boldsymbol{\nabla}\cdot[c\mathbf{p}^{(1)}], \tag{5.11}$$

$$0 = -\beta_0\boldsymbol{\nabla}c^{(2)} - \beta_1\boldsymbol{\nabla}c + \nabla^2\mathbf{p}^{(1)} - \mathbf{p}^{(2)} + 4\kappa_c c\boldsymbol{\nabla}c. \tag{5.12}$$

One would first solve the second equation for $\mathbf{p}^{(2)}$, and substitute it together with $\mathbf{p}^{(1)}$ into the first equation. It turns out that terms in $c^{(2)}$ exactly cancel out, leaving an expression for the dynamics of the large scale fluctuations in the field $c(\mathbf{r}; t)$ alone,

$$\partial_t c = \sigma_1\nabla^2 c - \nabla^4 c - 2g\boldsymbol{\nabla}\cdot(c\boldsymbol{\nabla}c) = \nabla^2\frac{\delta\mathcal{F}}{\delta c}, \tag{5.13}$$

with the nonlinear coefficient $g = 2\kappa_c(\alpha_0 + \beta_0) > 0$. This completes the equivalence with a Cahn-Hilliard formulation of a classical scenario of phase separation once one identifies an effective free energy functional,

$$\mathcal{F}[c] = \int\left[\frac{1}{2}|\nabla c|^2 + f(c)\right]d\mathbf{r}, \tag{5.14}$$

and $f(c)$ given by $f(c) = \frac{1}{2}\sigma_1 c^2 - \frac{1}{3}gc^3$. It is worth noticing that the expression for $f(c)$ lacks the conventional stabilizing c^4 term. This is justified by the authors in the original reference [373]. In any case, the onset of the phase separation process is, as claimed, appropriately described.

5.3   Active Turbulence

Active turbulence, sometimes referred to as well as *mesoscale turbulence* [431], is probably one of the most transversal concepts in active systems, whether in the form of swimmer dispersions or as active fluids [399]. I have mentioned it already in relation to magnetic spinners in Chapter 3 (see Sect. 3.2.1.2). Its importance was much emphasized in Chapter 4 when referring to dedicated experiments for active nematic fluids (see Sect. 4.2.1). Still in this latter context, I will address this issue theoretically in two sections of next chapter (see Sects. 6.2.3 and 6.3).

Prior to enumerate those scenarios where this topic has been mostly analyzed, I would like to justify the interest for active turbulence. From the very beginning, it is essential to recognize that most active systems display patterns of motion at large scales that, at first sight, are completely deprived of any spatio-temporal coherence. This motion can either refer to the moving specimens themselves or the fluid that embeds them. In any case, these erratic flow patterns look similar to what one qualifies as turbulent flows in classical hydrodynamics. Under turbulent conditions, highly enough energized laminar flows get destabilized into a system of chaotic streams and whirling eddies that span a wide range of length scales, conforming a regime that has been customary called *inertial turbulence*.

It seems thus natural to try to establish a connection between the two disparate scenarios of classical and active hydrodynamics. Disparate is a term used here deliberately in a twofold sense. First because, inertial turbulence is characteristic of high Reynolds number conditions, i.e. when inertial effects are prevalent, whereas, as mentioned several times in this text, active systems perform oppositely at low Reynolds, i.e. when viscous forces are largely dominant. The second difference is still deeper from a conceptual point of view. In classical turbulence, energy is injected at large scales and is progressively transferred into smaller scales where it is dissipated. This is totally opposite to what one can expect in the context here, where energy is furnished at the level of the active individual specimens that compose the studied system. Inherent chaotic flows self-organize, however, at much larger length scales. This principle is in fact the landmark of active systems regardless of the considered source of activity.

The existence of an inertially boosted energy cascade is thus totally meaningless in active systems. Nonetheless, one is still tempted to characterize active flows similarly to what has been canonical in the analysis of classical turbulence. Among these standard tools the likely most useful is the characterization of classical chaotic flows in terms of their *(kinetic) energy spectra*. In an intermediate range of length scales, where inertial effects get indeed dominant, flows look self-similar, i.e. their attributes are scale invariant. This further translates into power law characteristics and scaling regimes of the velocity autocorrelation functions over a wide range of length scales.

What is most celebrated in classical turbulence, and mostly sought for in parallel scenarios, are eventual features of universality of the exponents that characterize such scalings. Universal is here understood as irrespective of the characteristics of the flows, of their spatial range and of the driving force that fuels them. This is precisely the great contribution due to Kolmogorov who proved more than eighty years ago the existence of a down-scale cascade $E(q) \sim q^{-5/3}$ in $d = 3$ [111].[2]

With these conceptual antecedents of classical turbulence in mind, I will devote the remaining of this section to briefly contextualize this issue as it applies to active systems in as many experimental and theoretical contexts as I am aware of. Naturally, I will limit myself to those scenarios that are outside the ambit of active nematics to what I have dedicated, and will still dedicate later on, a very special attention. Along the line of the last two sections, I first consider experimental studies. Just for historical reasons in the study of this particular topic I start by referring to living microswimmers, to continue with artificial microswimming, and finish with driven colloidal systems.

As a matter of fact, the context where the concept of active turbulence first appeared in the literature was that of bacterial suspensions, and for this reason it was originally termed *bacterial turbulence* [86] (see Fig. 5.5). Later on, this topic got fully recognition as a generic emblem of active systems, and turned its denomination into active turbulence, the form actually mostly accepted. Such a flow regime in bacterial baths has received considerable attention during years, and from different perspectives, mainly from the R. E. Goldstein's and I. S. Aranson's groups for *B. subtilis*, both looking either at cortical or bulk suspensions [255, 86, 67, 367, 176, 431, 366, 96, 226, 304].

Among this lengthy, yet likely incomplete, list, I highlight the paper by Wensink et al. [431]. These authors reported experiments, particle simulations, and a continuum theory to identify the statistical properties (energy spectra and structure functions) of self-sustained turbulence in dense *B. subtilis* suspensions in quasi-2d and 3d closed geometries. In particular, the analysis of energy spectra in two-dimensional baths renders exponents different from classical: $E(q) \sim q^{+5/3}$ for small q, and $E(q) \sim q^{-8/3}$ for large q. This scaling was attributed to an apparent viscoelastic response of highly concentrated bacterial suspensions.

Moving to synthetic swimmers, the issue of active turbulence was addressed much more recently by Bourgoin et al. [41]. The considered system consists of a relatively small number of self-propelling disks, performing as Marangoni surfers at an air-liquid interface (see Sect. 3.1.2). These authors reported that

---

[2]In $d = 2$, due to the suppression of vortex stretching, there can be both an energy upward cascade, rendering again the exponent $-5/3$ at length scales larger than those of forcing, and an enstrophy-transfer downward cascade providing an exponent -3 in the energy spectrum at smaller scales [152].

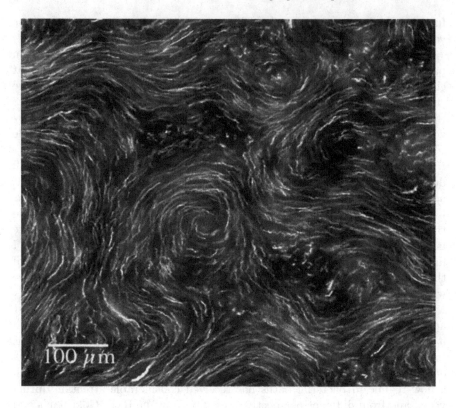

FIGURE 5.5
Bacterial turbulence. Dense suspension of swimming *Escherichia coli* displaying spontaneous collective motion at high concentration. The image is taken in the middle plane of two glass plates separated by 220 μm. Fluorescent tracers are dispersed in the suspension. Images are accumulated over few seconds to render the trace of the collective motion. Image provided by E. Clement PMMH-ESPCI, Paris / FAST-Univ. Paris Orsay / ICMCS, Univ. Edinburgh.

the statistical properties of the particles' velocities display a turbulent-like behavior, that is compatible with Kolmogorov's predictions. Interestingly, the analogy is found in the two representations that are commonly employed in classical turbulence: a *Lagrangian picture*, based on temporal correlations computed along particle trajectories, and following, as well, the more conventional *Eulerian picture* built from spatial correlations between equal-time instantaneous velocities.

A few more experimental scenarios are worth commenting, both pertaining to the category of driven systems. The first is based on a system of flocking ferromagnetic colloids (see Sect. 3.2.1.2) as analyzed by Kokot et al. [201]. The system is again a two-dimensional interfaced realization, but this time consists of a gas of self-assembled spinners rotating in either direction, while energized by a uniform uniaxial (perpendicular) alternating magnetic field. Authors demonstrate that spinners generate vigorous vortical and chaotic-like flows at the interface that are analyzed both experimentally and through computer simulations. It is worth noticing that the flow field refers in this case to the fluid flows elicited by the active units, rather than the swimming specimens themselves as was considered in the experiments by Bourgoin et al. mentioned previously. On what respects to turbulent spectra, authors report the inverse energy-scaling with exponent $-5/3$, consistent with high-Reynolds number two-dimensional turbulence, albeit $Re \approx 30$ in the spinners gas. A unique characteristics of this system is that activity originates from rotations only and is not associated with self-propulsion.

Still within the context of driven colloids, and, more precisely, by tuning the dynamic modes of Quincke rollers (see Sect. 3.2.2) under a pulsed electric driving, Karani et al. [191] were able to achieve controlled sequences of repeated particle runs and random reorientations. Authors find that a population of these random walkers exhibit behaviors reminiscent of bacterial suspensions such as mesoscale turbulent-like flows, even rendering similar exponent for energy spectra. Finally, using self-propelling Janus particles under an AC electric field based on the ICEO mechanism (see Sect. 3.2.4), trends of mesoscopic turbulence were reported by Nishiguchi et al. [282]. A broad peak in the energy spectrum of the velocity field appears at the spatial scales where the polar alignment and the cluster formation are observed. Exponents are extracted in this case in the range of small wave numbers.

I turn now attention into theoretical approaches. It is beyond the scope of this text to discuss in detail the different sorts of analyses that have been devoted during these last years to active turbulence. I limit myself to mention some representative references, and invite interested readers to go directly to the original reports to have a more comprehensive picture.

All the papers that are going to be mentioned in this section were more or less directly addressing the issue of bacterial turbulence. Hydrodynamic models, although not looking specifically at spectral properties, were early proposed by Aranson [10] and Wolgemuth [439]. Großmann et al. [144] reported a simple (dry) self-propelled particle model with short-range alignment and antialignment at larger distances. It is able to produce orientationally ordered states, periodic vortex patterns, and mesoscale turbulence. In the turbulent regime, these authors were able to reproduce quite consistently the experimental results for dense bacterial baths, although authors recognize themselves that obtained exponents were not universal, but dependent on the specific choice of parameters.

Other continuum-like models extend the Toner–Tu equations for polar flocks, thus incorporating nonlinear alignment and polarity self-advection terms (see Sect. 7.1.1). As a matter of fact, the original model proposed by

Wensink et al. ([431]) belong to this class. Non universality was again claimed and attributed by Bratanov et al. [42] to the coupling between the cubic non-linearity of this particular model with the quadratic advective one. This series of theoretical approaches is continued by the paper by James et al. [180] who employed conventional techniques borrowed from turbulence theory to propose a quantitative characterization of spectra. More recently, the perspective was widened by modeling not the swimmers themselves but the incompressible solvent that disperses them. In particular, Slomka et al. [360] predicted the existence of inverse energy cascades in three-dimensional active fluids, while Linkmann et al. [230] reached similar conclusions on two-dimensional systems. For a recent review on the topic of active turbulence, encompassing both polar and nematic symmetries, as well as systems with and without momentum conservation, interested readers might consult the review paper published by Alert et al. [4].

5.4   Thermodynamic Concepts in Active Matter

Finally, I would like to bring a new aspect into discussion before finishing this chapter that may help to better appraise the particularities of the study of collective effects in active systems. I refer to the recent attempts to extend to this context the powerful framework of thermodynamics. On what follows, I briefly comment on the two most characteristic concepts of what has been termed **active thermodynamics**. More details can be found in the review published by Takatori and Brady [390] dedicated entirely to this subject.

5.4.1   Active Temperature

Contrarily to what is standard in thermodynamic equilibrium, it is generally not possible to assign state variables to active systems in a meaningful way. This is already apparent on what respects to temperature.

Experimentally the question was addressed, as commented earlier, in relation to sedimentation profiles of active colloids [291, 131] (see Sect. 3.1.1.3). More recently, experiments by M. Han et al. considered the option of applying external electric and magnetic fields to orbitally circulating Janus particles, and an effective temperature was measured in terms of long time diffusion kinetics [153].

Turning to theoretical approaches, Loi et al. [238] considered an ensemble of interacting self-propelled particles in contact with an equilibrated thermal bath, and applied fluctuation-dissipation relations that allowed a definition of an **active temperature**. It was further shown that the calculated value was compatible with the results obtained using a tracer particle as a thermal probe. Such effective temperature takes a value that is higher than the temperature of the bath, and that is intrinsically controlled by the activity of the system. A more recent approach involving both the notion of an active temperature and a swim pressure (see below) was proposed by Takatory and Brady [389].

5.4.2 Active Pressure

Just because active pressure appeared in the specific literature as a pure conceptual development, here I start the discussion with reference to theoretical approaches, and later on I will refer to experiments addressing the same issue.

The notion of a **swim pressure** that is generated entirely through self-motion was put forward originally by Takatori et al. [391]. More precisely, it was formulated as the trace of the swim stress, i.e. the first moment of the self-propulsive force, being entirely athermal in origin, and characteristic of all active matter systems. As a matter of fact, the concept of active pressure is implicitly rooted on the notion that an active body would swim away in space unless confined by boundaries. The same authors proposed a non-equilibrium equation of state with pressure-volume phase diagrams that resemble Van der Waals loops for gas-liquid coexistence.

Simultaneously, Mallory et al. published numerical simulations to compute the equation of state of a suspension of spherical self-propelled nanoparticles in two and three dimensions [244]. A non monotonic dependence of pressure on temperature was found and interpreted with scaling arguments. The concept of pressure appears also under different formulations in simulations of self-propelled particles confined in two and three dimensions [105, 433], or in the presence of obstacles [280]. The notion of swim pressure in relation to a system of ABPs is also tackled in the more recent review by Marchetti et al. [246].

Other aspects related to the concept of active pressure are also worth mentioning. For instance, a sort of depleting effect, addressing the question of whether active matter can generate unexpectedly effective interactions between large objects immersed in a bath of self-propelled colloidal hard spheres and direct their further assembly, was analyzed by Ni et al. [279]. Moreover, a numerical model of active suspensions was used to compute a negative tension in a configuration of two coexisting, dilute and dense, phases [32]. In addition, osmotic effects arising from active solutes were considered by Lion et al. [232].

More recent comprehensive discussions on the concept of active pressure were published in a pair of papers by Solon et al. [369, 368]. At this point it is worth recalling that in thermodynamic equilibrium, the pressure, defined (mechanically) as the force per unit area exerted by a fluid on its containing vessel, coincides with alternative definitions either as a thermodynamic state function, obtained from the derivative of a free energy, or, finally, as the trace of a bulk stress tensor in the context of hydrodynamics. This equivalence is not longer true in general for active systems as emphasized in [368]. In [369], it is indeed demonstrated that mechanical pressure is a state function, i.e. it is independent of the wall-particle interactions, but only for the particular class of spherical ABPs with isotropic repulsions.

The key result is that in this case the mechanical pressure is the sum of an ideal gas contribution, and a non-ideal term stemming from interactions. The latter combines two contributions: one is a standard "direct" term (the density of pairwise forces acting across a plane), while the "indirect" term, absent in passive systems, accounts for the reduction in momentum flow caused by collisional slow-down of the particles, the idea put forward already in relation to MIPS. This indirect terms dominates at intermediate densities and it is responsible for the motility-induced phase separation phenomenon discussed earlier. The authors further show that adding up the ideal and the indirect terms one recovers the swim pressure originally introduced by Takatori et al. via a force-moment integral [391].

As a matter of fact, Solon et al. proved in [368] that for a wide class of systems, pressure depends on the precise interactions between active particles and confining walls, and thus, in general, pressure is not a state variable. The lack of an equation of state was also claimed. To get a taste on this issue, and, in particular, since it is relatively straightforward to obtain an explicit expression for the swim pressure, I briefly sketch the treatment in [369].

It refers to a finite density system of ABPs while neglecting hydrodynamic interactions and thus strictly applies to dry active systems (no momentum conservation). Realizations may correspond typically to motile entities that use a solid support close contact to get rid off linear momentum. Different contributions to the mechanical pressure are explicitly calculated and their meaning discussed on what follows. In particular, one recovers the notion of swim pressure, as the basic hallmark of the treatment.

Let's consider a 2d system of ABPs in the presence of walls that introduce a conservative wall-particle force $\mathbf{F}_w$. One assumes confining walls parallel to the y-axis and periodic boundary conditions along them. Interparticle forces are also retained and generically denoted $\mathbf{F}$. Taking for simplicity unit mobility one then have,

$$\dot{\mathbf{r}}_i = V_0 \mathbf{p}_i + F_w(x_i)\hat{\mathbf{x}} + \sum_{j \neq i} \mathbf{F}(\mathbf{r}_j - \mathbf{r}_i) + \sqrt{2D}\boldsymbol{\xi}_i, \qquad (5.15)$$

$$\dot{\mathbf{p}}_i = \sqrt{2D_r}\boldsymbol{\xi}_{r_i} \times \mathbf{p}_i, \qquad (5.16)$$

where in two dimensions $\mathbf{p}$ is expressed as usual in terms of an angle θ: $\mathbf{p} = (\cos\theta, \sin\theta)$. Following standard procedures, revisited several times in this text, one can convert this set of (single-particle) Langevin-like equations into an equation for the fluctuating distribution function $\hat{\psi}(\mathbf{r}, \theta; t)$, whose zeroth, first and second angular (d=2) harmonic components correspond, respectively, to the fluctuating particle density $\hat{\rho}$, polarization $\hat{\mathcal{P}}$ and nematic order $\hat{\mathcal{Q}}$

normal to the wall, i.e. $\hat{\rho} = \int \hat{\psi} d\theta, \hat{\mathcal{P}} = \int \hat{\psi} \cos\theta d\theta, \hat{\mathcal{Q}} = \int \hat{\psi} \cos 2\theta d\theta.$[3] Invoking particle conservation for impermeable boundaries, and integrating over noise terms (moments are then denoted without the hat notation), the total current in steady state is null,

$$0 = V_0\mathcal{P} + F_w\rho - D\partial_x\rho + I_1(x), \tag{5.17}$$

with $I_1(x) = \int F_x(\mathbf{r}' - \mathbf{r})\langle \hat{\rho}(\mathbf{r}')\hat{\rho}(\mathbf{r})\rangle d^2\mathbf{r}'$. The same procedure can be repeated for the polarization variable,

$$D_r\mathcal{P} = -\partial_x \left[\frac{V_0}{2}(\rho + \mathcal{Q}) + F_w\mathcal{P} - D\partial_x\mathcal{P} + I_2(x) \right], \tag{5.18}$$

the left integral being $I_2(x) = \int F_x(\mathbf{r}' - \mathbf{r})\langle \hat{\rho}(\mathbf{r}')\hat{\mathcal{P}}(\mathbf{r})\rangle d^2\mathbf{r}'$. Both integrals are x-functions only, given the translational invariance assumed along the y-axis.

The mechanical pressure on the wall is computed from an integral of the force exerted by the particle density. The wall is confining, i.e. $F_w(x)\rho(x) \to 0$ for $x >> 0$. It is further assumed that the wall force is conservative, i.e. derives from a potential whose range extends for $x > 0$, while its effect fades away at $x = \Lambda << 0$ well in the bulk of the fluid. Thus the reciprocal pressure exerted by the motile particles is just $P = -\int_\Lambda^\infty F_w(x)\rho(x)dx$. Using Eq. 5.17,

$$P = V_0 \int_\Lambda^\infty \mathcal{P}(x)dx + D\rho(\Lambda) + \int_\Lambda^\infty I_1(x)dx. \tag{5.19}$$

The integral involving $\mathcal{P}$ can be evaluated invoking Eq. 5.18 and noticing that $\mathcal{P}$ and $\mathcal{Q}$ vanish in the bulk and all terms vanish at infinity. Then,

$$P = \frac{V_0}{D_r} \left(\frac{V_0}{2}\rho(\Lambda) + I_2(\Lambda) \right) + D\rho(\Lambda) + \int_\Lambda^\infty I_1(x)dx. \tag{5.20}$$

The final result for the mechanical pressure can be written as a sum of three terms of very different nature,

$$P = P_0 + P_I + P_D. \tag{5.21}$$

The first contribution P_0,

$$P_0 = \left(D + \frac{V_0^2}{2D_r} \right) \rho(\Lambda), \tag{5.22}$$

is a pressure term that does not involve explicitly any sort of particle interaction (it is called "ideal" in the original Solon's paper ([369]). In fact, it is nice to express it in terms of the effective diffusion coefficient defined in relation to phoretic swimmers (see Sect. 3.1.1) i.e. $P_0 = D_{eff}\rho(\Lambda)$ (having set mobility

---

[3]In this particular section, I use $\mathcal{P}$ to denote a polarization variable and not to confuse it with the notation P employed here to indicate the total pressure, including active terms. In the next chapter, the polarization will be generically denoted in its vectorial form **P**, and reserve p for the pressure variable as in Navier-Stokes's like equations.

equal to 1, implies that units should be checked accordingly). The second term is an "indirect" pressure contribution with no passive counterpart,

$$P_I = \frac{V_0}{D_r} I_2(\Lambda). \tag{5.23}$$

Finally, the last term does not contain explicitly the activity, though indirectly activity appears through the two-point correlation function, and it can be considered a "direct contribution" formally identical for any classical system,

$$P_D = \int_{x>\Lambda} dx \int_{x'>\Lambda} F_x(\mathbf{r'} - \mathbf{r}) \langle \hat{\rho}(\mathbf{r'}) \hat{\rho}(\mathbf{r}) \rangle d^2 \mathbf{r'}. \tag{5.24}$$

The important point here is that the mechanical pressure P calculated in this way for interacting ABPs is a state function in the sense that can be calculated from bulk correlations but is independent of the particle-wall force $F_w(x)$.

The added contribution $P_0 + P_I$ can be alternatively written as,

$$P_0 + P_I = \left(D + \frac{V_0 V(\rho)}{2D_r} \right) \rho, \tag{5.25}$$

with a density-dependent velocity, a concept previously introduced in the discussion of MIPS,

$$V(\rho) = V_0 + 2I_2/\rho. \tag{5.26}$$

Moreover if one neglects translational diffusion, i.e. $D = 0$, the sole contribution is the swim pressure introduced by Takatory et al. [391] ($P_S = \frac{\rho}{2} \langle \mathbf{r} \cdot \mathbf{F}_a \rangle$, and $\mathbf{F}_a = V_0 \mathbf{e}$ being the propulsion force),

$$\frac{V_0 V(\rho)}{2D_r} \rho = P_S. \tag{5.27}$$

Thus for $D = 0$, one alternatively has,

$$P = P_S + P_D, \tag{5.28}$$

that confirms that P_S defined as a (bulk) propulsion force determines together with P_D the pressure on a confining wall.

For an experimental analysis of the concept of active pressure one could consult the paper by Junot et al. [185] dealing with vibrated disks ([81] (see Fig. 5.6). Authors study the mechanical pressure exerted by a set of respectively passive isotropic and self-propelled polar disks onto two different flexible unidimensional membranes (different number of beads with different diameters and separation). In the case of the isotropic disks, the mechanical pressure, inferred from the shape of the membrane, is identical for both membranes and follows the equilibrium equation of state for hard disks. On the contrary, for the self-propelled disks, the mechanical pressure strongly depends on the

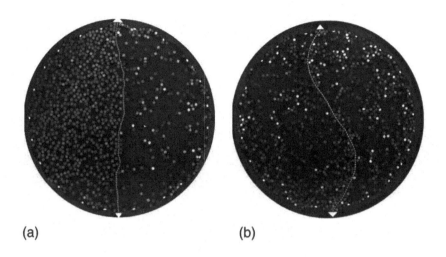

(a)   (b)

FIGURE 5.6
Mechanical pressure for a system of active hard disks. (a) Mechanical equilibrium between a small number of self-propelled polar disks (right) and a large number of isotropic passive disks (left). (b) Membrane instability when self-propelled disks are distributed in equal number on both sides. Image and caption text adapted from Junot et al. [185].

membrane in use and, thus, is not a state variable. When self-propelled disks are present on both sides of the membrane, the membrane develops an instability akin to the one predicted theoretically for active Brownian particles against a soft wall following the analysis by Nikola et al. [280].

6

Modeling Active Fluids

In the two remaining chapters, I aim at providing the reader with a self-contained, albeit brief, presentation of the theoretical corpus applying to active fluids. From the very beginning, I declare myself more interested in continuum, i.e. coarse-grained descriptions, rather than in more microscopically, and likely more system-dependent, motivated approaches. Singularly in relation to the former, active systems are classified in terms of their basic symmetries, and the eventually existing conservation rules that constrain their observables. In particular, such a conservation principle in relation to linear momentum is advantageously exploited to separate fluid-like systems discussed next, from systems where friction is dominant that are postponed to the next chapter.

The theoretical description of active fluids has been extremely successful during the last couple of decades. Two parallel, and very generic, versions of this theoretical doctrine correspond to what could be termed as *vectorial and tensorial descriptions of active fluids*. As a matter of fact, both perspectives share a common spirit, i.e the aim to adapting to active anisotropic fluids what was pursued more than fifty years ago to theoretically describe the hydrodynamics of Liquid Crystals.

The vectorial scheme minimally extends the *Leslie-Ericksen* nematodynamics [77] approach to active fluids. This classical body of study in the theory of Liquid Crystals can be consulted by interested readers in the book by De Gennes [77] that was already mentioned in relation to the basic notions introduced in the first chapter. On the other hand, the tensorial level of modeling is rooted in the *Beris-Edwards* treatment, originally aimed at obtaining constitutive equations for polymeric liquid crystals [97]. In relation to active nematics, a nematodynamic approach supposes to concentrate on the dynamics of the (nematic) director n̂, including an actives source of stress, but neglecting eventual spatial inhomogeneities of the scalar (nematic) order parameter. This latter approximation can be naturally relaxed by employing the second approach based on the tensorial (nematic) order parameter Q (see Sect. 2.2.2).[1]

---

[1]A caution remark is worth making at this point. Both, Leslie-Ericksen and Beris-Edwards methodologies suppose continuum descriptions aimed at describing nematic fluids in three-dimensional environments. Here both models will be employed in relation to nematic suspensions, most typically in two-dimensional preparations. Thus, strictly speaking neither model has a first-principle justification for this latter situation. In this respect, an important issue is whether or not the usually invoked incompressibility condition is fully justified when applied to interfacial flows. Another perspective from which to address the

DOI: 10.1201/9781003302292-6

The nematodynamic-like approach has the advantage that in some cases can be worked out even analytically, at the expenses of neglecting the proper consideration of singularities (defects). Conversely, analytics is practically discarded in the tensorial approach towards pure numerically-based solutions. The first two sections in this chapter are devoted respectively to each one of the just introduced schemes, introducing them first and consideting in each case, a few worked out examples of application. The remaining sections address more specific questions, for instance confinement or constraining effects on active fluids.

Before ending these introductory remarks, and to be honest with other approaches outside the pair of strictly liquid crystal-like theories just mentioned, I would like to recognize a few theoretical efforts to describe active fluids in reasonably more specific ways. With a completely subjective criterion, I only highlight in this introduction four of these contributions that have the merit, each on its own way, to build up continuum descriptions starting from microscopic models. Ordered following a loosely chronological perspective, I quote the already cited paper published by Simha et al. [1], devoted to the analysis of instabilities in ordered suspensions, and the similarly aimed contribution by Saintillan et al. [335]. More oriented towards rheological aspects of active fluids I cite Liverpool et al. [237], and the paper by Baskaran et al. [19] that is motivated by the study of bacterial suspensions.

6.1   Linearized Leslie-Ericksen Theories for Active Polar Fluids

The ultimate goal of this theoretical corpus is to generate hydrodynamic equations for active fluids, assuming that the latter operate close to thermodynamic equilibrium. This justifies the use of linear approximations for the constitutive relations between fluxes and forces within the spirit of the Onsager formulation of the *linear irreversible thermodynamics*, as elaborated by de Groot et al. [78].

---

same question concerns the density of the microtubules suspension. It is commonly taken as constant in most of the applications, although this is not apparent at all in real experiments, singularly considering the unavoidable presence of topological defects, regions in the assembled material which are void of microtubules. From this point of view, a major theoretical advancement would be to consider the active suspension as, at least, a two component fluid, and retain fully the concentration of active specimens as a state variable, with its corresponding dynamics incorporated into the scheme of basic equations, extending in this way the Leslie-Ericksen and Beris-Edwards approaches. A few phenomenological approaches have been formulated following these directions, like Voituriez et al. [424] in relation to Leslie-Ericksen formulation, or Giomi et al. [138, 133] in relation to Beris-Edwards scheme, but, still, none of them can be strictly considered first-principles formulated.

The justification of this hypothesis is far from obvious for the most part of the realistic protein-based fluid systems one can prepare in the laboratory and that I have referred to in Chapter 4. As several times mentioned, their activity is based on the chemical conversion of chemical energy suppliers. This is mediated by a chemical reaction whose driving force comes from an imbalance of chemical potentials between products and reactants (for instance ATP and ADP). Whether this driving is, or under what conditions it may be considered, "weak", justifying in turn this strategy, is a question difficult to answer *a priori*, and likely very much system-dependent.

Nevertheless, the use of such Onsager-like formulation is conceptually very powerful and elegant, while it permits to systematically address quite a different number of scenarios both for polar and nematic symmetries, as I will comment on what follows. In fact, the nematodynamic approach proposed by Leslie-Ericksen, mentioned in the introductory remarks to this chapter was constructed exactly under the same spirit [77]. The idea now is to extend this perspective from classical to **active liquid crystals**, introducing the minimal necessary complexity to account for the new ingredient, i.e that of the activity of the liquid crystal material.

I consider very instructive to present on what follows a detailed, albeit brief, treatment of the theory of active liquid crystals (theory of active gels as it is sometimes known in the literature). I invite the interested reader to consult the review papers by Kruse et al. [206], Joanny et al. [182] and Julicher et al. [184] to find a more comprehensive derivation of this theory. For the sake of simplicity, I will focus on the simplest scenario, consisting in a single component system. I thus admit that I am neglecting permeation effects pertinent to real gels [50]. Viscoelastic effects are not going to be explicitly considered either, though I will minimally comment on them later on. Moreover, I neglect as well extensions of the basic scheme that appeared later on in the literature, and specifically aimed at assessing the role of fluctuations [21], or incorporating chiral effects [116].

6.1.1 General Scheme of Equations

As in any hydrodynamic theory, we need to retain only coarse-grained variables of our system, i.e. those varying on large spatial scales and long times. Within this category, we consider the number density ρ, the **polarization vector p**,[2] and the momentum density $\mathbf{g} = \rho m \mathbf{v}$, where m is an effective mass of the active constituent and $\mathbf{v}$ represents the local velocity of the active fluid. The basic idea underlying the whole approach is to derive an expression

---

[2] Although this piece of theory will be mostly used later in relation to the nematic director and its corresponding symmetries, I prefer here to respect the classical formulation planned for a polarization vector. This equivalence is justified later at the end of this subsection. Also notice that without any possibility of confusion, I employ the same notation to indicate here the polarization vector as a coarse-grained variable, and that employed earlier to designate the orientation variable when I introduced the ABP model in Sect. 3.1.1.4.

for the entropy production that further would permit to identify fluxes and forces [78], and establish in this way constitutive equations for the system variables.

Considering the system at a constant temperature, the rate of production of entropy is related to the rate of change of its total free energy, i.e $T\dot{S} = -d\mathcal{F}_t/dT$. For a passive (conventional) liquid crystal-like system there are three essential contributions to the free energy. The first one arises from the system flow velocity, and corresponds to the density of kinetic energy $(1/2)\rho m\mathbf{v}^2$. The pair of remaining contributions are better understood when writing them in differential form for local variations of the free energy density. One of the contributions arises from the (scalar) chemical potential, as the field associated to the density, and the last one is a (vectorial) **molecular (orientational) field h** ([77]) conjugated to the polarization, i.e. $df = \mu d\rho - h_\alpha dp_\alpha$ (with the usual convention of summing over repeated indices).

Distinctively for active liquid crystals, we add to this set of free energy contributions the active part written in terms of the rate of ATP hydrolysis and the associated imbalance of chemical potential. This particular contribution to the rate of change of free energy per unit volume is expressed simply in a linearized way as $-r\Delta\mu$. Adding all the contributions we have,

$$T\dot{S} = \int \left\{ -\frac{\partial}{\partial t}\left(\frac{1}{2}\rho m\mathbf{v}^2\right) - \mu\frac{\partial \rho}{\partial t} + h_\alpha \dot{p}_\alpha + r\Delta\mu \right\} d\mathbf{r}. \qquad (6.1)$$

The next step is to write the corresponding conservation laws. The two conserved quantities are the density and linear momentum,[3]

$$\frac{\partial \rho}{\partial t} + \boldsymbol{\nabla} \cdot (\rho\mathbf{v}) = 0, \qquad (6.2)$$

$$\frac{\partial g_\alpha}{\partial t} + \partial_\beta \Pi_{\alpha\beta} = 0, \qquad (6.3)$$

where the momentum flux is written as the sum of two terms: $\Pi_{\alpha\beta} = \rho m v_\alpha v_\beta - \sigma^t_{\alpha\beta}$. The first term is the so-called *Reynolds stress*, while $\sigma^t_{\alpha\beta}$ denotes the *total stress*. Under the usual assumption of small Reynolds number the first contribution can be safely neglected.[4]

Using the conservation laws and after integration by parts, the entropy production can be rewritten as,

$$T\dot{S} = \int \left\{ \sigma_{\alpha\beta}v_{\alpha\beta} + P_\alpha h_\alpha + r\Delta\mu \right\} d\mathbf{r}, \qquad (6.4)$$

in terms of the *symmetric deviatoric stress tensor* $\sigma_{\alpha\beta}$. In anisotropic systems, the stress is not symmetric, since there are contributions associated to torques.

---

[3]Notice that conservation of linear momentum qualifies this approach as wet. For a discussion on the difference between "wet" and "dry" approaches dominated by friction, I direct the reader to the introductory paragraphs of Chapter 7.

[4]This assumption has been questioned in the context of active (nematic) turbulence in a recent paper published by Koch et al. [198].

Such a term was calculated in Chapter 5 of de Gennes's book [77], and amounts to a contribution $\sigma_{\alpha\beta}^a = 1/2(p_\alpha h_\beta - h_\alpha p_\beta)$. Thus the total stress tensor is expressed as,

$$\sigma_{\alpha\beta}^t = \sigma_{\alpha\beta} + \sigma_{\alpha\beta}^a - P\delta_{\alpha\beta}, \tag{6.5}$$

containing the usual pressure term. The velocity gradient tensor has been separated into its symmetric (*strain rate*) and antisymmetric (*vorticity*) components,

$$v_{\alpha\beta} = (1/2)(\partial_\alpha v_\beta + \partial_\beta v_\alpha) \qquad \omega_{\alpha\beta} = (1/2)(\partial_\alpha v_\beta - \partial_\beta v_\alpha), \tag{6.6}$$

the latter entering into the definition of the comoving and corotational derivative of the polarization written as,

$$P_\alpha \equiv \frac{Dp_\alpha}{Dt} = \frac{\partial p_\alpha}{\partial t} + v_\beta \partial_\beta p_\alpha + \omega_{\alpha\beta} p_\beta. \tag{6.7}$$

Finally, the molecular field **h** is calculated by taking functional derivatives of the *polarization free energy*, here considered in its pure elastic form. Its explicit form follows what is usual in nematic liquid crystals, i.e the Frank-Oseen form (see Sect. 2.2.3), incorporating here a spontaneous splay term permitted by symmetry. The modulus of the polarization is assumed to be fixed far from the isotropic/nematic transition, and taken arbitrarily at unit value. Otherwise, as commented earlier in Sect. 2.2.3, the polarization free energy would contain additional terms in a *Landau-de Gennes*–like expansion in powers of the polarization magnitude.

We write the polarization free energy in its more general form appropriate to three-dimensional systems, and free from the condition of equal elastic constants. Particular applications of this formalism that will be worked out later on will make the usual assumption of equal elastic constants, and particularized to 2d systems we will omit the second contribution associated to twist distortions. The polarization free energy reads,[5]

$$\mathcal{F}_p = \int \left\{ \frac{K_1}{2} (\mathbf{\nabla} \cdot \mathbf{p})^2 + \frac{K_2}{2} [\mathbf{p} \cdot (\mathbf{\nabla} \times \mathbf{p})]^2 + \frac{K_3}{2} [\mathbf{p} \times (\mathbf{\nabla} \times \mathbf{p})]^2 \right. \tag{6.8}$$
$$\left. + k\mathbf{\nabla} \cdot \mathbf{p} - \frac{1}{2} h_\parallel^0 p^2 \right\} d\mathbf{r}.$$

After integration by parts, the spontaneous splay term can be associated to boundary effects and it is normally not important. Finally the $h_\parallel^0$ term, parallel to the polarization, is interpreted as a Lagrange multiplier that permits to guarantee a fix unit modulus for the polarization.

---

[5]Notice that the here used variable **p** for the polarization field is totally equivalent to the director field denoted n̂ in Sect. 2.2.2. In doing this, I respect what is the conventional notation employed in these different contexts.

The entropy production given above in Eq. 6.4 is written as the sum of three contributions, each of them expressed as a product of a *flux* and a *force*. More precisely $v_{\alpha\beta}$, h_α and $\Delta\mu$ are identified as forces, and the associated fluxes are, respectively, $\sigma_{\alpha\beta}$, P_α and r. For what follows it is important to identify the time-reversal symmetries of the forces: $v_{\alpha\beta}$ is odd under time-reversal, while h_α and $\Delta\mu$ are even.

In the spirit of the linear irreversible thermodynamics, the next step is to use *Onsager relations* applied to the most general linear relations that express fluxes in terms of forces. The only limitation is to respect the translational and rotational symmetries. On the other hand, the rate of change of the free energy written above in terms of the time derivative of the entropy is clearly a dissipative contribution, and thus only the dissipative contribution of each flux contributes to the entropy production. Therefore, we need to separate fluxes into *dissipative* and *reactive* contributions, the former with the same symmetry as the corresponding force, while reactive parts have opposite time signature to the forces they come from. The two basic different orientational measures of the system are represented by the unit vector **p** and the unit trace nematic tensor **Q**, i.e. $Q_{\alpha\beta} = p_\alpha p_\beta - (1/3)\delta_{\alpha\beta}$.[6] Both, together with the strain and vorticity tensors, will play a fundamental combined role on what follows.

We now write the Onsager linear equations which in this context are known as *constitutive equations*. Those relations involve contributions with equal time-signature of fluxes and forces, and will be symmetric when coupling dissipative flux contributions and antisymmetric for reactive flux contributions.

One final remark is in order. Handling of the different terms will be simpler after splitting the tensor terms into diagonal and traceless components. This will permit later on to easily implement, for instance, the normally invoked condition of flow incompressibility. Thus we write $\sigma_{\alpha\beta} = \sigma\delta_{\alpha\beta} + \tilde{\sigma}_{\alpha\beta}$, with $\sigma = (1/3)\sigma_{\alpha\alpha}, \tilde{\sigma}_{\alpha\alpha} = 0$. Similarly $v_{\alpha\beta} = (u/3)\delta_{\alpha\beta} + \tilde{v}_{\alpha\beta}$, with $u = \partial_\alpha v_\alpha$ which is the divergence of the flow velocity field. Finally we add reactive and dissipative contributions,

$$\sigma_{\alpha\beta} = \sigma_{\alpha\beta}^r + \sigma_{\alpha\beta}^d, \tag{6.9}$$

$$P_\alpha = P_\alpha^r + P_\alpha^d, \tag{6.10}$$

$$r = r^r + r^d. \tag{6.11}$$

We start by writing the equations in the symmetric scheme involving the dissipative contributions of the fluxes (fluxes and forces with with the same time-reversal symmetry),[7]

---

[6] In relation to the generic definition given in Sect. 2.2.2, the prefactor $Q(T)$ is taken here arbitrarily unity.

[7] On writing next the set of constitutive equations, I follow the symbol convention that was employed in the papers where the theory of active gels was originally introduced, for instance those of Kruse et al. [207] and Voituriez et al. [423]. Also for the same reason, I change the notation for the activity parameter to $\zeta\Delta\mu$ instead of using a single parameter α as was done in Chapter 4.

$$\sigma^d = \bar{\eta}u, \tag{6.12}$$

$$\tilde{\sigma}^d_{\alpha\beta} = 2\eta\tilde{v}_{\alpha\beta}, \tag{6.13}$$

$$P^d_\alpha = \frac{h_\alpha}{\gamma_1} + \lambda_1 p_\alpha \Delta\mu, \tag{6.14}$$

$$r^d_\alpha = \lambda\Delta\mu + \lambda_1 p_\alpha h_\alpha. \tag{6.15}$$

Notice than to simplify the scheme we have neglected the tensorial nature of the viscosity, and we have retained simply two components, with η, $\bar{\eta}$, denoting, respectively, the conventional shear and longitudinal components. γ_1 itself stands for a rotational viscosity.

Now we turn into the antisymmetric scheme involving the reactive contributions (fluxes and forces with opposite time-reversal symmetry),

$$\sigma^r = -\bar{\zeta}\Delta\mu + \bar{\nu}_1 p_\alpha h_\alpha, \tag{6.16}$$

$$\tilde{\sigma}^r_{\alpha\beta} = -\zeta\Delta\mu Q_{\alpha\beta} + \frac{\nu_1}{2}(p_\alpha h_\beta + p_\beta h_\alpha - \frac{2}{3}p_\gamma h_\gamma \delta_{\alpha\beta}), \tag{6.17}$$

$$P^r_\alpha = -\bar{\nu}_1 p_\alpha \frac{u}{3} - \nu_1 p_\beta \tilde{v}_{\alpha\beta}, \tag{6.18}$$

$$r^r = \bar{\zeta}\frac{u}{3} + \zeta Q_{\alpha\beta}\tilde{v}_{\alpha\beta}. \tag{6.19}$$

In the above equations, ν_1 and $\bar{\nu}_1$ denote the flow-alignment parameters respectively associated to shear and compressional flows. The rest of parameters associated to the active terms in both the set of symmetric and antisymmetric Onsager relations have been introduced in a complete phenomenological way. The sign of the activity term $\zeta\Delta\mu$ is taken negative for contractile systems as in the actin cytoskeleton. A positive value corresponds, conversely, to an extensile stress, as observed for the most part of realizations of the microtubule/kinesin widely discussed in the previous chapter.[8]

In the most common situation when the fluid is assumed incompressible, the symmetric part of the stress tensor can be adsorbed into the pressure, and the scheme of equations get slightly simplified since the corresponding coefficients of the general scheme can be taken null, i.e. $\bar{\eta} = \bar{\nu} = \bar{\zeta} = 0$.

It is interesting to notice that the Onsager relations just written display an invariance under the transformation $\mathbf{p}$ to $-\mathbf{p}$.[9] This means that the description

---

[8]A couple of comments are worth doing here. First, an additional term appearing in the stress tensor as an Ericksen-like contribution, typical of classical nematodynamics (see [77]), and written in the form $\frac{\delta F}{\delta(\partial_\beta p_\gamma)}\partial_\alpha p_\gamma$ [182] has been neglected since it involves higher order contributions in gradients of the polarization field, assumed implicitly small in the most common applications of the theory of active gels. Second, in several worked out examples of application of the scheme just presented, the coefficient λ_1 appears later in a particular parameter combination, (see [207] for instance), that redefines the activity parameter.

[9]Remember that the molecular field $\mathbf{h}$ is defined in terms of a functional derivative with respect to $\mathbf{p}$ of a free energy (purely reorientational) that is functionally quadratic in $\mathbf{p}$, once the spontaneous, normally irrelevant, splay term is neglected.

above applies similarly to a system with nematic symmetry by assimilating **p** to the orientational nematic field n̂.

On what follows, I proceed to apply this scheme to particular scenarios of the dynamics of active gels to better appraise the power of the just derived theoretical framework.

6.1.2   Analysis of +1 Defects: Asters, Vortices, and Spirals

One of the first applications of the theory of active gels I am aware of was worked out by Kruse et al. [207]. The aim of the paper was to assess the stability of asters and vortices considered as the basic self-organizing patterns in a system of microtubules and kinesins. I remind the reader that these patterns followed from the observations reported by Nédélec et al. [273] (see Sect. 4.1.2).

In fact, the scenario analyzed by Kruse et al. extends the scheme that has been just presented in the sense that it includes viscoelastic effects. Viscoelasticity is incorporated through a Maxwell model that introduces a *elastic/viscous relaxation time* τ, defined as the ratio between the viscous and elastic modulus E, i.e. $E = \eta/\tau$. The system thus properly becomes a viscous fluid only at time scales longer than τ.

Grouping the different terms, the previously obtained generalized flux-force Onsager relations extended to include viscoelastic effects permit to obtain a pair of separate equations for the velocity and polarization fields,[10]

$$2\eta v_{\alpha\beta} = \left(1 + \tau\frac{D}{Dt}\right)\left\{\sigma_{\alpha\beta} + \zeta\Delta\mu p_\alpha p_\beta - \frac{\nu_1}{2}(p_\alpha h_\beta + p_\beta h_\alpha)\right\}, \qquad (6.20)$$

$$\frac{Dp_\alpha}{Dt} = -\nu_1 p_\beta v_{\alpha\beta} + \frac{1}{\gamma_1}\left(1 + \tau\frac{D}{Dt}\right)h_\alpha + \lambda_1 p_\alpha \Delta\mu, \qquad (6.21)$$

$$r = \zeta p_\alpha p_\beta v_{\alpha\beta} + \lambda\Delta\mu + \lambda_1 p_\alpha h_\alpha. \qquad (6.22)$$

The convective derivative of the stress tensor is defined here as,

$$\frac{D\sigma_{\alpha\beta}}{Dt} = \frac{\partial\sigma_{\alpha\beta}}{\partial t} + (v_\gamma\partial_\gamma)\sigma_{\alpha\beta} + [\omega_{\alpha\gamma}\sigma_{\gamma\beta} + \omega_{\beta\gamma}\sigma_{\gamma\alpha}]. \qquad (6.23)$$

Notice that the previous equations obviously reduce to those written down for the generic scheme presented in the previous subsection in the limit of a zero value of τ.

These equations were solved by Kruse et al. in relation to singular solutions of the polarization field, i.e. defects, of topological charge +1.[11] To describe

---

[10]In Kruse's original reference there is no term with τ appearing in the equation for the polarization field. The form here employed appears in [184], although is different from what is expressed in [206]. However, this omission has no effect on the final results discussed on what follows regarding defect stability since we will be looking for long-time stationary solutions.

[11]It is worth remarking that the treatment of polar systems justifies to deal here with defects of integer charge, as defects in nematic systems in $d = 2$ would have associated $\pm 1/2$ values, as extensively commented in Chapter 4.

them one uses polar coordinates (r, θ), while reserving ψ for the polarization angle, i.e. $p_r = \cos \psi, p_\theta = \sin \psi$. Adapting the set of equations in steady state to the polar coordinates framework, one obtains,

$$\nu_1 v_{r\theta} \sin 2\psi = h_\parallel / \gamma_1 + \lambda_1 \Delta\mu, \tag{6.24}$$

$$v_{r\theta}(1 - \nu_1 \cos 2\psi) = -h_\perp / \gamma_1, \tag{6.25}$$

from the equation for the polarization, and,

$$2\eta v_{r\theta} = \sigma_{r\theta} + \frac{\zeta}{2}\Delta\mu \sin 2\psi - \frac{\nu_1}{2}(h_\parallel \sin 2\psi + h_\perp \cos 2\psi), \tag{6.26}$$

from the equation for the stress.

This system of equations would be closed using the equations for the molecular field and the condition of momentum conservation. The field $h_\perp = -\delta \mathcal{F}_p / \delta\psi$ is calculated from the polarization free energy, reduced to splay and bend terms given the $d = 2$ dimensionality,

$$h_\perp = (K + \Delta K \cos^2 \psi)\left[\psi'' + \frac{\psi'}{r}\right] - \frac{\Delta K}{2} \sin 2\psi \left[\frac{1}{r^2} + \psi'^2\right], \tag{6.27}$$

where primes denote derivative with respect to r. K stands for the splay constant K_1, while $K + \Delta K$ denote the bend elastic modulus. Force balance leads on the other hand,

$$\partial_\beta \left(\sigma_{\alpha\beta} + (p_\alpha h_\beta - p_\beta h_\alpha)/2 - P\delta_{\alpha\beta}\right) = 0, \tag{6.28}$$

which contains also the antisymmetric part of the stress tensor.[12]

Without active terms (i.e. for ferroelectric nematics), solutions of these equations correspond to topological singularities, in absence of flow, in the form pf *asters, vortices and spiral defects*. Under appropriate boundary conditions, asters correspond to splay-favored solutions with $\psi = 0$ or $\psi = \pi$, while vortices correspond to bend-favored solutions with $\Psi = \pm\pi/2$. Spirals appear as mixed solutions for $\Delta K = 0$ with constant angle ψ_0. Under the usual single constant approximation the three solutions are degenerated. For sufficiently active systems in a contractile regime, (rotating) spiral defects, accompanied by flow patterns of the corresponding symmetry, are obtained, under the equal constant approximation, as the most stable solutions, with a polarization orientation and velocity given as,

$$\cos(2\psi_0) = 1/\nu_1, \tag{6.29}$$

$$v_\theta(r) = \omega_0 r \log(r/r_0), \tag{6.30}$$

---

[12]This latter term was missing in the original paper by Kruse et al. [207], but corrected later in a published erratum [208]. As quoted by Kruse et al. in the erratum note, the stability diagram (see main text and Fig. 6.1 in [207]) does not change qualitatively.

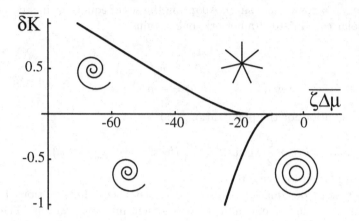

FIGURE 6.1
Analysis of defects in polar active gels. The panel depicts the stability
diagram of asters, vortices and spirals. Image and caption text adapted from
Kruse et al. [207].

where flow-aligning conditions have been naturally assumed ($|\nu_1| > 1$), to
justify the stable polarizarion orientation, similar to conventional nematics
under shear flow. The magnitude of the angular velocity is, as expected, a
linear function of the activity term. The explicitly quoted result is,

$$\omega_0 = \frac{2\sin 2\psi_0}{4\eta + \gamma_1 \nu_1^2 \sin^2 2\psi_0} \tilde{\zeta}\Delta\mu, \qquad (6.31)$$

written in term of an effective activity coefficient $\tilde{\zeta} = \zeta + \nu_1\gamma_1\lambda_1$.

The boundaries of stability of asters and vortices are calculated explicitly
in the original Kruse et al. paper [207] (see Fig. 6.1). This example is also
worked out up to the minimum detail in the main text and appendices of
[206].

6.1.3   Activity-Induced Flows from Aligned States

Another early and very interesting application of the theory of active polar
fluids was the analysis of the spontaneous flow transition induced by activity
on an, otherwise, aligned and resting configuration. This feature has been
several times mentioned earlier in the manuscript (see for instance the papers
by Simha et al. [1] and Saintillan [335] quoted above in the introductory
paragraphs of this chapter, or the reported observation by Martínez-Prat et
al. [253] in Sect. 4.2.1 for an experimental realization in microtubule-based

active nematic systems). The detailed original treatment was published by Voituriez et al. [423].

The bottom line conclusion is that aligned active polar gels are prone to undergo a *bend-like instability for extensile systems, or a splay-like torsion under contractile conditions*, with flows that are brought about accompanying the reorientation of the aligned material. This renders this scenario equivalent to a driven nematic liquid crystal, orientationally rotated from a uniformly aligned configuration (i.e the popular Freederickz transition of electric or magnetic origin in conventional liquid crystals [77]), although now the final state displays active motion. Although strictly speaking the analysis of Voituriez refers to a channeled system, and thus could be as well commented later on in relation to confined active fluids (see Sect. 6.4), I prefer to do it here since the underlying concept is itself that of the intrinsic instability of the aligned and motionless fluid. As a matter of fact, very strong confinement conditions suppress the instability as it will become clear on what follows, a scenario that is also confirmed by the experiments reported in Sect. 4.4.2.

I illustrate the treatment for an active gel in contact with a boundary that imposes strong planar anchoring at $y = 0$, and at a free surface at $y = h$. Stationary states are examined in this slab considered translationally invariant along the x-axis. I look for solutions $p_x = \cos\theta(y)$, $p_y = \sin\theta(y)$. Velocity is along the x-direction, thus the single component of the vorticity tensor is given by $\omega = (1/2)\partial_y v_x$. Force balance is simply written as $\partial_y \sigma_{yx} = 0$. Since at the free surface the shear stress is null $\sigma_{yx} = 0$ everywhere. By transforming the components of the molecular field ($h_\parallel = h_x \cos\theta + h_y \sin\theta$, $h_\perp = -h_x \sin\theta + h_y \cos\theta$), the first two constitutive equation lead to the result for the velocity field gradient,

$$w = \frac{(\zeta\Delta\mu/\gamma_1)\sin 2\theta}{4\eta/\gamma_1 + 1 + \nu_1^2 + 2\nu_1 \cos 2\theta}, \tag{6.32}$$

and an equivalent equation for the perpendicular component of the molecular field,

$$h_\perp = \frac{\zeta\Delta\mu(1 + \nu_1 \cos 2\theta)\sin 2\theta}{4\eta/\gamma_1 + 1 + \nu_1^2 + 2\nu_1 \cos 2\theta}. \tag{6.33}$$

Retaining the lowest order in θ in this last expression, and since $h_\perp = K\partial_y^2\theta$ under the hypothesis of equal elastic constants, one is lead to a differential equation for the reorientation angle of the form,

$$\frac{d^2\theta}{dy^2} + \frac{\theta}{L^2} = 0, \tag{6.34}$$

written in terms of a characteristic length L,

$$\frac{1}{L^2} = -\frac{2\zeta\Delta\mu(1 + \nu_1)}{K[4\eta/\gamma_1 + (1 + \nu_1)^2]}. \tag{6.35}$$

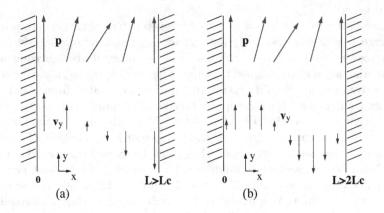

FIGURE 6.2
Spontaneous flow transition of active polar gels. (a) With free slip
domain walls the system displays a spontaneous flow transition for $L > L_c$. (b)
With no-slip domain walls the system displays a spontaneous flow transition
for $L > 2L_c$. Notice that the axes prescription in this figure is reverse to that
prescribed in the main text. Image and caption text adapted from Voituriez
et al. [423].

For a contractile system $\zeta \Delta \mu$ is negative and assuming $1 + \nu_1 > 0$, this
characteristic length is real. Note that in general the onset of spontaneous flow
is controlled by the sign of the product $\zeta \Delta \mu (1 + \nu_1)$. The flow-alignment coefficient ν_1 can in general have both positive and negative values and depends
both on the shape of the active units and the degree of nematic order. Deep in
the nematic state, $\nu_1 < -1$ corresponds to elongated rod-like particles, while
$\nu_1 > 1$ to disk-like.

The final outcome of this analysis is that a polarizarion field $\theta = \theta_0 \sin(y/L)$ is possible whenever h is larger than $L_c = \pi L$. In this case, the
state of quiescent uniform alignment gets unstable in front of a polarized and
sheared configuration of amplitude given by θ_0 (see Fig. 6.2). This prediction
for a shear flow regime is confirmed with the experimental observations by
Hardoüin et al. [156] mentioned in Sect. 4.4.2.

Voituriez et al. [424] published a little bit later an extension of this approach reporting a phase diagram in the case of two-dimensional compressible
active polar films. The gel is assumed to display a weakly fluctuating density
$c(\mathbf{r}) = c_0 + \rho(\mathbf{r})$. The theoretical scheme corresponds to the set of linear hydrodynamic equations just reviewed with a couple of differences that directly

come from the assumption of compressibility. One difference is at the level of the polarization free energy (see Eq. 6.8) that is now expressed as,

$$
\mathcal{F}_p = \int \left\{ \frac{K_1}{2} (\boldsymbol{\nabla} \cdot \mathbf{p})^2 + \frac{K_3}{2} [\mathbf{p} \times (\boldsymbol{\nabla} \times \mathbf{p})]^2 + k \boldsymbol{\nabla} \cdot \mathbf{p} - \frac{1}{2} h_\parallel^0 \mathbf{p}^2 \right.
$$

$$
\left. + \omega \rho \boldsymbol{\nabla} \cdot \mathbf{p} + \frac{\beta}{2} (\nabla \rho)^2 + \frac{\alpha}{2} \rho^2 \right\} d\mathbf{r}. \tag{6.36}
$$

The three added terms can be easily interpreted. The first one is the linear splay term now additionally coupled to the variable density field. The last term with positive coefficient α accounts for the compressibility, and the contribution with positive coefficient β accounts for the density fluctuations whose length scale is given by $((\beta/\alpha)^{1/2})$. The second difference with the traditional scheme is found at the level of the force balance equation (see Eq. 6.28) now rewritten in terms of a pressure which is given by $P = \frac{\delta F}{\delta \rho} = \omega \boldsymbol{\nabla} \cdot \mathbf{p} + \alpha \rho - \beta \Delta \rho$. Modulated flowing phases and a macroscopic phase separation at high activities are predicted. I refer the reader to the original reference for more details.

6.1.4 Minimal Version for a Two-Dimensional Active Nematic in Absence of Flow-Alignment

An even simpler form of the just introduced scheme can be employed under the assumption of negligible effects of flow-alignment for a two-dimensional unbounded active nematics. This is the scenario analyzed by Alert et al. [5] in relation to the issue of active (nematic) turbulence, several times mentioned earlier (see Sects. 4.2 and 5.3). Moreover, since Alert et al.'s description is defect free, the Leslie-Ericksen–based framework developed in this section in fully in order.[13] To render the notation more conventional to the analysis of flows in nematics, we just replace formally the polarization field $\mathbf{p}$ by the nematic orientation vector $\hat{\mathbf{n}}$ in the remaining of this section.

The equation for momentum conservation reads,

$$
0 = -\partial_\alpha P + \partial_\beta (\sigma_{\alpha\beta} + \sigma_{\alpha\beta}^a). \tag{6.37}
$$

The symmetric part of the stress tensor, in absence of flow-alignment effects, is expressed simply as,[14]

$$
\sigma_{\alpha\beta} = 2\eta v_{\alpha\beta} - \zeta Q_{\alpha\beta}, \tag{6.38}
$$

---

[13]I remind the reader that this approach is equivalent to accept that the scalar nematic order parameter S adopts a fixed value throughout the sample. At the same time the solely part of the free energy playing a role comes from long range spatial distortions of the director field, with no reference to the bulk part constructed from the invariants of the $\mathbf{Q}$ tensor.

[14]I incorporate the $\Delta\mu$ term in the activity coefficient denoting it simply as ζ. Moreover, in this particular 2d set up, $Q_{\alpha\beta} = n_\alpha n_\beta - (1/2)\delta_{\alpha\beta}$ (see Sect. 2.2.2).

while the antisymmetric part $(\sigma_{\alpha\beta}^a = 1/2(n_\alpha h_\beta - h_\alpha n_\beta))$ is written in terms of the molecular field,

$$h_\alpha = -\frac{\delta\mathcal{F}}{\delta n_\alpha} = K\nabla^2 n_\alpha, \tag{6.39}$$

under the conventional one-constant approximation for the pure elastic, i.e. orientational, part of the free energy. Finally, the system is closed in terms of the dynamical equation for the director field,

$$\partial_t n_\alpha + v_\beta\partial_\beta n_\alpha + \omega_{\alpha\beta} n_\beta = \frac{1}{\gamma_1} h_\alpha. \tag{6.40}$$

One further renders the equations dimensionless by rescaling length by the system size L, time by the active time $\tau_a = \eta/|\zeta|$, pressure by the active stress coefficient $|\zeta|$, and the molecular field by K/L^2.

The pressure of the momentum conservation equation is eliminated by taking its curl and, thus, converting it into a Poisson-like equation for the vorticity. Further, this equation is transformed into the corresponding one for the *stream function*, i.e. $v_x = \partial_y\psi, v_y = -\partial_x\psi$, that reads,

$$\nabla^2\omega = -\nabla^4\psi = s(\mathbf{r};t), \tag{6.41}$$

where $s(\mathbf{r};t)$ denotes a vorticity source,

$$s(\mathbf{r};t) = \frac{1}{2}\frac{R}{A}\nabla^4\theta + S\left[\frac{1}{2}(\partial_x^2 - \partial_y^2)\sin 2\theta - \partial_{xy}^2\cos 2\theta\right]. \tag{6.42}$$

The last equation, coupling the spatial variations of the angle θ of the nematic director with the vorticity pattern, is written in terms of three dimensionless parameters: the activity parameter $A = L^2/l_a^2$, the dimensionless viscosity ratio $R = \gamma_1/\eta$, and the sign of the active stress $S = \zeta/|\zeta| = \pm 1$, respectively for extensile/contractile systems. l_a is the active length scale commented largely in Sect. 4.2.1, slightly different written here as $l_a = (K/|\zeta|R)^{1/2}$. The above equation clearly identifies two contributions to the vortical stirring, the first one arising from director relaxations and the second being purely active. Finally, the dynamics of the director is simply expressed as,

$$\partial_t\theta + (\partial_y\psi)(\partial_x\theta) - (\partial_x\psi)(\partial_y\theta) + \frac{1}{2}\nabla^2\psi = \frac{1}{A}\nabla^2\theta. \tag{6.43}$$

Uniformly oriented and quiescent solutions $(\psi = 0, \theta = \theta_0)$ are identified as unstable to orientational fluctuations. This is nothing but another manifestation of the scenario commented in the previous subsection. The growth rate of small perturbations of wave number $\mathbf{q}$ forming an angle ϕ with the director reads in dimensionless form,

$$\Omega(\mathbf{q}) = \tau_r^{-1}\left[\frac{SA}{2}\cos 2\phi - \left(1 + \frac{R}{4}\right)(qL)^2\right], \tag{6.44}$$

with the time-scale $\tau_r = (\gamma_1 L^2/K)$. Depending on the extensile/contractile nature of the systems the fastest growing modes will be different: for contractile (extensile) stress with $S = -1(+1)$, the most unstable mode is transverse (longitudinal) to the reference director orientation (i.e. $\phi = \pi/2$ for contractile and $\phi = 0$ for extensile). A proxi of the length scale at the onset of the instability is given by the critical wave number (maximum wave number for null growth rate) q_c. The corresponding wavelength, λ_c, can be expressed in terms of the active length scale as $\lambda_c = 2\pi l_a(2 + R/2)^{1/2}$. If one reasonably takes $R = 1$, assuming rotational and shear viscosities are similar, this sets a threshold for the activity number A_c of the order of 100.

Numerical simulations show that the first instability gives rise to a stripe pattern that undergoes a cascade of orthogonal instabilities very much similar to what was reported experimentally by Martínez-Prat et al. [253], as reviewed in Sect. 4.2.1. What is more interesting is the analysis of the fully developed turbulent state in terms of the different energy contributions.

In this respect, Alert et al. [5] performed a spectral analysis of the energy balance at the steady turbulent regime. In Fourier space, this energy balance is expressed as,

$$\dot{F}(q) = -D_s(q) - D_r(q) + I(q) + T(q) = 0, \tag{6.45}$$

where the l.h.s. term denotes the rate of change of the orientational energy.[15] The latter must be equilibrated by the shear viscous dissipation rate D_s, plus the rotational viscous dissipation after adding eventual injecting (active energy) I, and transference (energy advected) T contributions. The striking result reported by Alert et al. is that energy injection has a broad spectrum peaked at a length scale $\lambda_i \sim l_a$ that also corresponds to the maximum of stored elastic energy, while there is no signature of energy transference, at odds with the classical result of inertial turbulence (see Sect. 5.3).

It is quite straightforward within the reduced scheme proposed by Alert et al. to obtain explicit expressions for the contributions to the energy balance above. One first identifies the free energy that has permitted us to derive the system of fluxes and forces with the orientational free energy of the active nematics. In other words, one writes,

$$-\frac{d\mathcal{F}_{elastic}}{dt} = T\frac{dS}{dt} = \int [\sigma_{\alpha\beta} v_{\alpha\beta} + P_\alpha h_\alpha] d\mathbf{r}. \tag{6.46}$$

In absence of the contribution from the chemical reaction of hydrolysis (the term $r\Delta\mu$), the above equation expresses a balance between the orientational and velocity fields that store the energy flowing through the system according

---

[15] In the spirit of this work the orientational energy is largely dominant with respect to the kinetic energy under the claimed hypothesis of very small Raynolds numbers.

to their respective degrees of freedom. From the explicit expression for σ and P, one has,[16]

$$-\frac{d\mathcal{F}}{dt} = \int [2\eta v_{\alpha\beta} v_{\beta\alpha} + (1/\gamma_1) h_\alpha h_\alpha - \zeta Q_{\alpha\beta} v_{\beta\alpha}] d\mathbf{r}, \quad (6.47)$$

which identifies the contributions by shear dissipation, the rotational dissipation and the injection terms. To get an explicit expression for the transference term one needs to work out explicitly the term $\frac{d\mathcal{F}}{dt}$. The procedure is better handled in Fourier space. One starts with the definition of the density of the orientational energy,

$$\mathcal{F} = L^2 \int_{\mathcal{R}^2} F(\mathbf{q}) d^2\mathbf{q} = \frac{K}{2} L^4 \int_{\mathcal{R}^2} q^2 |\theta_\mathbf{q}|^2 d^2\mathbf{q}, \quad (6.48)$$

from which one obtains,[17]

$$\partial_t F(\mathbf{q}) = q^2 Re[\theta_\mathbf{q}^* \partial_t \theta_\mathbf{q}]. \quad (6.49)$$

Finally one uses the Fourier version of the dynamical equation 6.43. Interested readers can find the detailed calculation in the original paper [5]. We simply write the final expressions for the dissipation, injection and transference contributions written in terms of the Fourier components of the orientation and velocity fields (s_a denotes the active contribution in Eq. 6.42, i.e $s_a = S\left[\frac{1}{2}(\partial_x^2 - \partial_y^2)\sin 2\theta - \partial_{xy}^2 \cos 2\theta\right]$),

$$D_s(\mathbf{q}) = \frac{A}{R} q^4 |\psi_\mathbf{q}|^2, \quad (6.50)$$

$$D_r(\mathbf{q}) = \frac{1}{A} q^4 |\theta_\mathbf{q}|^2, \quad (6.51)$$

$$I(\mathbf{q}) = -\frac{A}{R} Re[s_{a,\mathbf{q}}^* \psi_\mathbf{q}], \quad (6.52)$$

$$T(\mathbf{q}) = \frac{q^2}{2\pi^2} \int_{\mathcal{R}^2} (\mathbf{k} \times \mathbf{q}) \cdot \hat{z} Re[\theta_\mathbf{k} \psi_{\mathbf{q}-\mathbf{k}} \theta_\mathbf{q}^*] d^2\mathbf{k}. \quad (6.53)$$

In the final part of [5], authors discuss scaling features of the spectra of the kinetic energy (per unit mass density) and enstrophy i.e $\mathcal{E} = (1/2)\int v^2 d\mathbf{r}$, $\Omega = (1/2)\int \omega^2 d\mathbf{r}$. Both spectra show characteristic peaks associated to the wavelength λ_i. At smaller wave numbers (left from the peak) $E(q) \sim q^{-1}$, while past the peak $E(q) \sim q^{-4}$. For enstrophy those scaling get trivially multiplied by q^2, as follows directly from a dimensional analysis. The q^{-4} scaling of the kinetic energy spectrum confirmed the results obtained previously by Giomi [132] that I will going to comment with more detail later on in this chapter (see Sect. 6.2.3).

---

[16]For simplicity, I omit the subindex of the free energy on what follows.
[17]One omits here L and K considered as pure dimensional factors.

I finish with an important remark: all the results quoted in this subsection correspond to a pure two-dimensional active nematic fluid described in absence of defects, flow-alignment contributions, and completely unbound, i.e. free of any contacting medium. This latter condition will be relaxed later on when discussing the modeling conditions of interfaced active fluids (see Sect. 6.3).

Extending the framework based on the linearized Leslie-Ericksen formulation of nematodynamics under the one-fluid approximation, I finish this section with a reference to the work of Giomi et al. [139] that describes a two-component (filament plus solvent) active (polar) suspension. According to the results of this paper, spontaneous flows in polar fluids come always accompanied by strong concentration inhomogeneities. In addition, spontaneously oscillating and banded flows, even at low activity, were also reported.

6.2 A Beris-Edwards Approach to Model Active Nematics

The previous section was devoted to the theory of polar active fluids. However, in many instances the polar nature of the fluid was not by itself a singular characteristics. As opportunely mentioned, this comes from the fact that the scheme of fundamental equations supports a natural inversion symmetry of the polarization field **p**. This means that nematic systems perfectly fit into such a description, as mentioned earlier, except for an important fact: the role of defects. Defects in polar and nematic systems have fundamentally different topological characteristics. Thus, to properly describe active nematic fluids, and singularly to compare with experiments, we must extend the simple consideration of a nematic director **n̂**, to fully incorporate a nematic tensor **Q**-based description.

6.2.1 General Scheme of Equations

The first attempt I am aware of in this context was published by Marenduzzo et al. [249]. It appeared more or less simultaneously to the classical papers on active gels I mentioned in the previous section. Marenduzzo et al. formulated a closed tensorial description for the nematic order **Q**, and further performed a numerical study based on the use of a lattice Boltzmann algorithm. On what follows I sketch the basic set of employed equations in their most generic formulation (3d). Later on I will consider either expanded or simplified forms of this scheme with special motivations.

In this case, I will not employ the original notation employed in [249], but rather I will largely follow that proposed in the review paper by Doostmohammadi et al. [87], and also adopt the most general definition of the **Q** tensor (see Sect. 2.2.2). The latter is here defined for uniaxial nematics as $\mathbf{Q} = \frac{3}{2}q(\hat{\mathbf{n}}\hat{\mathbf{n}} - \mathbf{I}/3)$, where q denotes the magnitude of the order parameter,[18] and the director $\hat{\mathbf{n}}$ is the characteristic nematic unit vector.

I start by writing the Landau-de Gennes free energy (density) of the nematic liquid crystal, similar to what I did for the polarization free energy discussed largely in the previous section, now incorporating terms that control the magnitude of the nematic order. We write it under the convention of equal elastic moduli as,

$$F = \frac{A}{2}\mathbf{Q}^2 + \frac{B}{3}\mathbf{Q}^3 + \frac{C}{4}\mathbf{Q}^4 + \frac{K}{2}(\nabla\mathbf{Q})^2, \tag{6.54}$$

where we have neglected the spontaneous splay term equivalent to having infinitely strong anchoring at the boundaries.[19] The next step is to write the dynamical equation for the tensor order parameter. It is obtained starting from the strain rate $\mathbf{E}$ and vorticity tensors $\mathbf{\Omega}$,[20]

$$(\partial_t + \mathbf{v} \cdot \nabla)\mathbf{Q} - \mathbf{S} = \gamma_1^{-1}\mathbf{H}, \tag{6.55}$$

in terms of a rotational diffusivity γ_1^{-1} and a corotation term,

$$\mathbf{S} = (\lambda\mathbf{E}+\mathbf{\Omega})\cdot(\mathbf{Q}+\mathbf{I}/3)+(\mathbf{Q}+\mathbf{I}/3)\cdot(\lambda\mathbf{E}-\mathbf{\Omega})-2\lambda(\mathbf{Q}+\mathbf{I}/3)(\mathbf{Q}:\nabla\mathbf{v}). \tag{6.56}$$

Notice that the last term with the double-dot product can be similarly written, as often appears in the literature, as the trace of the matrix product. The molecular (orientational) field $\mathbf{H}$, here of a corresponding tensorial nature,

---

[18] I use here q instead of the more conventional S notation for the scalar nematic order parameter because the same symbol in tensorial form, i.e. $\mathbf{S}$, will be employed later on to refer to a corotation of the $\mathbf{Q}$ tensor.

[19] This is the standard form of the free energy density in this context, adding the phase and elastic contributions, adapted from what has been stated in Sects. 2.2.2 and 2.2.3. In this respect, I denote the single elastic constant as K to comply with the convention in the $\mathbf{Q}$ theory of active nematics, identifying it with the parameter L employed in Eq. 2.32, although it is then clear that K here can not be any longer identified strictly with the single Frank-Oseen elastic constant. Also notice that the shorthand notation $(\nabla\mathbf{Q})^2$ for the conventional elastic term must be interpreted as $(\partial_\gamma Q_{\alpha\beta})^2$, where Greek indices denote Cartesian coordinates, and summation over repeated indices is implied.

[20] In the context of active nematics formulated in the $\mathbf{Q}$ representation, the flow-alignment parameter is denoted λ as proposed here, or ξ in the original reference [249]. The relation of this parameter with the Leslie-Ericksen nematic alignment parameter employed in the precedent section can be found in [248] for $d = 3$, and for $d = 2$ in [158], and depends on the magnitude of the order parameter. For instance for $A = 0$, $\xi > 0.6$ corresponds to flow-aligning and flow-tumbling otherwise. Also an active coefficient introduced in terms of a parameter λ in the dynamical equation for $\mathbf{Q}$ that was retained in the original equations in Marenduzzo et al. [249] turns out to be effectively adsorbed in an effective activity parameter, and thus can be ignored, very much like what I did when considering explicit examples of application of the theory of active gels. Finally, to homogenize the notation with the one used earlier in this chapter, I have preferred to keep the notation γ_1 for the rotational viscosity, although in many formulations of the $\mathbf{Q}$ theory a symbol Γ is employed whose definition strictly involves again the magnitude of the order parameter [248].

is defined, similarly as for polar systems, in terms of derivatives of the free energy functional,

$$\mathbf{H} = -\frac{\delta \mathcal{F}}{\delta \mathbf{Q}} + \frac{\mathbf{I}}{3} Tr(\frac{\delta \mathcal{F}}{\delta \mathbf{Q}}). \tag{6.57}$$

Assuming constant density, the velocity field obeys the incompressible Navier-Stokes equation,

$$\rho(\partial_t \mathbf{v} + \mathbf{v} \cdot \boldsymbol{\nabla} \mathbf{v}) = \boldsymbol{\nabla} \cdot \sigma, \tag{6.58}$$

with σ a global stress tensor, which includes the pressure, together with the viscous, elastic and active contributions,

$$\sigma_{viscous} = 2\eta \mathbf{E}, \tag{6.59}$$

$$\sigma_{elastic} = - P\mathbf{I} + 2\lambda(\mathbf{Q} + \mathbf{I}/3)(\mathbf{Q}:\mathbf{H}) - \lambda \mathbf{H} \cdot (\mathbf{Q} + \mathbf{I}/3) - \lambda(\mathbf{Q} + \mathbf{I}/3) \cdot \mathbf{H}$$
$$- \boldsymbol{\nabla} \mathbf{Q} \frac{\delta \mathcal{F}}{\delta \boldsymbol{\nabla} \mathbf{Q}} + \mathbf{Q} \cdot \mathbf{H} - \mathbf{H} \cdot \mathbf{Q}, \tag{6.60}$$

$$\sigma_{active} = - \zeta \mathbf{Q}. \tag{6.61}$$

In absence of the active term, this scheme reduces to the conventional nematodynamic equations written in tensorial form. As in the previous section $\zeta > 0$ refers to extensile systems, and $\zeta < 0$ to contractile.

In the original Marenduzzo et al. paper [249], authors analyzed the active flows elicited in a slab of both, flow-aligning and flow-tumbling materials. The transition to an active flowing phase is confirmed. In the second part of the paper, the rheological response to a Poiseuille flow was considered. The just presented scheme was extended to include a concentration variable by Giomi at el. [138], very much like what had been done earlier by these authors for polar fluids [139]. In this case, the interplay between nonuniform nematic order, activity, and flow results in spatially modulated relaxation oscillations, similar to those seen in excitable media.

6.2.2 A Simplified Analysis of Defect Dynamics

Reduced to strictly two-dimensional versions, the $\mathbf{Q}$-based set of equations admits considerable simplifications. This has permitted to address several scenarios while reducing the complexity of the numerical work. An interesting example is the study of defect annihilation and proliferation as published by Giomi et al. [133, 134]. Since this study is particularly illustrative, I will comment it on what follows with some detail. The bottom line of the study is the observed different behavior shown by pairs of closely interacting defects in active systems, depending on their extensile or contractile nature. In relation to conventional liquid crystals, back-flow currents are largely modified by active stresses in such a way that defect annihilation can be speed up, slowed down or even suppressed, according to the relation between the level of activity and the time scale of the orientational relaxation.

The considered situation assumes incompressibility in a system of driven filaments whose fundamental variables are the flow velocity $\mathbf{v}$, and the nematic alignment tensor $\mathbf{Q}$. The reduced set of equations read,[21]

$$\rho\frac{Dv_i}{Dt} = \eta\nabla^2 v_i - \partial_i P + \partial_j\sigma_{ij}, \tag{6.62}$$

$$\frac{DQ_{ij}}{Dt} = \lambda S v_{ij} + Q_{ik}\omega_{kj} - \omega_{ik}Q_{kj} + \gamma_1^{-1}H_{ij}. \tag{6.63}$$

In the previous system of equations, the pure comoving derivative is used without vorticity terms, i.e. $D/Dt = \partial_t + \mathbf{v}\cdot\nabla$, and S stands for the scalar magnitude of the nematic order, as commonly used. The Landau-de Gennes free energy is written as in Eq. 6.54, except for the absence of the cubic term in the bulk free energy part (see Sect. 2.2.2), and finally the stress tensor, whose viscous part has been already sorted out, is expressed in terms of an elastic and an active part denoted respectively as: $\sigma_{ij} = -\lambda S H_{ij} + Q_{ik}H_{kj} - H_{ik}Q_{kj}$, and $\sigma_{ij} = \alpha_2 c^2 Q_{ij}$, with c denoting the concentration variable. According to the authors the c^2 dependence of the active stress is appropriate for systems where activity comes form pair interactions, as is the case for paired filaments interacting through kinesin motors. The α coefficient stands here for the activity coefficients, reproducing the notation employed in Sect. 4.3.1. Notice that here the activity coefficient α_2 is taken positive for contractile and negative for extensile, and I will adopt this notation hereafter.

One considers a pair of half-integer defects with respective positions denoted x_+ and x_-, and $x = x_+ - x_-$ accounts for the corresponding separation apart. Neglecting first back-flow effects, the pair dynamics is purely relaxational and given by a force balance. The attractive force is given by $\mathbf{F_{pair}} = -\nabla E_{pair}$, with $E_{pair} \sim K\log(x/a)$, where K stands for the elastic constant and a denotes the defect core radius. The drag force is expressed as usual, i.e. $F_{fric} = \mu\dot{\mathbf{x}}$, with $\mu \sim \gamma_1$ (i.e similar values of translational and rotational friction coefficients). Clearly, the distance separating defects decreases as the square root of t, with equal velocity for defects of either sign that follow symmetric trajectories. The case with back-flow is solved numerically. In absence of activity, it is known [415] that positive defect tend to speed up and negative to slow down.

Activity provides a new contribution to be added to pure back-flow and has opposite effects in contractile and extensile systems. Active back-flow drives positive defects to move in the direction of their tail in contractile systems, and following their head in extensile, while negative defects do not move due to its

---

[21] Giomi et al. original paper includes a (filament) concentration variable as well, with its corresponding conservation equation. I am not going to retain it here since it is not fundamental for the analysis it follows.

triangular-like symmetry and flow compensation. I mentioned this clear distinction when referring to experimental observations in the original reference on experimental active nematics by Sanchez et al. [337].

In active systems, Giomi et al. in [133] considered a pair of opposite charge aligned defects with the negative placed left of the positive, and confirm that in this relative simple orientation extensile systems lead to an apparent counterintuitive defect unbinding with the $+1/2$ unit moving away from its negative counterpart, as also observed experimentally [337]. At large values of activity, the asymmetry in defect dynamics is very pronounced.

The dynamics of pair annihilation is in this case effectively represented by the balance equation,

$$\mu[\dot{x}_{\pm} - v_b(x_{\pm})] = \mp \frac{K}{x_+ - x_-}, \tag{6.64}$$

where $v_b(x_{\pm})$ denotes the back-flow velocity at the position of the respective defect. Giomi et al. retain only the active contribution to the back-flow and replace the flow distribution by the velocity at the defect core, with $v_b(x_+) = v_\alpha$ and $v_b(x_-) = 0$, with v_α positive (respectively negative), for extensile (respectively contractile) systems. This gives a differential equation that can be readily integrated to,

$$x(t) = x_0 + v_\alpha t - 2(\tilde{\kappa}/v_\alpha) \ln\left[\frac{x(t) - 2\tilde{\kappa}/v_\alpha}{x_0 - \tilde{\kappa}/v_\alpha}\right], \tag{6.65}$$

with $\tilde{\kappa} = K/\mu$. The pair annihilation time is calculated in a straightforward way, confirming that for the particular alignment considered in the paper, activity speeds up annihilation in contractile conditions, whereas slows it down for extensile active systems. This study was extended by the same group with a precise calculation of the back-flow distribution around the defect cores [135].

Interestingly enough this simple phenomenological expression differs only by the logarithmic factor from the result obtained by Pismen [309] from a detailed calculation of a system of sparse defects populating an active layer that is confined within a no-slip Hele-Shaw cell with strong wall friction. A more realistic scenario in relation with experiments with MT-based (interfaced) active nematics was considered a few years later [310], while a theoretical analysis of the particular scenario of defect unbinding was independently published by Shanlar et al. [355].

In addition to the works by Giomi et al., Pismen et al., and Shankar et al., the issue of the defect dynamics in active nematics based on the **Q**-based scheme has deserved a considerable attention in the last decade, mainly from the group of J. M. Yeomans. I singularly mention a series of papers by Thampi et al. [397, 398].

In [397], the relation between the defect density and the length scale of the velocity field was investigated in terms of the elastic constant and the activity coefficient. In the second paper [398], the question of active (nematic) turbulence was addressed, by showing how defects are first nucleated from an hydrodynamic instability-generated wall. These walls appear as regions of very strong bend deformations. Further these walls relax the accumulated elastic tension by defect unbinding. This rationale is in complete agreement with the original observations on microtubule-based active nematics by Sanchez et al. [337], or with the more recent characterization by Martínez-Prat et al. [253].

6.2.3   Theoretical Description of Active Nematic Turbulence

The question of active nematic turbulence deserves a special attention since it has motivated several experimentally dedicated works, as commented earlier (see Sect. 4.2.1), while at the same time has justified a few theoretical efforts (see Sect. 6.1.4). The central attention will be focused here on the paper published by Giomi [132] that I have quoted already several times in the manuscript singularly in the just mentioned sections. Giomi's contribution reports mean-field arguments and numerical results, including the proper consideration of defects and flow-alignment characteristics, for the full set of equations that describe active nematics in terms of the **Q** tensor.[22]

The system is considered again two-dimensional, and the free energy (see Eq. 6.54) is written with the particular choice of coefficients ($B = 0, C = -2A$). The numerical study follows a direct integration scheme applied to the dynamic equations for the velocity and nematic fields, and it is, essentially, devoted to extensile systems (contractile conditions behave similarly as quoted in the original reference) (see Fig. 6.3).

The first characterized observable is the distribution of vortex sizes. I already referred to such a distribution (see Sect. 4.2.1 in relation to the work by Guillamat et al. [149] and Fig. 4.4). The results obtained by Giomi indicate a prominent exponential distribution in a range qualified as an active range ($A_{min} < A < A_{max}$),

$$n(A) = \frac{N}{Z} \exp(-A/A^*), \tag{6.66}$$

where N and Z represent, respectively, the total number of vortices, and a normalization constant for the exponential distribution. Conversely, the mean vorticity remains roughly constant among scales. As activity is increased, vortices become smaller and this leads naturally to identify the characteristic vortex size A^* with the active length scale $A_{min} \approx A^* \sim l_a^2$. The activity coefficient α appears in this analysis as the direct coefficient of the **Q** tensor entering into the stress as the active contribution. Since **Q** is taken dimensionless, this means that α here bears units of stress.[23] From a balance of

---

[22] A multi-scale statistical analysis of two- and three-dimensional turbulent active (nematic) flows was published a couple of years later by Urzay et al. [418], claiming fundamental differences with high-Reynolds number turbulence. Singularlu, authors conclude that convection plays a minor role in transporting momentum across scales.

[23] For extensile systems Giomi's analysis considers α negative and thus, strictly this coefficient should be considered in absolute value in this subsection.

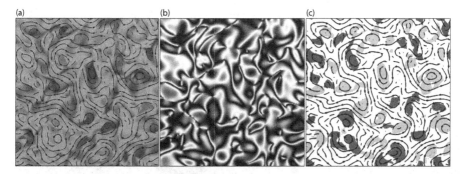

FIGURE 6.3
A simulated two-dimensional active nematic. (a) Flow velocity and vorticity fields. (b) Schlieren texture for the director field, with indicated positive and negative semi-integer defects. (c) Clockwise rotating and counterclockwise vortices (characterized with different greyscale levels) detected by means of the Okubo-Weiss indicator (colored image in original version). Image and caption text adapted from Giomi [132].

active and viscous stress, one might expect that the typical mean vorticity would be given as $\omega_{vor} \sim \alpha/\eta$, as confirmed. The multiscale organization and large-scale isotropy of the flow is captured by the velocity components that are Gaussian distributed, whereas, similarly to classic turbulence, the vorticity does not obey a Gaussian distribution, but show long tails.

Giomi also evaluated the average kinetic energy $\langle v^2 \rangle/2$ and enstrophy $\langle \omega^2 \rangle/2$ per unit area, considering different values of activity. These quantities respectively display a linear and quadratic dependence on the activity parameter that are easily interpreted. Accumulating enstrophy on the different vortices, one has $\Omega_{tot} = \frac{1}{2} \int \omega^2(\mathbf{r}) d\mathbf{r} = \frac{1}{2} \int n(A) \omega_{vor}^2 dA = \frac{1}{2} N \overline{A} \omega_{vor}^2$ since ω_{vor} is assumed independent of the vortex size. The mean vortex size is taken as $\overline{A} = L^2/N$ with L the typical linear dimension of the system. Thus $\langle \omega^2 \rangle/2 = \Omega_{tot}/L^2 \approx \omega_{vor}^2 \sim \alpha^2$. Similarly, using the fact that the characteristic vortex area is given by l_a^2, and invoking the scaling of the latter with the activity coefficient, Giomi concluded that $\langle v^2 \rangle/2 \sim \alpha$.

The spectra of the total kinetic energy and enstrophy were also extracted from the velocity fields. Non-monotonous behaviors are clearly evidenced with peaks around the inverse of the active length scale, and $E(q) \sim q^{-4}$ for higher wave numbers. This result was reproduced later on in Alert et al. [5] (see Sect. 6.1.4). Results for smaller wave numbers are less conclusive in Giomi's analysis. Finally, Giomi derived a mean-field (statistical) approach that reproduces the spectral structure, as well as the short-scale velocity and vorticity correlations. The theory is based on assuming a distribution of vortices statistically independent, and extends to this case the approach used for decaying

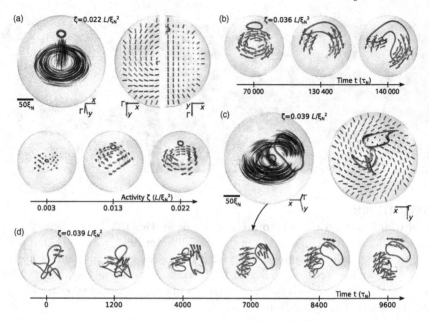

FIGURE 6.4

A three-dimensional active nematic confined within droplets. (a) Active nematic at low activities with a single point defect (small ring) displaced from the droplet center, and flow vortex (gray streamlines). The right panel shows director profiles in the plane of the vortex and perpendicular to it. The bottom panel shows how, with increasing activity, the point defect moves further away from the droplet center, while flow magnitude increases. (b) Transition of the point-defect regime into the turbulent regime upon an increase of activity. (c) Turbulent active nematic regime. Defect loops are shown as isosurfaces of reduced degree of order. (d) Selected timeline of active turbulence within the droplet (colored image in original version). Image and caption text adapted from Čopar et al. [69].

two-dimensional self-similar coherent structures in non-active turbulence (see Giomi's paper [132] for the original references). Pieces of this theory will be employed in next section when referring to interfaced ANs.

Just to finish this section, and restricting to those approaches that use **Q**-based descriptions to address active nematic turbulence, I cite the recently published paper by Čopar et al. [69]. This report extends the dimensionality of the problem to encompass an scenario of encapsulated active nematics for regimes of droplet radius comparable but slightly larger than the active length scale. No-slip, and both homeotropic and planar-degenerated alignment anchoring conditions were separately considered (see Fig. 6.4).

More precisely, an structure with a center-separated stationary point defect is observed for homeotropic anchoring and low activity, whereas for higher activities, a regime of three-dimensional active turbulence with spatially varying and time-varying defect loops is observed (see Fig. 6.4). These two regimes are separated by a (hysteretic) structural transition. Under degenerate planar anchoring, the defect lines within the bulk are not necessarily closed into loops but can also terminate at the surface. Indeed, in addition to closed defect loops observed for droplets with perpendicular surface alignment, defect lines with surface-to-surface spanning ends, and surface boojum defects are observed, resulting in the turbulent defect dynamics. In the bulk, this dynamics is generally similar to the dynamics in homeotropic droplets, but it is additionally coupled to the topological events at the droplet surface.

6.3 Modeling Interfaced Active Fluids

One of the problems inherent to the existing theories on active fluids, singularly when trying to compare their predictions with experiments, is that, for the most part, studies published so far, and certainly those commented in the previous two sections, actually refer to (ideally) unbounded situations, i.e. the active fluid is considered in these modeling approaches as completely isolated from the medium where it resides. Obviously this is never realized in the laboratory, and at most one wishes that the effects of limiting interfaces be subdominant.

This feature is particularly sensible in relation to active nematics based on microtubules and motor proteins. I remind the reader at this point that the conventional preparation refers to an active layer interfaced with two contacting immiscible fluids. I already reported in Sect. 4.3 striking effects on the active textures and flows that arise from the viscosity contrast established at the passive/active interface. From this perspective, the presence of interfaces needs to be properly accounted for in any realistic theory of active fluids, and, in particular, the role played by viscosity contrast must be appropriately assessed in any modeling aimed to be applied to two-dimensional active nematics.

I could argue similarly on what respects to the role played by confining environments, to which I devoted a specific (experimental) section in 4.4. I dedicate the present section to address the former question, i.e interfacial effects, while I reserve the following one (see Sect. 6.4) to comment on modeling approaches that describe encapsulated and confined active fluids.

I start by recalling the observations reported by Guillamat et al. [150] on the effects of changing the viscosity contrast on an active nematics residing at a conventional oil/water interface (see Sect. 4.3.1). I emphasized at that point the important observed effects on the density of defects, as well as in the velocity of the active flows. Specifically, I will examine a model proposed by Shankar and Marchetti ([150]) that enabled to obtain the dependence of the (positive) defect velocity on the viscosity of the contacting passive oil. Moreover, this permitted to extract the first estimation, at the time the paper was published, of the active nematic viscosity.

The starting point is the Stokes equation for an active monolayer, supposed to evolve in steady state as a single and incompressible phase, and endowed with a two-dimensional shear viscosity η_n,

$$\eta_n \nabla^2 \mathbf{v} - \boldsymbol{\nabla} P + \mathbf{f}_w + \mathbf{f}_o + \mathbf{f} = 0, \tag{6.67}$$

where the subindices w/o refer to water and oil force (surface) densities exerted respectively by the corresponding accompanying phases on the nematic film. The left term, $\mathbf{f}$, denotes the force density that results from stresses intrinsically associated to the nematic film. P stands for the two-dimensional pressure.

The idea of the calculation is to separate the three forces appearing in the Stokes equation, in such a way that the force pair $\mathbf{f}_i$ arising from the coupling with the bulk passive phases will enter directly into the Green operator, while $\mathbf{f}$ denotes a source force. Thus, we write,

$$v_\alpha(\mathbf{r}) = \int G_{\alpha\beta}(\mathbf{r} - \mathbf{r}') f_\beta(\mathbf{r}') d\mathbf{r}', \tag{6.68}$$

or, equivalently, in Fourier space,[24]

$$\tilde{v}_\alpha(\mathbf{q}) = \tilde{G}_{\alpha\beta}(\mathbf{q}) \tilde{f}_\beta(\mathbf{q}). \tag{6.69}$$

The next step is to obtain the viscous forces exerted by the oil and water phases on the nematic film,

$$\mathbf{f}_w(\mathbf{r}) = -\eta_w \left. \frac{\partial \mathbf{v}_{w\parallel}(\mathbf{r}, z)}{\partial z} \right|_{z=0^-}, \tag{6.70}$$

$$\mathbf{f}_o(\mathbf{r}) = -\eta_o \left. \frac{\partial \mathbf{v}_{o\parallel}(\mathbf{r}, z)}{\partial z} \right|_{z=0^+}, \tag{6.71}$$

with the convention that the nematic, strictly two-dimensional, film is placed at $z = 0$. Bulk flows are supposed to fulfill Stokes equations, respectively within $-h_w < z < 0$ and $0 < z < h_o$ for the water and oil phases.[25] Following

---

[24] I prefer to track explicitly here each quantity when expressed in real or, alternatively, in Fourier space, and, thus I reserve a special notation in this latter case. In any case, the Fourier decomposition is introduced here as $\mathbf{v}(\mathbf{r}) = \int \frac{d^2\mathbf{q}}{(2\pi)^2} \tilde{\mathbf{v}}(\mathbf{q}) e^{i\mathbf{q}\cdot\mathbf{r}}$, and equivalently for the other quantities.

[25] I remind the reader that in the experiments aqueous phase is placed at the bottom, and the oil phase on thetop of the cell.

Lubensky et al. [240], these flows are demonstrated planar, evolving under a uniform pressure, and with planar Fourier modes satisfying,

$$\eta_w(\partial_z^2 - q^2)\tilde{\mathbf{v}}_{w\|} = 0, \tag{6.72}$$

$$\eta_o(\partial_z^2 - q^2)\tilde{\mathbf{v}}_{o\|} = 0, \tag{6.73}$$

in their respective domains.

These back-flows are driven by the hydrodynamic coupling with the nematic film, i.e,

$$\mathbf{v}_{o\|}(\mathbf{r}, 0^+) = \mathbf{v}_{w\|}(\mathbf{r}, 0^-) = \mathbf{v}(\mathbf{r}). \tag{6.74}$$

The remaining boundary conditions are prescribed according to the experimental design. The water layer is in contact with a solid substrate, where a no-slip condition is assumed, i.e. $\mathbf{v}_{w\|}(\mathbf{r}, -h_w) = \mathbf{0}$. Correspondingly, the oil layer is open to air, and, thus, introduces a no-shear stress boundary condition, i.e. $\partial_z \mathbf{v}_{o\|}(\mathbf{r}, z = h_o) = \mathbf{0}$.

Implicit solutions of the bulk flows in terms of the nematic flow field read,

$$\tilde{\mathbf{v}}_{w\|}(\mathbf{q}, z) = (\cosh(qz) + (1/\tanh(qh_w))\sinh(qz))\tilde{v}(\mathbf{q}), \tag{6.75}$$

$$\tilde{\mathbf{v}}_{o\|}(\mathbf{q}, z) = (\cosh(qz) - \tanh(qh_o)\sinh(qz))\tilde{v}(\mathbf{q}), \tag{6.76}$$

from which we obtain trivially the expressions for the surface force densities in their Fourier representation,

$$\tilde{\mathbf{f}}_w(\mathbf{q}) = -\eta_w q(1/\tanh(qh_w))\tilde{v}(\mathbf{q}), \tag{6.77}$$

$$\tilde{\mathbf{f}}_o(\mathbf{q}) = -\eta_o q\tanh(qh_o)\tilde{v}(\mathbf{q}). \tag{6.78}$$

Finally, one obtains the Green function in the form, $\tilde{\mathbf{G}}(\mathbf{q}) = \tilde{G}(q)(\mathbf{I} - \mathbf{q}\mathbf{q}/q^2)$, where $\tilde{G}(q)$ is the transverse part of the Green function expressed as,

$$\tilde{G}(q) = \frac{1}{\eta_n q^2 + \eta_o q \tanh(qh_o) + \eta_w q(1/\tanh(qh_w))}. \tag{6.79}$$

Notice that this last expression introduces two viscous length scales respectively associated to the water and oil layers,

$$l_w = \eta_n/\eta_w; l_o = \eta_n/\eta_o. \tag{6.80}$$

The meaning of these two length scales is straightforward: At scales larger than either viscous length, dissipation in the external fluid layer, either oil or water, dominates over dissipation within the nematic film.

In the context of the calculation by Shankar et al., one assumes for the sake of simplicity that $h_w \sim h_o = h$.[26] In terms of the latter, one splits the Green function in two spatial ranges separating long and short length scales. Explicitly,

$$\tilde{G}(q) \simeq G_>(q) = \frac{1}{\eta_n q^2 + \eta_o q}, \tag{6.81}$$

$$\tilde{G}(q) \simeq G_<(q) = \frac{1}{(\eta_n + \eta_o h)q^2 + \eta_w/h} = \frac{1}{\eta_r(q^2 + l_\Gamma^{-2})}, \tag{6.82}$$

as asymptotic forms, respectively valid for $qh \gg 1$ and $qh \ll 1$. In the first expression, the contribution from the water viscosity is disregarded in front of the oil viscosity associated term. In the second, $\eta_r = \eta_n + \eta_o h$ stands for a renormalized 2d viscosity,[27] and $l_\Gamma = (\eta_r h/\eta_w)^{1/2}$ represents a frictional screening length.

In the spirit of Shanhar and Marchetti's calculation [150], one is interested in calculating the flow field due to the force distribution created by a positive semi-integer defect located at the origin. In this situation, the force left at the level of Eq. 6.67 is expressed as $\mathbf{f}(\mathbf{r}) = |\alpha| \nabla \cdot \mathbf{Q}$, where $\mathbf{Q}$ is written in terms of the scalar order parameter $S = S(\mathbf{r})$. Assuming a defect core radially symmetric, one has $\mathbf{f}(r) = |\alpha|(\frac{dS}{dr} + \frac{S}{r})\hat{e}_x$, where x denotes the symmetry axis of the positive parabolic defect. The classical solution $S(r) \simeq \frac{r}{2\tilde{\zeta}_Q}$ vanishes linearly as $r \to 0$ ($\tilde{\zeta}_Q$ denoting here the core radius), and tends to a constant (normalized here to unity) outside the core. This permits to approximate the only non-vanishing component of the force as $f_x = |\alpha|\tilde{\zeta}_Q$ for $r < \tilde{\zeta}_Q$, and $f_x = \frac{|\alpha|}{r}$ for $r > \tilde{\zeta}_Q$. Finally, the velocity of the flow elicited by the defect motion is expressed as $v = \int' \tilde{G}(q)(\mathbf{I} - \mathbf{qq}/q^2) \cdot \mathbf{f}(\mathbf{q})_x = \frac{1}{2}\int' dr G(r) f_x(r)$, where the prime denotes conveniently chosen large and small cutoffs.

The most obvious choice for these cutoffs are, respectively, the defect core radius $\tilde{\zeta}_Q$ and the system size L. Thus $v = v_< + v_>$. This leads to corresponding expressions for the velocities, written in terms of a yet intermediate cuttof l,

$$v_< = \frac{1}{2}\int_l^L G_< \frac{|\alpha|}{r} = \frac{|\alpha|l_\Gamma}{\eta_r}F_<\left(\frac{L}{l_\Gamma}, \frac{l}{l_\Gamma}\right), \tag{6.83}$$

$$v_> = \frac{1}{2}\int_{\tilde{\zeta}_Q}^l G_> \frac{|\alpha|}{r} = \frac{|\alpha|l}{\eta_n}F_>\left(\frac{\eta_o l}{\eta_n}, \frac{l}{\tilde{\zeta}_Q}\right). \tag{6.84}$$

---

[26] To be precise, in most of the experiments by Guillamat et al. [150] this is not quite so. Typically, h_w c.a. 40 μm, while h_o c.a. 10^3 μm

[27] Given the context of this specific calculation, there should be no possible confusion with the use of this notation that could lead us to think of a rotational viscosity.

Explicit expressions for the left functions $F_>$ and $F_<$ are given in terms of integrals of special functions in the original reference [150]. Approximate expressions keeping the dominant contributions, are more illustrative,

$$v_< \simeq \frac{|\alpha|(L-l)}{\eta_n} \log\left(\frac{2l_\Gamma}{l}\right), \qquad (6.85)$$

$$v_> \simeq \frac{|\alpha|l}{\eta_n} \log\left(\frac{\eta_n}{\eta_o l}\right). \qquad (6.86)$$

In the original paper, the contribution $v_<$ is considered subdominant. This permits to extract a clear logarithmic dependence of the flow velocity on the oil viscosity as predicted by the experiment.

The active nematic viscosity could be evaluated from a fitting to the experimental results. To do so one needs to specify the left cutoff l. The first obvious choice is to take l as the thickness of the most viscous fluid, i.e. the upper oil layer, of the order of 1 mm in a conventional experiment. This leads to $\eta_n \sim 13 \times 10^{-3}$ Pa s m. However, one could alternatively argue that the single-defect calculation should be cutoff at the typical defect mean inter-defect separation (of the order of 100 µm in the considered experiments). The latter estimation in fact does depend also in η_0 but marginally at the effects of this evaluation. In this case, the extracted value of the viscosity of the active nematic layer is one order of magnitude smaller, $6.5 - 13 \times 10^{-4}$ Pa s m.

More recently, these results were revisited. It turns out that the contribution $v_<$ can not be completely ruled out as was initially proposed [150]. Taking it into account, and assuming a value of l according to the second choice mentioned above, the value of the viscosity is reduced by nearly two orders of magnitude. This latter estimation is closer to the value extracted from the analysis of the turbulent energy spectra, as commented on what follows, and also agrees in orders of magnitude with the value reported in a scenario of a sheared conventional active nematic preparation by Rivas et al. [327].

The full structure of the flows of active nematics is certainly best appraised in terms of the kinetic energy spectrum, as mentioned several times in this context. From a modeling point of view, I have commented this issue in some detail in Sects. 6.1.4 and 6.2.3. In both cases, these studies referred to unbounded fluids, without considering explicitly the role of the interface. This is the pending problem to be explored on what follows. More in particular I will refer to the experiments by Martínez-Prat et al. [252] presented earlier (see Sect. 4.3.1), where striking effects at the level of the kinetic energy spectra were reported as arising from the viscosity contrast between the active and the passive contact fluids.

I will briefly sketch on what follows the theoretical analysis published together with experiments in the mentioned reference [252]. The treatment starts from the consideration of the coupling of active and passive fluids through a Green function, as it was elaborated earlier in this section. Next, one seeks a formal relation between the velocity field and the source force density that appears at the level of the Stokes equation for the active fluid, paralleling the approach sketched in Sect. 6.1.4 valid in absence of defects and neglecting flow-alignment effects as well. Finally, one uses results for the vorticity field derived earlier by Giomi ([132]) using mean-field assumptions (see Sect. 6.2.3).[28] The kinetic energy per unit mass density is defined as,

$$\mathcal{E} = \frac{1}{2} \int \mathbf{v}^2 d^2 \mathbf{r}, \tag{6.87}$$

or using the Fourier decomposition employed earlier in this section,

$$\mathcal{E} = \frac{1}{2} \int \frac{d^2 \mathbf{q}}{(2\pi)^2} |\tilde{\mathbf{v}}(\mathbf{q})|^2 = A \int_0^\infty E(q) dq, \tag{6.88}$$

where $A = L^2$ stands for the system area, and $E(q)$ denotes the *kinetic energy spectral density*, assuming the isotropic nature of the system. This latter quantity is the central piece of the analysis for which we seek an explicit expression in terms of the parameters of the active layer, and those of the passive contact bulk fluids. We next use Eqs. 6.69 and 6.79 to obtain,

$$|\tilde{\mathbf{v}}(\mathbf{q})|^2 = \left(\delta_{\alpha\beta} - \frac{q_\alpha q_\beta}{q^2} \right) \left(\delta_{\alpha\gamma} - \frac{q_\alpha q_\gamma}{q^2} \right) \frac{\tilde{f}_\beta(\mathbf{q}) \tilde{f}_\gamma^*(\mathbf{q})}{\Lambda^2(q)}, \tag{6.89}$$

$$= \frac{1}{\Lambda^2(q)} \left(|\tilde{\mathbf{f}}(\mathbf{q})|^2 - \frac{q_\alpha q_\beta}{q^2} \tilde{f}_\alpha(\mathbf{q}) \tilde{f}_\beta^*(\mathbf{q}) \right), \tag{6.90}$$

where a shorthand notation $\Lambda(q)$ has been introduced for the denominator in the expression of the Green function (Eq. 6.79),

$$\Lambda(q) = \eta_n q^2 + \eta_o q \tanh(q h_o) + \eta_w q (1/\tanh(q h_w)). \tag{6.91}$$

Using Eq. 6.67 in absence of external fluids, and employing a scheme similar to what is proposed in [5], it can be shown that the vorticity and source force density fields are related through the relation,

$$- q^2 \tilde{\omega}(\mathbf{q}) = \frac{i}{\eta_n} (q_y \tilde{f}_x - q_x \tilde{f}_y), \tag{6.92}$$

so that the velocity and vorticity fields get directly connected through the relationship

$$|\tilde{\mathbf{v}}(\mathbf{q})|^2 = \frac{\eta_n^2 q^2}{\Lambda^2(q)} |\tilde{\omega}(\mathbf{q})|^2. \tag{6.93}$$

---

[28]Notice that this whole strategy differs from that of Shankar et al. just commented since now we take into consideration in a self-contained way the whole structure of the flow field, albeit neglecting the presence of defects and flow-alignment. In this particular respect, the use of Giomi's results in [132], that are free of these assumptions, could be in principle questioned. However, this pragmatic way of doing is later justified by a favorable comparison with the experimental results, at least within a range of intermediate conditions.

At this point we use Giomi's result for the vorticity field as was obtained in [132], based on decomposing the vorticity field into a superposition of N uncorrelated vortices

$$|\tilde{\omega}(\mathbf{q})|^2 = \frac{N\omega_{vor}^2(R^*)^4}{8\pi^2}e^{-q^2(R^*)^2/2}\left[I_0(q^2(R^*)^2/2) - I_1(q^2(R^*)^2/2)\right], \quad (6.94)$$

where ω_{vor} and R^* denote, respectively, a typical mean vorticity and characteristic vortex size (see Sect. 6.2.3). This leads to our final result for the spectral density of kinetic energy,

$$E(q) = B\frac{q(R^*)^4e^{-q^2(R^*)^2/2}\left[I_0(q^2(R^*)^2/2) - I_1(q^2(R^*)^2/2)\right]}{[q + \eta_o/\eta_n\tanh(qh_o) + \eta_w/\eta_n(1/\tanh(qh_w))]^2}, \quad (6.95)$$

where, I_0 and I_1 are modified Bessel functions, and B is a prefactor related to the total enstrophy, independent of both the wave number q and the mean vortex radius.

A direct comparison of this theoretical analysis with the experimental results published in [252] is presented in Fig. 6.5. The first important conclusion is that of the two contact fluids the water subphase does not play any significant role, its viscous length scale being larger than the largest length scale of the experimental system. Thus, as expected, the system is largely dominated by the more viscous upper oily phase. Consistent with theoretical predictions, we experimentally observe a scaling of the spectral energy density $E(q) \sim q^{-4}$ at small scales, and $E(q) \sim q^1$ at large scales, for all oil viscosities [panels (a) and (b)]. For intermediate oil viscosities, we also observe signatures of a $E(q) \sim q^{-1}$ regime at intermediate length scales, larger than the vortex size R^* but smaller than the viscous length l_o [panel (c)]. In our measurements, the scaling regimes are limited in extension, and the data alone do not provide conclusive evidence for the scaling laws. However, the agreement with the theoretical predictions confirms that the active turbulent flows in our system indeed follow underlying scaling laws.

The q^{-4} and q^{-1} scaling behaviors can be interpreted as intrinsic properties of an active nematic film, respectively characterizing the small-scale flows inside vortices and the large-scale flows due to hydrodynamic interactions in the film [132, 5]. In contrast, the q^1 scaling stems from the hydrodynamic coupling to an external fluid. All these scaling relations turn out to be independent of the properties of the fluids. Consistently, by varying the oil viscosity, we tune the range of the different regimes without changing their scaling exponents. In other words, the values of the exponents are properties of the equations of active nematics and not of their parameters. The range of each scaling regime, however, depends on parameters.

Having demonstrated the scaling regimes, one can fit the theoretically obtained full energy spectrum (Eq. 6.95) to the experimental data. Knowing the

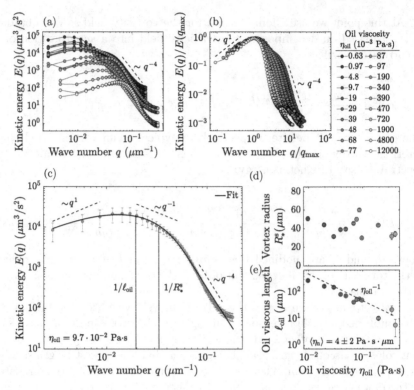

FIGURE 6.5

Kinetic energy spectra in interfaced active nematics. (a) Kinetic energy spectra of turbulent flows in an active nematic film in contact with a layer of oil, for different oil viscosities. (b) Rescaled spectra by its maximum and its corresponding wave number displaying the large-scale scaling regime. (c) Fit to a representative spectrum at intermediate oil viscosity. The spectrum features three scaling regimes, separated by two crossover lengths: the mean vortex size and the viscous length $l_o = \eta_n/\eta_o$ (vertical dashed lines). Error bars are standard deviations. Mean vortex radius (d) and oil viscous length (d) obtained from the spectral fits in the range of intermediate oil viscosities in which the theory fits the data well. Error bars are standard errors of the mean (colored image in original version). Image and caption text adapted from Martínez-Prat et al. [252].

values of η_w, η_o, h_w, and h_o, we use B, η_n, and R^* in Eq. 6.95 as fitting parameters. Despite the made assumptions, the theory fits the data remarkably well for a wide range of intermediate oil viscosities ($9.7 \times 10^{-3} < \eta_o < 0.39$ Pa s). Visual inspection of the experiments suggests smaller vortices as η_o increases (see Fig. 4.7). Yet, the mean vortex radius obtained from the fit, in the intermediate range where it is supposed to work, is independent of oil viscosity [panel (d)]. Finally, the oil viscous length $l_o = \eta_\eta/\eta_o$ decreases as $1/\eta_o$ [panel (e)], indicating that η_n does not vary with oil viscosity in the range of validity of the model. Hence, the fits allow us to estimate the viscosity of the active nematic: $\eta_n = 4 \pm 2$ Pa s µm. The present estimate is of the same order as the value obtained by Rivas et al. [327], as mentioned earlier in this section.

6.4 Modeling Confined Active Fluids

Several models applied to encapsulated and confined active fluids will be revised in this section. Some of them will be examined as a pure application of the theory of active fluids. In other instances, they were motivated by specially designated experiments, to which I have referred previously in Chapter 4. As I proposed there, I first deal in this subsection with active fluids under what was referred to as "soft confining" conditions. By this, I remind the reader that I am referring to active fluids prepared either as thin films, or, in most of the situations, when they come encapsulated within droplets of different topologies. In the next subsection, I comment on "hard confining" situations, represented by specific experimental designs that implement a variety of two- or three-bounding geometries.

6.4.1 Modeling Active Flows in Thin Films and Droplets

6.4.1.1 Thin Active Films

With a remote original motivation rooted in the analysis of the spreading of bacterial films or on the dynamics of lamellipodium in cell motion, Sankararaman [339] proposed a treatment that applies to a thin active film enclosed between a substrate and a deformable interface. In my opinion, there are a couple of reasons that qualify it for being commented here with some detail. First, since, apart from the paper by Voituriez et al. [423] mentioned several times in this chapter, it represents one of the earliest applications of the theory of active gels under bounding conditions. Second, because it incorporates the dynamics of a geometric variable that directly encodes the support perimeter. This latter instance allows to specifically formulate a classical stability analysis of the free interface when facing deformation stresses.

In Sankararaman et al.'s analysis, the role of the substrate is simply taken as a bounding surface without introducing any sort of friction damping. More precisely, the purposed idea is to formulate the thin-film hydrodynamics of a suspension of polar self-driven particles, and show how the latter is prone to undergo a (geometric) destabilization. Such instability comes from the interplay of activity, polarity, and the existence of a free surface. Anchoring is supposed to prescribe planar boundary conditions on both the substrate and the free surface.

Sankararaman et al. assume that the velocity of the polar particles relative to the background fluid follows the polarization variable, i.e. one prescribes it as $V_0\mathbf{p}$. The fields in which we are interested are the concentration of active units c, the polarization, written in its horizontal and vertical components i.e. $\mathbf{p} = (\mathbf{p}_\perp, p_z)$ with $\|\mathbf{p}\| = 1$, and the film height h. Dynamical equations are written for $c(\mathbf{r}_\perp; t), \mathbf{p}(\mathbf{r}_\perp; t)$ and $h(\mathbf{r}_\perp; t)$. The kinematic boundary condition $\dot{h} = v_z - \mathbf{v}_\perp \cdot \mathbf{\nabla}_\perp h$ expresses the evolution of the height h in terms of the velocity field $\mathbf{v} = (\mathbf{v}_\perp, v_z)$ evaluated at the free surface. Incompressibility leads to volume conservation, and the height dynamics converts into a local conservation law,

$$\partial_t h + \mathbf{\nabla}_\perp \cdot (h\overline{\mathbf{v}}_\perp) = 0, \tag{6.96}$$

where $\overline{\mathbf{v}}_\perp$ denotes the in-plane velocity averaged over the thickness of the film. One first solves for the velocity field using the Stokes equation under the *lubrication approximation* $v_z = 0$, $|\mathbf{\nabla}_\perp \mathbf{v}| << |\partial_z \mathbf{v}|$ to obtain,

$$\eta \partial_z^2 \mathbf{v}_\perp - \mathbf{\nabla}_\perp P - \hat{z}\partial_z P + \mathbf{\nabla} \cdot \sigma^a = 0, \tag{6.97}$$

with the active stress given by $\sigma^a = -\zeta\Delta\mu c(\mathbf{r})\mathbf{p}(\mathbf{r})\mathbf{p}(\mathbf{r})$. Following the notation introduced in Sect. 6.1.1, $\zeta\Delta\mu < 0$ corresponds to contractile systems, and $\zeta\Delta\mu > 0$ to extensile. Notice that elastic stresses have been disregarded as they are considered subdominant (see also [247]).

Clearly, to obtain a closed scheme we need to write equivalent equations for the remaining variables, i.e. the polarization and the concentration fields. Generically they read,

$$\partial_t c = -\mathbf{\nabla} \cdot [c(\mathbf{v} + V_0\mathbf{p})], \tag{6.98}$$

and,

$$\frac{Dp_\alpha}{Dt} + \nu_1 v_{\alpha\beta} p_\beta + \lambda_1(\mathbf{p} \cdot \mathbf{\nabla})p_\alpha = -\frac{\delta F_p}{\delta p_\alpha} + \frac{C}{h}\nabla_\alpha^\perp h, \tag{6.99}$$

with D/Dt defined as in Eq. 6.7, and ν_1 denoting the flow-alignment parameter. Two additional comments are worth making in relation to this last equation. First, a term proportional to λ_1 appears as an advective nonlinearity in the equation for the polarization that is not present in the conventional theory of active gels, since one must remember that the latter is formulated under the linear Onsager framework. Its precise meaning will be commented later on in Sect. 7.1. Second, the coefficient C introduces a term whose ultimate justification can be found in the original paper [339]. The simplest way to interpret it is to think on the way a curved interface accommodates polar particles. A positive value of this coefficient indicates that a tilt of the interface tries to make $\mathbf{p}_\perp$ to point uphill.

Active currents are directed along $\mathbf{p}$ and particles can not leave the film. Moreover planar anchoring is assumed everywhere on the bounding surfaces, as mentioned above. These conditions imply $p_z(z = 0) = 0$, and $\mathbf{p} \cdot \hat{n} = 0$ at $z = h$, where $\hat{n} = (-\nabla_\perp h, 1)/\sqrt{1 + (\nabla_\perp h)^2}$ denotes here the outward normal to the free interface. This means that $p_z \approx \partial_x h$ at the free surface (at first order in h variations), while its z-profile is given by $p_z \approx (z/h)\partial_x h$. For thin films one can take z-averaged values $p_z \approx (1/2)\partial_x h$ and the gradient $\partial_z p_z \approx h^{-1}\partial_x h$.

One takes as a reference state a configuration of the thin film with uniform concentration, height and x-directed polarization, c_0, h_0, p_0, and perform a linear stability analysis. For it one assumes small deviations in the polarization field, i.e. $\mathbf{p}_\perp = \hat{x} + \theta\hat{y}$ with $\theta << 1$. Explicit expressions for the active stress gradients read,

$$\nabla_i \sigma_{ix}^a = \sigma_0(\partial_y\theta + \partial_x c/c_0 + h^{-1}\partial_x h), \tag{6.100}$$

$$\nabla_i \sigma_{iy}^a = \sigma_0\partial_x\theta, \tag{6.101}$$

$$\nabla_i \sigma_{iz}^a = \sigma_0\partial_x h^2/2. \tag{6.102}$$

One finally eliminates the pressure by invoking continuity of the stress at the free surface and uses boundary conditions for the in-plane velocity field and its gradient $\partial_z \mathbf{v}_\perp(h) = 0$ and $\mathbf{v}_\perp(0) = 0$, to obtain a final equation for the vertical profile of the velocity in a z-averaged approximation to active stresses,

$$\mathbf{v}_\perp(z) = \frac{hz - z^2/2}{\eta}\left(\tilde{\gamma}\nabla_\perp\nabla_\perp^2 h - \frac{1}{2}\sigma_0 h\partial_x^2\nabla_\perp h - \mathbf{f}_\perp\right), \tag{6.103}$$

where $\tilde{\gamma}$ denotes the surface tension of the liquid phase, and $\mathbf{f}_\perp$ is a combination written in terms of the small deformations in the three fields, i.e. polarization, concentration and height,

$$\mathbf{f}_\perp = \sigma_0\left[(\partial_y\theta + \frac{\partial_x c}{c_0} + h^{-1}\partial_x h)\hat{x} + \partial_x\theta\hat{y}\right]. \tag{6.104}$$

Inserting this result into Eq. 6.96 we get finally a closed equation relating distortions of the three fields. This relation is better analyzed in its linearized version for $h = h_0 + \delta h$ and $c = c_0 + \delta c$ and resolved in its in-plane Fourier components (i.e. the basis of functions $\exp(i\mathbf{q} \cdot \mathbf{r} - iwt)$),

$$\partial_t \delta h_\mathbf{q} = -\frac{\sigma_0 h_0^2}{3\eta}\left[2h_0 q_x q_y\theta_\mathbf{q} + h_0 q_x^2\frac{\delta c_\mathbf{q}}{c_0}\right.$$
$$\left. +\left(1 - \frac{1}{2}h_0^2 q^2\right)q_x^2\delta h_\mathbf{q}\right] - \frac{\tilde{\gamma}h_0^3}{3\eta}q^4\delta h_\mathbf{q}. \tag{6.105}$$

One would proceed now to obtain dynamical equations for the corresponding Fourier components of the left fields. We do not write them here and only refer to the dispersion relation that follows when considering non-motile particles, i.e. $V_0 = 0$, but retaining extensile or contractile stresses. In this case, deviations in the concentration field are null at the lowest order, while the dispersion relation looks like,

$$w = \pm \frac{1 + isign(q_x C\sigma_0)}{\sqrt{2}} \left(\frac{h_0^2}{3\eta}\right)^{1/2} |C\sigma_0 q_x|^{1/2}|q_y|, \qquad (6.106)$$

so the relative signs of C and $\zeta\Delta\mu$ determine the propagation of the planar Fourier modes. More details on the implications of this predicted instability can be found in the original reference [339].

6.4.1.2 Active Droplets

The issue of active fluids encapsulated in spherical droplets has been quite extensively examined in the literature, not only from an experimental point of view as considered in Chapter 4, but also from the point of view of modeling. I start by considering bulk active fluids, and refer at the end to cortical (i.e. two-dimensional) flows.

In relation to encapsulated bulk active fluids, I only make here a brief mention to a few original papers, adding some specific comments that may be useful to interested readers. The first reference I am aware of in this context was published by Joanny et al. [183] relative to a three-dimensional active drop placed on a substrate. Planar anchoring is assumed, i.e., no normal component of the polarization at any bounding surface. The main conclusion is that active stresses sensibly modify the wetting properties, leading to new static shapes and spreading laws. The shape of the standing drop is determined primarily by the interplay of active stresses and surface tension, with the liquid crystal elasticity of the ordered fluid playing a role near the contact line.

Different from the perspective of a sessile droplet considered by Joanny et al., the question of activity-induced motility of droplets was also addressed by Tjhung et al. [406]. In this latter example, the addressed situation is that of a droplet of a polar fluid (actomyosin solution) confined by interfacial tension and surrounded by a Newtonian host fluid.

The main conclusion, applying to contractile as well as extensile-like systems, is that when the activity parameter exceeds some threshold an initially circular or spherical droplet spontaneously breaks the fore-aft inversion symmetry, leading to unidirectional motion. This spontaneous symmetry-breaking process is akin to a supercritical *Hopf bifurcation*, or alternatively can be considered as a continuous non-equilibrium phase transition (see Fig 6.6).

Still, the model was extended to mimic the behavior of three-dimensional crawling cells by Tjhung et al. [408] in a subsequent paper. The basic ingredients of the model used there were actin tread milling, and a variable effective friction representing focal adhesions to account for basic mechanisms that are crucial for cell motility on a substrate. The study of the onset of motility of three-dimensional droplets of active fluid on planar surfaces was also

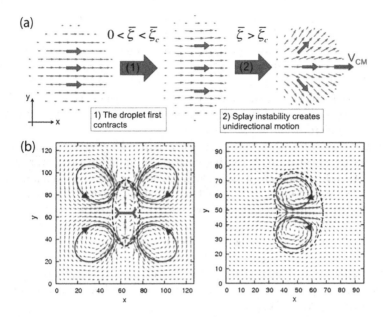

FIGURE 6.6

Spontaneously induced motion of a contractile active droplet. (a) Steady-state configuration without self-advection turns into an elongated shape perpendicular to the polarization, and further to a splayed motile droplet above a critical activity. (b) Left and right panels display respectively flow patterns: a quadrupolar one typical of contractile systems, and that based on a pair of vortices, respectively below and above the motility threshold. Image and caption text adapted from Tjhung et al. [406].

undertaken by Khoromskaia et al. [196]. Motility patterns are analyzed in this latter report in relation to different types of defects and their location, the droplet shape and the surface friction with the substrate.

Leaving polar active fluids, a similar study considering this time a nematic symmetry was published by Blow et al. [39]. The aim was to analyze a two-dimensional active nematic in coexistence with an isotropic fluid, under invariant amounts of each of them. A scalar parameter ϕ measures their relative quantity at a given point. No anchoring is included at the level of the free energy, but the active stress alone turns out to generate a preferential orientation, a phenomenon authors term "active anchoring".[29]

---

[29] A numerical investigation of the morphology of active deformable 3d droplets was recently published by Ruske et al. [331] from the same group. Interestingly, for extensile activity finger-like protrusions were reported to appear at points were disclination lines intersect droplet surfaces. For contractile systems, the activity field drives cup shaped droplet invaginations and the formation of surface wrinkles.

Still addressing the question of self-driven droplets powered by active nematics a paper by T. Gao et al. [123] formulates another proposal. These authors consider a confined suspension of nonmotile particles that can, however, elongate and create active stresses. The variable concentration field is encapsulated into a 2d droplet immersed in a Newtonian fluid. The field on the entire domain is assumed incompressible, and the corresponding Stokes equation incorporates a term related to the surface tension. Different types of locomotion and rotation modes are analyzed.

Next, I dedicate a brief attention to a paper by Giomi et al. [136] (see Fig 6.7) in the same context. My justification to highlight this contribution is again twofold. First because it, originally, introduces an interesting new aspect, i.e. that of spontaneous division of active nematic droplets.[30] Second, because the model applies to a multiphase flow, taking a perspective slightly different from considering permeation effects that are pertinent to real active gels. In this respect, Giomi et al.'s paper is similar to the just mentioned contribution by Blow et al. [39]. Both combine in an elegant way the methodology widely reviewed here to describe active fluids with a classical *phase-field formulation*, very popular in the study, a few decades ago, of a plethora of non-equilibrium pattern forming scenarios [70].

In [136], the two phases are described by a scalar ϕ, whose values are $\phi = -1$ in the isotropic and $\phi = 1$ in the nematic phase. The region with $\phi \approx 0$ would denote the diffuse interface. The capillarity of the interface is represented by a free energy density of a Ginzburg-Landau form, i.e. $\frac{1}{2}\kappa\left[|\nabla\phi|^2 + \frac{1}{2\epsilon^2}(\phi^2 - 1)^2\right]$. The surface tension of the interface is given in terms of the parameters κ and ϵ as $\Sigma = \sqrt{8}/3(\kappa/\epsilon)$. This interfacial tension gives rise to a body (capillary) force that permits to introduce a chemical potential $\mathbf{f}_{cap} = -\phi\nabla\mu$ with $\mu = \delta\mathcal{F}_{cap}/\delta\phi = -\kappa[\Delta\phi - \phi(\phi^2 - 1)/\epsilon^2]$.

Incompressibility of the fluid phases imply $d/dt \int \phi dA = 0$ and $\nabla \cdot \mathbf{v} = 0$. There are now three fields $(\phi, \mathbf{Q}, \mathbf{v})$ and we need to write hydrodynamic equations for all of them in the two-dimensional reduced form of the $\mathbf{Q}$-based theory,

$$\frac{D\phi}{Dt} = M\kappa\left[\Delta\phi - \frac{\phi(\phi^2 - 1)}{\epsilon^2} + \xi(\phi)\right], \tag{6.107}$$

$$\rho\frac{Dv_i}{Dt} = \eta\Delta v_i - \partial_i p - \phi\partial_i\mu + \partial_j\sigma_{ij}, \tag{6.108}$$

$$\frac{DQ_{ij}}{Dt} = \lambda S v_{ij} + Q_{ik}\omega_{kj} - \omega_{ik}Q_{kj} + \gamma_1^{-1}H_{ij}. \tag{6.109}$$

---

[30]Experimental realization of droplet division through the concerted effect of encapsulated constructs of actin and myosin motors was reported in a paper published by Weirich et al. [430].

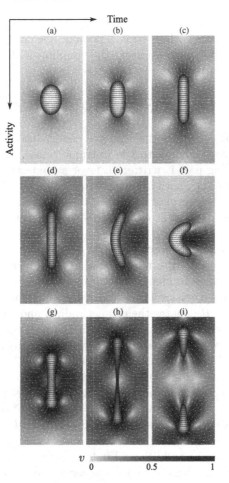

FIGURE 6.7
Induced motion and division of a contractile active droplet with homeotropic anchoring. Panels (a) to (c): For small activity, the droplet stretches under the effect of the pair of $+1/2$ disclinations. Panels (d) to (f): For intermediate activities, the droplet is unstable to splay, deforms and, further, it moves. Panels (g) to (i): For very large activity, the capillary forces are not strong enough to prevent cell division (colored image in original version). Image and caption text adapted from Giomi et al. [136].

In the previous equations, M denotes a mobility coefficient, using here a different notation to distinguish it from the chemical potential, η the unique viscosity, and the function ξ represents a Lagrange multiplier to guarantee mass conservation,

$$\xi(\phi) = |\phi^2 - 1| \frac{\int (\phi^2 - 1)dA}{\int |\phi^2 - 1|dA}. \tag{6.110}$$

Finally, we need to write the free energy density. In its simplest form, it contains a bulk plus anchoring terms. The bulk free energy density is written as the usual quadratic, incorporating the coupling with the phase variable, and quartic terms, in the form,

$$F = \frac{K}{2}\left[|\nabla \mathbf{Q}|^2 + \frac{1}{\delta^2}Tr\mathbf{Q}^2(Tr\mathbf{Q}^2 - \phi)\right]. \qquad (6.111)$$

This last equation implies that the free energy has a minimum for $S = 0$ where $\phi = -1$ corresponding to the isotropic phase, and for $S = 1$ when $\phi = 1$. The anchoring part is written to favor a homeotropic alignment $F_a = 1/2W_a Tr(|\nabla \phi|^2 \mathbf{Q} - \mathbf{A})^2$, with the tensor $A_{ij} = \partial_i \phi \partial_j \phi - |\nabla \phi|^2 \delta_{ij}/2$. The net effect of this anchoring term is to favor a director field parallel to $\nabla \phi$, thus normal to the interface. The stress tensor is written as the normal sum of the elastic and active parts.

In absence of activity, nematic droplets display two interior $+1/2$ defects. When activity is introduced, similarly to what I commented above in relation to the paper by Tjhung, droplets splay, turn motile and even may break under high enough activity. Division is controlled by a capillary number defined as $Ca_\alpha = \alpha R/\Sigma$ where R denotes the droplet radius, and I keep the convention that $\alpha > 0$ denotes contractile systems. For the particular set of simulation parameters considered in the paper, the threshold for division is estimated at a critical value $11.4 < Ca_\alpha^{div} < 16$ (see Fig. 6.7).

In the remaining of this section, I will refer to cortical, rather than bulk, flows that display active nematic characteristics. I already mentioned the paper by R. Zhang et al. [455] analyzing the dynamics of defects in active nematic shells, motivated by the experimental results by Keber et al. [193] (see Sect. 4.4.1). The scheme of Zhang et al.'s calculation is essentially that of the general $\mathbf{Q}$ theory with two additions. First, an extra term is introduced at the level of the free energy density to account for a planar degenerate anchoring of the nematic field. The corresponding contribution penalizes out-of-plane distortions of the $\mathbf{Q}$ tensor. An additional equation is introduced for the dynamics of the $\mathbf{Q}$ tensor at the droplet surface. Details can be found in the original reference. Inspired also by Keber et al.'s work, I also mention the paper by Metselaar et al. [256] addressing self-deformation patterns of active shells. Authors stress the role of defect in the morphodynamics of the shell.

A more complicated piece of theory was dedicated by Pearce et al. [301] to the experiments on toroidal shells published by Ellis et al. [99] (see Sect. 4.4.1). The idea is to adapt a $\mathbf{Q}$-based theory to a curved substrate taking into account the topology corresponding to the outer surface of a torus (see Fig. 6.8). I simply quote the main conclusion of the paper and refer the interested reader to the original reference for details. Using a combination of hydrodynamic and particle-based simulations, authors demonstrate that the fundamental structural features of the fluid, such as the topological charge density, the defect number density, the nematic order parameter, and defect creation and

(a) 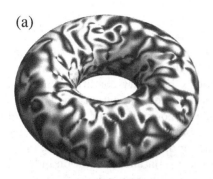   (b)

FIGURE 6.8
An active nematic cortex on a torus obtained from numerical integration of the Q tensor formalism. Left and right panels correspond respectively to Schlieren textures for the nematic field and to the vorticity distribution (colored image in original version). Image and caption text adapted from Pearce et al. [301].

annihilation rates, are approximately linear functions of the substrate Gaussian curvature, which then acts as a control parameter for the chaotic flow.

6.4.2   Modeling Active Flows under Geometric Confinement

The first paper I am aware of in this context was published by Fürthauer et al. [115]. The idea was to study the dynamics of an active polar fluid in a Taylor-Couette geometry, i.e. with the fluid confined between two coaxial rotating cylinders. As easily anticipated, this system is found to be able to generate flow and, as a consequence, to set the two cylinders into relative motion, either by spontaneous symmetry breaking or via asymmetric boundary conditions on the polarization field at the cylinder surfaces.

The addressed situation corresponds to two impermeable concentric cylinders of radii (outer) R_+ and (inner) R_-. One can choose to keep the inner cylinder fixed without loss of generality. The outer cylinder rotation rate is designed $\Delta\omega$. The model further enforces azimuthal symmetry and translational invariance along the cylinder long axis. In addition, the fluid is considered incompressible. The polarization field is assumed to be confined to the section perpendicular to the long axis and assumed to have fixed magnitude $\mathbf{p}^2 = 1$. Torques Γ_+ and Γ_- per unit axial length can be assumed to apply at the outer and inner cylindrical surfaces, such that $\Gamma_+ + \Gamma_- = 0$.

The analysis follows the same lines I commented previously in relation to the work by Kruse et al. [207] on aster or vortex-like defects (see Sect. 6.1.2). No slip boundary conditions are prescribed, i.e. $v_\theta(R_-) = 0, v_\theta(R_+) = \Delta\omega R_+$, and both asters and vortex-like solutions are considered: $\psi(R_+) = \psi(R_-) =$

ψ_0, with $\psi_0 = 0$ for asters and $\psi_0 = \pm\pi/2$ for vortices. Without applied torques, the analysis of the solutions and their stability is essentially what was published by Kruse et al. in [207]. In the present geometry, the flows that naturally appear past an activity threshold make the outer cylinder to rotate. Explicit solutions were worked out for vortex boundary conditions in [115]. Past the activity threshold, stable solutions for the polarization angle display a rather symmetric (concave) profile, while is monotonous for the azimuthal velocity. Increasing the activity, this stable branch is supplemented with unstable branches displaying a successive number of nodes at intermediate radii.

The stalled system is also considered, since spontaneous rotation indicate that the system can behave as a motor. Its stall corresponds to $\Delta\omega = 0$. A corresponding instability is found with a bifurcation diagram similar to the previous case. The difference is that in this case curved profiles are found but with opposite curvature: (convex) for the polarization and (concave) for the azimuthal velocity.

More or less simultaneously to the paper by Fürthauer et al. [115], another contribution was published on a similar geometric design by Ravnik et al. [322]. These authors addressed a single cylinder geometry, and rather than a polar fluid, a nematic state was considered for both planar and homeotropic boundary conditions with no-slip velocity characteristics. Solving the continuum equations of motion in the flow-aligning extensile regime, these authors found that active flows emerge not only along the capillary axis but also within the plane of the capillary, where radial vortices are formed.

Figure 6.9 shows the active flow profile in a capillary with surfaces imposing homogeneous alignment, parallel to the capillary axis. Upon increasing the activity one observes a transition to spontaneous flow. Above the transition, the leading flow is along the axis of the capillary, directed left or right depending on the random noise in the initial condition. At higher activities, the flow evolves into a bidirectional pattern, with the velocity in opposite halves of the capillary in opposite directions.

Also in-plane secondary flow fields within the capillary cross-section (Fig. 6.9 (b) and (c), second panels) were identified whose magnitude is two orders of magnitude smaller than the primary currents. The secondary flow consists of a circularly symmetric pattern of distinct vortices. The vortices are pairwise counter rotating, and thus exhibit no net angular momentum.

The geometries I have referred to so far strictly correspond to three-dimensional realizations. Since the most common experimental realizations of active fluids, specially those based on microtubules and kinesins, have been mostly prepared as two-dimensional active layers it is not surprising that theoretical attention turned soon into the study of two-dimensional forms of geometric confinement.

The simplest choice is a channel geometry as studied first in a couple of papers published more or less simultaneously by Doostmohammadi et al. [88] and Shendruk et al. [358]. In the first of the mentioned papers, the interest was to look at the onset of active turbulence under confinement, while the second

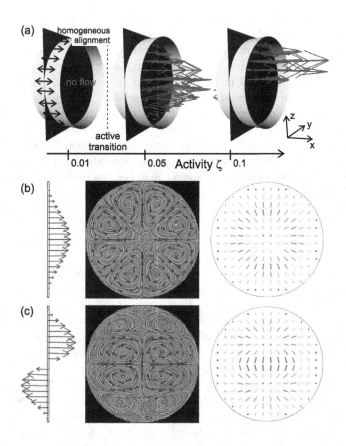

FIGURE 6.9
Active flows in a capillary with planar alignment parallel to the capillary axis. (a) Flow regimes upon increasing activity: basic state with no flow (left), first excited state with unidirectional flow (middle), and second excited state with bidirectional flow (right). Panels (b) and (c); Primary flow across the capillary diameter, secondary flow in the yz plane, and director profile of the active nematic (short lines, out of plane; longer lines, in plane) for (b) the unidirectional flow and (c) for the bidirectional flow (colored image in original version). Image and caption text adapted from Ravnik et al. [322].

predicted the existence of a dynamical organized regime known as "dancing disclinations" to what I referred earlier with occasion of its experimental observation by Hardoüin et al. ([156] in Sect. 4.4.2). Let's comment separately on both of them.

Doostmohammadi et al., analyze the transition between a regime of vortex lattice and a situation of chaotic flows. The transition takes place through the splitting of vortex pairs into smaller non-ordered vortices. Authors refer to these localized domains of non-ordered vorticity as "active puffs", in analogy with inertial puffs observed in experiments on turbulence in pipe flows [14]. Active puffs can themselves split into new puffs or decay into the ordered vortex-array state. Moreover, active puffs span the entire system when decay is slower than generation, and under these conditions of high enough activity the decay time far exceeds the splitting time. The turbulence fraction is measured as a function of the activity level and results indicate that it grows with a power-law dependence whose exponent matches the universal critical exponent of the (1+1) directed percolation model ($\beta = 0.276$). Strikingly enough the found exponent is practically identical to the value measured for inertial turbulence in Couette flow (see [88] for the original references). Other exponents that were measured correspond to the spatial and temporal distributions of vortex-lattice gaps.

In the second work, Shendruk et al. analyze the defect dynamics accompanying the vortex-lattice state just mentioned. The results indicate that this defect dynamics displays a dancing disclination state. These dancing disclinations are positively charged, long lived, and continuously navigate through the channel. As they move past each other on the vortex lattice, the positive disclinations form short-lived pairs, that permit the authors to qualify such a state under the term *topological Ceilidh dynamics* (named after some traditional Gaelic dances). A complete study reveals in fact that the dancing disclination regime appears as a state intermediate between unidirectional flow and the conventional turbulent-lile active flows. Employed parameters correspond to flow-tumbling and extensile, although, apparently, Ceilidh-like dynamics is also observed for contractile and other conditions of flow-alignment. Channel walls enforce no-slip boundary conditions and strong homeotropic anchoring of the orientation field.

If we remember from Sect. 4.4.2, circular confinement of two-dimensional active nematics was studied in experiments performed by Opathalage et al. [285], and by Hardoüin et al. in [157]. It thus seems natural to refer here to modeling approaches published in relation to this particular scenario of geometric confinement. I will briefly refer on what follows to a couple of simultaneously published papers. The first report considers confinement on circular disks by Norton et al. [283], and the second one on annuli by S. Chen et al. [63].[31]

Norton et al. [283] consider specifically the role played by boundary conditions on an extensile active nematics under no-slip conditions. Depending on disk radius and activity parameter three basic states are identified: a minimally distorted state totally equivalent to an equilibrium configurations (two

---

[31] A modeling scheme similar to that used in the paper by S. Chen et al. had been employed earlier by T. Gao et al. in [122] to analyze confinement is discs.

antipodal half-integer positive defects), a defect circulating regime, and finally the turbulent scenario typical of unconfined active nematics. What is more interesting from the work by Norton et al., is that, differently from the equilibrium situation, anchoring seems to be largely screened and playing a minor role in the active liquid crystal dynamics.

The second paper is from S. Chen et al. [63], simulating again no-slip boundary conditions on annuli. Both dilute and concentrated suspensions of extensors (i.e. non-motile but elongating specimens producing extensional flows) are considered separately in a Bingham-Q tensor scheme (an alternative to the most conventional **Q** theory). For dilute conditions, regimes of no-flow lead to circulation currents by increasing activity. Increasing further, circulating flow pattern still dominates but the streamlines exhibit periodic bending deformations in the radial direction to form traveling waves, leading to counter-clockwise (CCW) and clockwise (CW) vortices near the inner and outer boundaries respectively. At still larger activities flow becomes rather chaotic. Thresholds for respective crossovers decrease with increase the annulus width. For concentrated suspensions, steady-state solutions corresponding to nematic order without flows were obtained at relatively small activity, and small gaps. Relaxing this constraint, the internal collective dynamics features motile disclination defects and flows at finite gap widths.

6.5 Brief Commented List of Selected Review Papers

The list here essentially refers to the contents of this chapter, although some of the reviews deal totally or partially with developments that apply indeed to dry active matter that is discussed in next chapter.

Active nematics.
A. Doostmohammadi, J. Ignés-Mullol, J. M. Yeomans and F. Sagués, Nature Communications 9, 3246 (2018).
*This review supposes an effort to bring together a comprehensive account of experimental scenarios and theoretical developments built around the well-known concept of active nematics. Special emphasis is devoted to analyze active flows patterns and defect dynamics. Experiments refer to the celebrated system based on bundled microtubules exercised with kinesin motors. Interfacial preparations, singularly those involving nematic/smectic liquid crystals, originally developed by two of the authors are emphasized. The theoretical part is based on the accumulated experience of the rest of the contributors in using the Beris-Edwards **Q** formalism.*

Generic theory of active polar gels: a paradigm for cytoskeletal dynamics.
K. Kruse, J. F. Joanny, F. Jülicher, J. Prost and K. Sekimoto, European Physical Journal E 16, 5 (2005).

This reference is one of the series of review papers that appeared from the Dresden-Paris collaboration, and that established the basis of the theory of active gels. Other similar contributions were published by J. F. Joanny et al., New Journal of Physics 9, 422 (2007) and F. Jülicher et al. Physics Reports 449, 3 (2007), that were several times mentioned when developing the contents of this chapter. The reviewed theory follows the same principles that guided the formulation of nematodynamics a few decades ago.

The mechanics and statistics of active matter.
S. Ramaswamy, Annual Review Condensed Matter Physics 1, 323 (2010).
This article reviews the progress accumulated during the first decade of the 2000's in applying the principles of non-equilibrium statistical mechanics and hydrodynamics to build a systematic theory of the behavior of collections of active particles. Theory and experiments are presented side by side, and discussed scenarios encompass both living systems and inanimate realizations.

Hydrodynamics of soft active matter
M. C. Marchetti, J. F. Joanny, S. Ramaswamy, T. B. Liverpool, J. Prost, M. Rao and R. Aditi Simha, Review of Moder Physics 85, 1143 (2013).
Ten years after publication, this is in my opinion the most comprehensive review paper on active matter, encompassing both wet and dry realizations, that has been published to date. Doubtless, it has become since its appearance the classical reference in the field. Apart from encompassing a wide variety of swimming-like active systems, from microorganisms to chemical and mechanical analogues, one of the merits of this contribution is its declared interest to integrate from an unified theoretical perspective several approaches, from semi-microscopic to phenomenological, that paved the way to the development of the theoretical corpus of this discipline.

Active gel physics
J. Prost, F. Jülicher and J. F. Joanny, Nature Physics 11, 111 (2015).
*This more specific review is mostly focused on the biological relevance of the theory of active gels, emphasizing the possibilities of its application into a cell context, from sub-cellular structures to tissues. Most prominently discussed topics refer to cell motility, division, wound healing or cortical flows. Similar in spirit, I could also mention an earlier review by D. A. Fletcher and P. L. Geissler, **Active biological materials**, Annual Review of Physical Chemistry, 60, 469 (2009).*

Emergent self-organization in active materials
M. F. Hagan and A. Baskaran, Current Opinion in Cell Biology 38, 74 (2016).
The authors choose to discuss in a unified way the two most characteristic classes of active matter systems: propelled colloidal particles and extensile rod-like particles. It is a clear example of the close similarities and overlapping subjects when coming to discuss topics in the two categories.

The dynamics of microtubule/motor-protein assemblies in Biology and Physics
M. J. Shelley, Annual Review Fluid Mechanics 48, 487 (2016).
More focused than the previous broader-aimed references, this contribution reviews recent modeling work applied to systems of microtubules and motors. The perspective adopted by M. J. Shelley is to treat them as multi scale complex fluids. The underlying motivation behind the choice of selected topics is singularly interesting for those that do not have a special education in Cell Biology, since this review aims at establishing the basic scenarios and biological functions of assembled microtubules within the cell milieu.

Collective hydrodynamics of swimming microorganisms: Living fluids.
D. L. Koch and G. Subramanian, Annual Review Fluid Mechanics, 43, 637 (2011).
Primarily dedicated to review computer simulation efforts for systems of hydrodynamically interacting self-propelled particles. Stability analysis and numerical solutions of averaged equations of motion for ensembles of microswimmers are also profusely reviewed.

Rheology of active fluids.
D. Saintillan, Annual Review Fluid Mechanics 50, 563 (2018).
This review summarizes recent experiments, models and simulations, highlighting the critical role played by the rheological response of active fluids. It encompasses a variety of contexts, from enhanced transport of passive suspended objects to the emergence of spontaneous flows and collective motion. From the same author, and coauthored with M. J. Shelley, there is a precedent reference reviewing colloidal suspensions published by D. Saintillan et al., **Active suspensions and their nonlinear models** *Comptes Rendus Physique 14, 497 (2013).*

Collective motion.
T. Vicsek and A. Zafeiris, Physics Reports 517, 71 (2012).
From one of the pioneers in the field of dry active matter, this paper reviews both observations and basic laws describing the essential aspects of collective motion ranging from macromolecules, to robots, groups of animals and people. A balanced discussion includes experiments, mathematical treatments and simulation models.

Hydrodynamics and phases of flocks.
J. Toner, Y. Tu and S. Ramaswamy, Annals of Physics 318, 170 (2005).
The classical theoretical paper on flocking: the collective motion of a large number of self-propelled units. Phases are classified in terms of their symmetries much like in equilibriums systems. Analogies and differences with their equilibrium counterparts are stressed.

Computational models for active matter.
M. R. Shaebani, A. Wysocki, R. G. Winkler, G. Gompper and H. Rieger, Nature Reviews Physics 2, 181 (2020) [354].
A very recent review of the diverse approaches that have been employed during these last decades for modeling active matter. It comprises both discrete, colloidal-like, systems and active fluids, either natural or artificially designed. The last sections are briefly dedicated to cells, tissues and animal groups.

To finish, I also mention here a special issue of Annual Review of Condensed Matter Physics appeared during the completion of this monograph and containing a couple of interesting review papers on modeling of active particles: **Dry aligning dilute active matter** by H. Chaté, Ann. Rev. Condens. Matt. Phys. 11, 189 (2020), and **Self-propelled rods: Insights and perspectives for active matter** by M. Bär, R. Großmann, S. Heidenreich and F. Peruani, Ann. Rev. Condens. Matt. Phys. 11, 441 (2020).

7

Concepts and Models for Dry Active Matter

All the systems considered so far in this text have fluid-based characteristics. Systems considered in Chapter 3, for instance, can be considered as self-propelling or driven counterparts of classical liquid-dispersed colloids. In other cases, this colloidal nature, although undeniable, is less apparent, mainly when thinking of the protein-based active fluids I referred to in Chapter 4. Nevertheless, in all the active systems I have described so far we can clearly identify an ambient fluid.

The consideration of a chapter devoted to dry active matter may thus appear, at first sight, rather bewildering. However, the commonly accepted distinction between **"wet" active matter** and **"dry" active matter** [247] teach us that not all active systems involving a fluid phase can be strictly considered wet, contrary to what could be thought at first glance. I feel thus opportune to finish this monograph including this last, albeit briefer, chapter dedicated to dry systems with two main arguments in mind. First, because of the just stated fuzzy and subtle distinction between wet and dry, briefly elaborated in the next paragraph. Second, for the sake of completeness, thinking of the large community of practitioners in active matter who are being or have been educated along this perspective.

From a fundamental point of view, active systems are qualified as *wet* whenever we impose momentum conservation at the level of the embedding fluid. This principle has been many times invoked, either in implicit or explicit ways, when modeling active fluids in the previous chapter. Conversely, *dry* realizations correspond to scenarios when this constraint does not apply, commonly under the assumption of the dominant role played by the friction drag that a bounding support exerts on the ambient fluid of the system under study. In practice, all active fluids are bounded one way or another, and, thus, one needs always to consider two parameters that describe as a whole the active fluid and its contacting boundaries, i.e. its viscosity η and friction coefficients γ (assumed both scalar parameters). Their ratio renders the (squared) *hydrodynamic screening length*, i.e $\tilde{\delta} = (\eta/\gamma)^{1/2}$. Naturally, a system we may be interested to looking at will be endowed with its own values of these two parameters, and in turn will have a well-defined screening length. Eventually, and depending on the length scales we are interested, we are allowed to

DOI: 10.1201/9781003302292-7

consider a system as dry or wet, as hydrodynamic effects will fade away on scales larger than $\tilde{\delta}$.

This kind of ambivalent description is more or less apparent when looking at experiments as well. If one thinks, for instance, of shaken granular ensembles of particles of whatever geometry [271, 209, 81] there is no doubt they can be safely classified as dry, but some dense bacterial baths [89, 431] are commonly classified in the same way. Even the same primary system in slightly two different presentations can be considered either wet or dry. For instance, extracts of filamentary protein performing in microtubule assays might be considered dry, but, more often, active microtubule/kinesin blends are presented as paradigmatic active (nematic) fluids, as thoroughly discussed in this text.

Once this question has been clarified, we can go over to the next concept, basic in our presentation of dry active matter. As I have much insisted in previous chapters, active fluids can be classified into polar or nematic according to the symmetries of their textures and dynamic flows. This distinction applies again for dry systems, and I dedicate either class a specific attention. The freedom of modeling permits, however, to consider as well systems of mixed symmetry, as it will become clear on what follows. This characterizes, apart from polar and nematic, a third class of dry active systems [247].

Differently from what was commented in the introductory remarks of the previous chapter, and considering the tradition that the theoretical study in this specific context has accumulated over the past years, I feel appropriate to comment here both the hydrodynamic and microscopic, i.e. active unit-based, approaches to dry active matter. Either perspective has its intrinsic advantages and drawbacks, and although my own experience is closer to the first ambit, I prefer here to take an eclectic perspective dedicating both of them separate sections at similar depths. In any case, my aim is to highlight the principal aspects and results of both methodologies, avoiding to get trapped in an endless enumeration of papers that have been published in either context over the last years.

To systematize the presentation I choose to follow a couple of detailed reviews, respectively related to either one of the just mentioned levels of description. On what respects to hydrodynamic theories I refer to the review by M. C. Marchetti et al. [247], that devotes a whole chapter to present in a systematic way the most important results for active dry systems, according to the three different categories mentioned above. Alternatively, a more recent and specific paper published by H. Chaté [58] will serve us to elaborate the basics of the microscopic approaches. I will not go deep into the way the two approaches are connected, i.e. on the (often complicate) procedures needed to derive the former (hydrodynamic) from the latter (microscopic). Some minimal ideas and a worked out example, though, will be commented later on in the introductory paragraphs of Sect. 7.2.

7.1 Hydrodynamic-like Theories

For all three categories treated in this section, i.e. polar, nematic and mixed symmetry, I follow a similar style of presentation. I first motivate, either from experiments or modeling, the interest for the particularly chosen class. Next, I introduce the corresponding theoretical scheme constructed for the chosen field-variables and the set of equations they satisfy. Most of the interest resides in justifying and interpreting rigorously the category-distinct displayed contributions. Finally, I comment on the main results that one can derive from this set of equations, singularly those that are specific to the discussed class. Details are kept to a minimum not to distract the main flow of the discussion. Readers interested in particular questions will find a wealth of original references in the above mentioned couple of review papers. References to experimental realizations, scarce to the best of my knowledge at the colloidal scale, will be quoted for each specific category at appropriate passages in the discussion.

7.1.1 Flocking of Active Polar Particles

The motivation for this particular level of modeling clearly goes back to the celebrated Vicsek (microscopic) model [421], published more than twenty-five years ago. This seminal paper proposed the existence of a novel type of phase transition observed for self-propelling particles with group characteristics. Swarming features were evidenced as resulting from collective interactions when increasing density (or reducing noise) (see Sect. 5.1 for references to swarming in wet systems).

In Vicsek's model, particles are supposed to move with constant velocity (V_0) while, at each time step, they assume the average direction of motion of the particles in their neighborhood, with some added random perturbation (ferromagnetic alignment). This simple scheme immediately captured the interest of theoreticians, principally J. Toner and Y. Tu [412, 413] (see also a more recent account with the participation of S. Ramaswamy another pioneer in the field [414]).

These authors derived a set of hydrodynamic equations by purely invoking symmetry considerations as they apply to the retained gradient terms that describe a (non-equilibrium) polar fluid in contact with a frictional substrate (the latter amounts to consider the system lacking Galilean invariance).[1] Under the usual hypothesis of fixed number of flowing particles, one prescribes the number density field as the only conserved (scalar) quantity. On the other

---

[1] In classical mechanics, Galilean invariance states that the laws of motion are the same in all inertial frames, i.e. in those frames that are not undergoing acceleration. In an inertial frame of reference, a physical body subjected to a zero mean force will move with constant velocity.

hand, we need to introduce the (coarse-grained) version of the polarization vector field $\mathbf{p}(\mathbf{r}; t)$ written this time, differently from the ad-hoc way as it was introduced in the previous chapter, in terms of unit-vector individual polarizations $\mathbf{p}_i$,

$$\rho(\mathbf{r}; t) = \sum_i \delta(\mathbf{r} - \mathbf{r}_i(t)), \tag{7.1}$$

$$\mathbf{p}(\mathbf{r}; t) = \frac{1}{\rho(\mathbf{r}; t)} \sum_i \mathbf{p}_i(t) \delta(\mathbf{r} - \mathbf{r}_i(t)). \tag{7.2}$$

The respective equations of motion are written in terms of a (polarization) free energy functional $\mathcal{F}_p$, incorporating the corresponding noise terms,

$$\partial_t \rho + V_0 \boldsymbol{\nabla} \cdot (\rho \mathbf{p}) = -\boldsymbol{\nabla} \cdot \left(-\frac{1}{\gamma_\rho} \boldsymbol{\nabla} \frac{\delta \mathcal{F}_p}{\delta \rho} + \boldsymbol{\xi}_\rho \right), \tag{7.3}$$

$$\partial_t \mathbf{p} + \lambda_1 (\mathbf{p} \cdot \boldsymbol{\nabla}) \mathbf{p} = -\frac{1}{\gamma_p} \frac{\delta \mathcal{F}_p}{\delta \mathbf{p}} + \boldsymbol{\xi}_p, \tag{7.4}$$

where γ_ρ, γ_p denote kinetic coefficients. The term λ_1, with dimensions of velocity (i.e related to a current for the polar field-variable) does not necessarily has to coincide with V_0 as a consequence of the lack of Galilean invariance.[2]

Derivations of these hydrodynamic equations from microscopic models, as published by Bertin et al. [29], assess as well this inequality between the coefficients in the convective terms for the density and polarization fields.[3]

As usual, the noise in the second equation, denoted $\boldsymbol{\xi}_p$, is assumed to enter additively, and be Gaussian with zero mean and δ correlations.[4] The right hand side of the conservation equation for the particle density is of diffusive nature and is sometimes neglected on some of the considerations that follow. Notice that the left hand sides of these equations clearly distinguishes between the density and polarization fields. The first is naturally advected by the intrinsic particle velocity, whereas the second has a dual nature, as a current itself, and, at the same time, as an orientational order parameter, i.e

---

[2]The use of the symbol λ_1 here, and λ later, follows the convention employed in the review by Marchetti et al. [247]. Their meaning has not to be confused with what was attributed to these symbols in Sect. 6.1.1, as an activity related coefficient, where the general framework of equations for active polar fluids was presented.

[3]A more recent derivation of Toner-Tu equation for mesoscale dynamics in suspensions of microswimmers was published by Reinken et al. [326].

[4]The assumption of additive noise is again the usual one. However, the possibility to choose a multiplicative-approximation, i.e to assume that the effective noise value entering into the dynamical equation depends on the local value of the dynamical variable, remains an open issue. See the review by Marchetti et al. for a note on this direction. Moreover, I implicitly assume on all what follows that the noise contributions have a thermal origin, and that the usual Stokes-Einstein relation connecting diffusion coefficients and friction apply.

it is considered a field distinctively advected by itself [247]. The distinctive coefficient that accounts for this self-advection of the polarization is precisely the one denoted λ_1.[5]

The left free energy functional is expressed as,

$$\mathcal{F}_p = \int_V \left[\frac{\alpha'(\rho)}{2} |\mathbf{p}|^2 + \frac{\beta'}{4} |\mathbf{p}|^4 + \frac{K'}{2} (\delta_\alpha p_\beta)(\delta_\alpha p_\beta) + \frac{\omega}{2} |\mathbf{p}|^2 \nabla \cdot \mathbf{p} - \omega_1 \nabla \cdot \mathbf{p} \frac{\delta\rho}{\rho_0} + \frac{A}{2} \left(\frac{\delta\rho}{\rho_0} \right)^2 \right],$$
(7.5)

written in terms of a mean density ρ_0 and density fluctuations $\delta\rho$ about its mean value.

Let's look in some detail at the contributions retained in this expression for the free energy of the system. The first two terms refer to the continuous (mean-field) order-disorder transition when α' crosses zero at $\rho_0 = \rho_c$, according to a phenomenological functional choice $\alpha'(\rho) = a_0(1 - \rho/\rho_c)$ (β is taken positive as usual for non active systems to assure stability). The next term is the conventional elastic contribution, this time written for a polar self-assembled material under the one-constant approximation. The last term accounts for a compressional penalty in terms of the modulus A. The left terms with ω-proportional factors will be easier to interpret later on when looking at their effect at the level of the explicit dynamical equation for the polarization field $\mathbf{p}$ that we are going to obtain next. The rotational diffusion rate $D_r = a_0/\gamma_p$ has dimensions of frequency, and it is chosen to set the time unit such that $D_r = 1$. The dynamical equation for the polarization field $\mathbf{p}$ thus reads (quantities without prime are obtained by dividing by γ_p),

$$\partial_t \mathbf{p} + \lambda_1 (\mathbf{p} \cdot \nabla) \mathbf{p} = -[\alpha(\rho) + \beta |\mathbf{p}|^2] \mathbf{p} + K\nabla^2 \mathbf{p} - v_1 \nabla \frac{\rho}{\rho_0} + \frac{\lambda}{2} \nabla |\mathbf{p}|^2 - \lambda \mathbf{p}(\nabla \cdot \mathbf{p}) + \boldsymbol{\xi}_p,$$
(7.6)

with two effective parameters with dimensions of velocity $v_1 = \omega_1/\gamma_p$ and $\lambda = \omega/\gamma_p$. These are in fact equilibrium terms, also present in describing ferroelectric effects in liquid crystals, that introduce contributions to polar ordering that may potentially arise either from gradients in the density or the magnitude of the polar order. Once the scheme has been set, let's comment on some important results from it.

We look first at homogeneous steady states in absence of fluctuations. By construction the system exhibits a phase transition between two homogeneous steady states: an isotropic and disordered (gas) configuration with zero-mean velocity ($\mathbf{p} = 0$) for $\alpha > 0$ ($\rho_0 < \rho_c$), and a moving (flocking) state ($\mathbf{V} = V_0 \mathbf{p}_0$) with $p_0 = \sqrt{(-\alpha(\rho_0) = -\alpha_0)/\beta}$ ($\rho_0 > \rho_c$), characteristic of a broken rotational symmetry. Notice in this respect that the propulsion direction is completely degenerated in this formulation. It is important to remark [247]

---

[5]A similarly denoted contribution, bearing the same meaning, was introduced in the treatment by Sankararaman et al. [339] of thin deformable active films in Sect 6.4.1.

that these mean-field results apply as well when fluctuations are allowed, as noted in the original Toner and Tu papers.

The characteristics of these homogeneous steady states are conventionally analyzed in terms of their (linear) stability properties with respect to virtual deviations, whose dispersion relations are easily obtained in terms of Fourier modes. The isotropic state turns out to be stable provided $v_1 > 0$. In microscopic derivations of these equations $v_1 = V_0/2$ at low densities ([29]), and thus the condition for stability is trivially established. Thus, as long as $\alpha_0 > 0$ the system displays a linearly stable gas configuration. However, even in this case interesting observations refer to the existence of anomalous, i.e. non-diffusive, modes of propagation of density inhomogeneities as $\alpha_0 \to 0^+$.[6]

More interesting are the results for the polarized state. The linear stability analysis is easier performed close to the order-disorder transition, $\alpha_0 \to 0^- (p_0 \to 0^+)$, predicting instability conditions. Nonlinear effects, already in absence of noise effects, manifest themselves as a replacement of the uniform polarized state by propagating bands aligned transverse to the direction of mean polarization, as also observed from direct simulations of Vicsek-like models [143, 59]. Other simply announced results in two-dimensional systems confirm the presence of *sound waves* deep in the ordered phase, and *long-range order* of a continuous order polarization parameter that violates Mermin-Wagner theorem.[7]

7.1.1.1 Giant Number Fluctuations

I reserve a special attention to another striking observation of the theoretical analysis of flocking of active polar particles: the presence of **giant density fluctuations**, a concept that was already announced in the introductory words of Chapter 5. I present in this subsection a simple treatment appropriate to linearized equations, including this time a diffusive current and noise contributions [247]. Later on, I will slightly refine this first obtained result.

Away from the mean-field transition, where the ordered state is linearly stable, one can calculate correlation functions of the density field. These density fluctuations are conveniently accounted for in terms of an important quantity in Condensed Matter Physics i.e., the so-called *static structure factor* $S(\mathbf{q}) = \frac{1}{\rho_0 V}\langle \delta\rho_{\mathbf{q}}(t)\delta\rho_{-\mathbf{q}}(t)\rangle = \int_{-\infty}^{\infty} S(\mathbf{q},\omega)\frac{d\omega}{2\pi}$, derived from the *dynamic structure factor*,

$$S(\mathbf{q},\omega) = \frac{1}{\rho_0 V} \int_0^\infty \exp(iwt)\langle \delta\rho_{\mathbf{q}}(0)\delta\rho_{-\mathbf{q}}(t)\rangle dt, \qquad (7.7)$$

---

[6]Since here Fourier components are taken as $\exp(i\mathbf{q}\cdot\mathbf{r} - iwt)$, this corresponds to real ω determinations in the dispersion relation for a range of intermediate wave numbers. In this case, density inhomogeneities propagate as sound waves [247].

[7]Mermin-Wagner theorem does not apply here as in conventional equilibrium systems, where it precludes the spontaneous breaking of a continuous symmetry in two-dimensional systems.

In the present situation, the leading order dependence of the structure factor corresponds to a singularity of second-order [247],

$$S(\mathbf{q}) \propto \frac{1}{q^2}. \tag{7.8}$$

It is worth remembering that in the limit of the smallest wave number, the static structure factor is connected to the number fluctuations in the system through the relation $\lim_{q \to 0} S(q) = \frac{(\Delta N)^2}{\langle N \rangle}$ where $(\Delta N)^2 = \langle (N - \langle N \rangle)^2 \rangle$ is the variance in the number fluctuations. In equilibrium, it is known that the standard deviation $\Delta N \sim \sqrt{\langle N \rangle}$ so that $\Delta N / \langle N \rangle \sim 1/\sqrt{\langle N \rangle} \to 0$ as $\langle N \rangle \to \infty$. We rewrite the variance in terms of the structure factor as $\Delta N \sim \sqrt{\langle N \rangle S(q \to 0)}$, and assume that the smallest wave number is taken in terms of the system size $V^{-1/d}$. Further, using Eq. 7.8 one has $S(q \to 0) \sim V^{2/d} \sim \langle N \rangle^{2/d}$, that finally leads to,

$$\Delta N \sim \langle N \rangle^{(1/2)+1/d}, \tag{7.9}$$

clearly showing the mentioned anomaly that yields, for instance, a unity exponent for $d = 2$.

As a matter of fact, the Toner and Tu original and more complete theoretical analysis of flocking [412], based on the application of renormalization techniques, had anticipated this striking behavior of the density inhomogeneities, although predicting quantitative a different result with respect to that just quoted. In spite of the fact that there is still some debate on the exact values of the scaling exponents, Toner and Tu's results are accepted as qualitative correct [58], and I feel worth to mention them to finish this subsection.

According to Toner and Tu, the two-point equal-time correlation is predicted to have (anisotropic) algebraic decay in the form $C(\mathbf{r}) = |\mathbf{r}_\perp|^{2\xi} f(r_\parallel / |\mathbf{r}_\perp|^\varsigma)$. Theoretically predicted values for the exponents were $\xi = -1/5$ and $\zeta = 3/5$. Such scaling translates into a dominance of transverse density fluctuations at small wave numbers of the form $q^{1-d-2\xi-\varsigma}$.[8] The final result then is expressed as $\Delta N \sim \langle N \rangle^{1+(2\xi+\varsigma-1)/(2d)}$. For $d = 2$ this renders an exponent $4/5$ instead of the mentioned unity value.

Canonical dry polar systems have allowed to detect such anomalous behavior of number fluctuations, as reported from experiments with vibrated granular polar discs by Deseigne et al. [81], where a value slightly smaller than unity but still larger than the conventional 0.5 determination, was obtained. More precisely the extracted value in these experiments is $(1.45 \pm 0.05)/2$. Another radically different scenario where such anomalies have been detected is that of bacterial colonies as published by Zhang et al. [451] (see Fig. 7.1) where a quite similar exponent was observed, i.e. 0.75 ± 0.03.[9] I will come

---

[8]Notice that this combination agrees with the quoted result above for the linearized theory through the equalities $\xi = 1 - d/2$ and $\zeta = 1$.

[9]Actually, the phenomenon of giant density fluctuations is even far more general since they apparently may exist even in entirely colloidal-based systems, as for example in

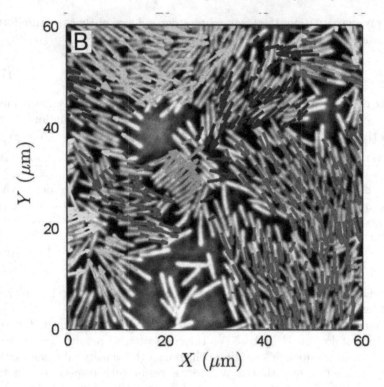

FIGURE 7.1
Collective motion in bacterial colonies. Configuration of a wild-type
B. subtilis colony showing correlated patterns of motion of the individual
specimens. Velocity vectors are overlayed on the raw images of bacteria. The
length of the arrows corresponds to bacterial speed, and nearby bacteria with
arrows of the same color belong to the same dynamic cluster (colored image
in original version). Image and caption text adapted from H. P. Zhang et al.
[451].

back to this topic in the following sections, as giant number fluctuations has
become a landmark in active systems of different sorts.

7.1.2   Particles Interacting Nematically on a Substrate

The best designed experimental realization of dry systems displaying nematic
order I am aware of was published by Narayan et al. [271] on a vibrated monolayer of apolar rods (see Fig. 7.2). Precisely this paper is probably best known

---

relation to living crystals [292], an autonomous microswimming scenario mentioned earlier in Sect. 3.1.1.

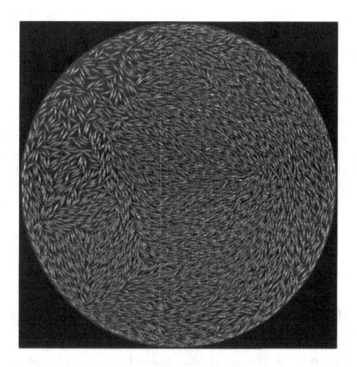

FIGURE 7.2
Giant number fluctuations in a system of active granular rods. Snapshot of the nematic order in a particle set vibrated sinusoidally perpendicular to the plane of the image. A large density fluctuation can be detected at the sparse region at the top left. Image and caption text adapted from Narayan et al. [271].

for its pioneering observation of anomalies in density fluctuations. These authors actually reported a scaling exponent of the standard deviation close to unity. Narayan et al.'s observations were extended a little bit later to spherical particles by Aranson et al. [9]. Other antecedents in this specific context are worth mentioning, principally the earlier theoretical work by Ramsawamy et al. [320, 259], or the simulation schemes of Vicsek-like models run by Chaté et al. [60].

The description of a system of elongated active particles interacting nematically on a substrate is totally parallel to what was introduced in the previous subsection by replacing this time the polarization vector field by a nematic tensor field. This is nothing but the particle-based version of the much used (traceless) tensor $\mathbf{Q}$ employed in the previous chapter to describe nematic active fluids. It is expressed in terms of the density field-averaged local degree of mutual nematic alignment of the axes of the particles (assuming head/tail inversion). The components of the tensor $\mathbf{Q}$ explicitly read,

$$Q_{\alpha\beta}(\mathbf{r};t) = \frac{1}{\rho(\mathbf{r};t)} \sum_i \left(\mathbf{p}_{i,\alpha}(t)\mathbf{p}_{i,\beta}(t) - \frac{1}{d}\delta_{\alpha\beta} \right) \delta(\mathbf{r} - \mathbf{r}_n(t)). \qquad (7.10)$$

The dynamics of the orientation encoded in the $\mathbf{Q}$ tensor is given as usual in terms of a free energy functional, apart from noise terms. Notice that advection-like contributions are not included in this case as distinctive of dry representations,

$$\partial_t \mathbf{Q} = -\frac{1}{\gamma_Q}\frac{\delta \mathcal{F}_Q}{\delta \mathbf{Q}} + \boldsymbol{\xi}_Q. \qquad (7.11)$$

The explicit expression for the free-energy functional reads correspondingly,[10]

$$\mathcal{F}_Q = \int_V \left[\frac{\alpha_Q(\rho)}{2}\mathbf{Q}:\mathbf{Q} + \frac{\beta_Q}{4}(\mathbf{Q}:\mathbf{Q})^2 + \frac{K_Q}{2}(\boldsymbol{\nabla}\mathbf{Q})^2 + C_Q\mathbf{Q}:\boldsymbol{\nabla}\boldsymbol{\nabla}\frac{\delta\rho}{\rho_0} + \frac{A}{2}\left(\frac{\delta\rho}{\rho_0}\right)^2 \right]. \qquad (7.12)$$

The appearing contributions look very much similar to the previous case with the necessary adaptation to the nematic rather than polar symmetry. For two-dimensional systems, it contains terms up to fourth order in Q (i.e $\mathbf{Q}:\mathbf{Q}$ and $(\mathbf{Q}:\mathbf{Q})^2$), the conventional elastic term, the compressibility penalty and a coupling of the density fluctuations and inhomogeneities in the nematic field.[11]

The most interesting new aspect, as proposed by Ramaswamy et al. [320], is the form activity enters into the dynamical equation for the density variable. Avoiding to consider explicitly the effects of fluctuations, one proposes a continuity equation in the form,

$$\partial_t \rho = -\boldsymbol{\nabla} \cdot \mathbf{J}, \qquad (7.13)$$

with,

$$\mathbf{J} = -\frac{1}{\gamma_\rho}\boldsymbol{\nabla}\frac{\delta\mathcal{F}_Q}{\delta\rho} + \mathbf{J}_{act}. \qquad (7.14)$$

After what I have commented in the previous chapter, the choice for the active contribution is straightforward. As a matter of fact, a simple argument

---

[10]Notice that this form, apart from the density terms, is totally equivalent to that postulated earlier to describe active fluids with nematic symmetry (see Eq. 6.54)

[11]I remind the reader, in relation to the Beris-Edwards approach to active fluid nematics commented in the previous chapter, that the shorthand notation $(\boldsymbol{\nabla}\mathbf{Q})^2$ for the conventional elastic term must be interpreted as $(\partial_\gamma Q_{\alpha\beta})^2$, where Greek letters denote Cartesian coordinates, and summation over repeated indices is implied.

[319] establishes that a curvature of the nematic field $\boldsymbol{\nabla} \cdot \mathbf{Q}$ (say of splay or bend type), even for apolar particles, is enough to induce a local polarity. Under non-equilibrium conditions sustained by autonomous or driven activity this is enough to generate a current,

$$\mathbf{J}_{act} = \zeta_Q \boldsymbol{\nabla} \cdot \mathbf{Q}, \tag{7.15}$$

with ζ_Q the phenomenological active parameter.[12] The final continuity equation for the density field, incorporating conventional and active terms, thus looks like,

$$\partial_t \rho = D\nabla^2 \rho + B\nabla^2 \boldsymbol{\nabla}\boldsymbol{\nabla} : \mathbf{Q} + \zeta_Q \boldsymbol{\nabla}\boldsymbol{\nabla} : \mathbf{Q} + \boldsymbol{\nabla} \cdot \boldsymbol{\xi}_\rho, \tag{7.16}$$

where the coefficients D and B derive from those denoted, respectively A and C_Q in the expression of the energy functional. Notice also that the kept noise terms, appearing through a gradient term, correspond to number conserving fluctuations.

I mention only the basic results from this scheme, while detailed results can be found in the Marchetti et al.'s review, or references there cited, and in the original paper [320]. An ordering transition (for $\alpha_Q < 0, \beta_Q > 0$) is predicted (continuous in d=2 according to the mean field approach of Ahmadi et al. [2]), and found numerically by Chaté et al. [60], giving rise to scaling of the (density) structure factor which is direction dependent, but of the form ($S(q) \sim q^{-2}$). The uniform nematic state just past onset displays a finite wave number instability into a perpetual evolving state that is reminiscent of the active nematic turbulent state much referred to in previous chapters. Giant density fluctuations are predicted as well, with results similar to those applying to systems with polar symmetry commented previously.

7.1.3 Self-Propelled Rods with Nematic Alignment

This is the kind of intermediate scenario announced previously in the introductory remarks to this section. It refers to systems of self-propelled units yet interacting nematically (for instance via hard-core collisions), rather than head-to-head/tail-to-tail as in Vicsek's original model. Although vaguely inspired by motility assays of filamentary actin proteins, as in the model by Kraikivski et al. [203], the direct antecedents in this context are the pure modeling approach of Baskaran et al, [18, 20], and the particle-based simulations of Ginelli et al. [129].

In particular, Baskaran et al. derived a Smoluchowski equation appropriate to self-propelled rods that was further coarse-grained into an hydrodynamic equation. The main conclusion is that, rods do not order in this

---

[12]I use here the notation ζ_Q for the activity parameter, instead of the simplest form ζ employed as scaling exponent in this chapter. The subindex also reminds us that activity is related to spatial inhomogeneities of the nematic tensorial order.

case into a propagating flock, but rather the nematic, Onsager-like, order is enhanced by self-propulsion. The quoted result [247] is that the density threshold for the isotropic/nematic transition is renormalized as $\rho_{iso/nem} = \rho_{Ons}/(1+V_0^2/2l^2D_r^2)$, where l stands for the filament length and D_r represents the rotational diffusion coefficient.

In spite of the fact that the only conserved quantity is the density field, hydrodynamic equations must be derived simultaneously for the pair polarization vector and nematic tensor fields. The whole set of equations now read,

$$\partial_t\rho + V_0\boldsymbol{\nabla}\cdot(\rho\mathbf{p}) = D\nabla^2\rho + \frac{1}{\gamma_\rho}\boldsymbol{\nabla}\boldsymbol{\nabla}:\frac{\delta\mathcal{F}_{pQ}}{\delta\mathbf{Q}} + \boldsymbol{\nabla}\cdot\boldsymbol{\xi}_\rho, \qquad (7.17)$$

$$\partial_t\mathbf{p} + \lambda_1(\mathbf{p}\cdot\boldsymbol{\nabla})\mathbf{p} - \delta_1\mathbf{p}\cdot\mathbf{Q} + \delta_2\mathbf{Q}:\mathbf{Qp} = -\frac{1}{\gamma_p}\frac{\delta\mathcal{F}_{pQ}}{\delta\mathbf{p}} + \boldsymbol{\xi}_p, \qquad (7.18)$$

$$\partial_t\mathbf{Q} + \lambda_1'(\mathbf{p}\cdot\boldsymbol{\nabla})\mathbf{Q} = -\frac{1}{\gamma_Q}\left[\frac{\delta\mathcal{F}_Q}{\delta\mathbf{Q}}\right]_{ST} + \boldsymbol{\xi}_Q, \qquad (7.19)$$

where the subscript ST indicates a symmetric traceless tensor, i.e. $T_{ij}^{ST} = 1/2(T_{ij} + T_{ji}) - (1/2)\delta_{ij}T_{kk}$. The explicit expression for the free energy now adds intrinsic $\mathbf{p}$-terms and $\mathbf{p}, \mathbf{Q}$-coupling contributions to the nematic free energy,

$$\mathcal{F}_{pQ} = \mathcal{F}_Q + \int_v\left[\frac{\alpha_{pQ}}{2}|\mathbf{p}|^2 + \frac{K}{2}(\delta_\alpha p_\beta)(\delta_\alpha p_\beta) + P(\rho,p,Q)\boldsymbol{\nabla}\cdot\mathbf{p} - \nu_2 Q_{\alpha\beta}(\delta_\alpha p_\beta)\right]. \tag{7.20}$$

The most striking feature to be noticed at the level of the set of above equations is the appearance of a wealth of "advection" terms in the dynamics of the polarization field that incorporate crossed terms between $\mathbf{p}$ and $\mathbf{Q}$. The kinetic theory of Baskaran et al. predicts $\alpha_{pQ} > 0$ for all densities, precluding an isotropic-polar transition. The only homogeneous steady solutions are the isotropic state with $\rho = \rho_0$ and $\mathbf{p} = \mathbf{Q} = 0$, and the nematic configuration corresponding to $\mathbf{p} = 0$ but $\mathbf{Q}$ finite. The transition occurs at $\rho_{iso/nem}$. As in the polar and nematic active systems, the isotropic state is stable, but may support finite wave vector propagating sound-like waves. Numerical simulations of collections of self-propelled rods with steric repulsion reveal a rich behavior, quite distinct from that of polar Vicsek-type models. As predicted by theory, self-propelled rods with only excluded volume interactions do not order in a macroscopically polarized state, but exhibit only nematic order, which appears to be long range in two dimensions [129].

7.2 Microscopic-like Theories

I abandon the treatment of hydrodynamic equations applied to dry active matter to enter the realm of particle-based modeling. As announced above I have prepared the contents of this section following the recent review by Chaté [58].

7.2.1 Particle-Based Models for Dry Systems

Chaté identifies the class of systems to be described in this section with a particularly illustrating acronym: DADAM, for **dry aligning dilute active matter**. Their most distinctive feature, that has already been implicit in the treatment of the previous section, is the reciprocal coupling between orientational order and density. From the very beginning it is useful to distinguish DADAM from the other most celebrated scheme of simulating active particles we referred to as the Active Brownian Particle (ABP) model in Chapter 3. Let's clarify this difference in the simplest way.

In the vast majority of (microscopically) modeled particle-based active matter systems, the main interactions between active particles are assumed local and consist of a combination of repulsion and velocity/polarity alignment. The two extreme cases, corresponding to pure repulsion or pure alignment. precisely identify the two just mentioned model categories: respectively ABP and DADAM.

In the latter context that will be discussed on what follows, modeling supposes to take point-like particles, with negligible or sub-dominant repulsion (justifying the assumption for diluteness), that self-propel, while mutually or collectively aligning within some distance. In continuous time, the appropriate description is based on a pair of equations,[13]

$$\dot{\mathbf{r}}_i = \pm V_0 \mathbf{p}(\theta_i), \tag{7.21}$$

$$\dot{\theta}_i = \frac{K}{n_i} \sum_{j \sim i} \sin k[(\theta_i - \theta_j)] + \xi_{\theta,i}, \tag{7.22}$$

where the sign $\pm$ changes at some prescribed rate a, K is a measure of the interaction strength on alignment, and k indicates the alignment symmetry: $k = 1$ reproduces ferromagnetic alignment, whereas $k = 2$ stands for nematic interactions. The original Vicsek model is easily identified with the pair $a = 0, k = 1$. We can readily assign pairs of $a - k$ values to the remaining

---

[13]To emphasize the role of stochastic effects in the alignment that are essential to DADAM systems, I consider the pure deterministic version of the dynamic equation for the position variable.

classes considered (macroscopically) in the previous section: Particle systems interacting nematically on a substrate correspond to $a \neq 0, k = 2$, and self-propelled rods with nematic alignment to $a = 0, k = 2$.

Most of the models analyzed in the literature correspond to the original or slightly modified versions of time discrete realizations of the previous scheme. I follow [58] and restrict to the most representative variants that consider unit modulus displacement, isotropic radius of interaction range, and uniform angular noise at each time step,

$$\mathbf{r}_i^{t+1} = \mathbf{r}_i^t \pm \mathbf{V}_i^{t+1}, \tag{7.23}$$

$$\theta_i^{t+1} = (\mathcal{R}_\eta \circ \theta) \langle \mathbf{V}_j^t \rangle_{j \sim i}, \tag{7.24}$$

where θ denotes an operator that resizes vectors to unit norm, i.e. $\theta(\mathbf{V}) = \mathbf{V}/\|\mathbf{V}\|$, $\mathcal{R}_\eta$ rotates a vector $\mathbf{V}$ a random angle drawn from a uniform distribution inside an arc length $2\pi\eta$ centered on $\mathbf{V}$, and the averaged velocities reproduce either the ferromagnetic or nematic alignment rules,

$$\langle \mathbf{V}_j^t \rangle_{j \sim i}^{ferro} = \sum_{j \sim i} \mathbf{V}_j^t, \tag{7.25}$$

$$\langle \mathbf{V}_j^t \rangle_{j \sim i}^{nem} = \sum_{j \sim i} sign[\mathbf{V}_i^t \cdot \mathbf{V}_j^t] \mathbf{V}_j^t. \tag{7.26}$$

The hydrodynamic equations that were considered in the previous section can be derived from this sort of microscopic schemes following different strategies.[14] I only give here some basic indications of the most canonical procedures

The first step consists in transforming the Langevin-like, i.e. noisy, equations for the elementary (particle-based) dynamical variables into a *kinetic equation* (Fokker-Planck/Smoluchowski or Boltzmann/Chapman-Enskog) written for the one-particle distribution function, similarly to what was done for ABPs in Sect. 3.1. Strong assumptions, familiar from classical Statistical Mechanics, are commonly invoked at this stage. Next, one uses such kinetic equation to derive equations for the moments of the one particle distribution function, either corresponding to conserved quantities (in dry systems the number-density of particles) and to broken symmetries (polar or nematic). As it is well-known such a procedure leads to a hierarchy of equations where higher-order moments are expressed in terms of lower-order ones. At this point this hierarchy must be truncated in an educated way to finally obtain the hydrodynamic equations for the field variables of interest. Since I consider this procedure a key point to obtain hydrodynamic-like theories for

---

[14]For a general discussion, interested readers can consult Sect. V in Marchetti et al.'s paper [247].

particle-based dry systems, I illustrate it in the remaining of this section as it applies to the derivation of continuum equations for aligning Vicsek-like particles on a substrate. I adapt arguments borrowed from the corresponding derivation, as can be found in the review by Marchetti et al. [247].

As the simplest scenario, I consider a two-dimensional system of self-propelled point-like particles with pure angular interaction. Each particle bears its individual scalar velocity V_0 (no reversals in this case) and orientation defined in terms of the unit vector $\mathbf{p}_i = (\cos\theta_i, \sin\theta_i)$. The system of dynamic equations for the position and angular variables are written as,

$$\dot{\mathbf{r}}_i = V_0\mathbf{p}_i - \gamma^{-1}\sum_j \frac{\partial U}{\partial \mathbf{r}_j} + \boldsymbol{\xi}_i(t), \tag{7.27}$$

$$\dot{\theta}_i = \gamma_r^{-1}\sum_j \frac{\partial U}{\partial \theta_j} + \xi_{\theta,i}. \tag{7.28}$$

The potential is assumed of strength Γ,

$$\gamma_r^{-1}U(\mathbf{r}_i, \theta_i; \mathbf{r}_j, \theta_j) = -\frac{\Gamma}{\pi R^2}\Theta(R - |\mathbf{r}_i - \mathbf{r}_j|)\cos(\theta_i - \theta_j), \tag{7.29}$$

and of limited range R as expressed by the Heaviside function. Since we are interested on large length scales, one can assume that the potential is of local nature. The system of coupled Langevin-like equations is transformed into a Smoluchowski equation for the one-particle distribution function. Notice that, albeit the context is substantially different, this is similar to the treatment that was proposed in Sect. 3.1.1 to describe a system of sedimenting self-phoretic particles.

The sought equation contains separate translation and rotation fluxes. Either one contains a drift term written in terms of the interaction potential, and eventually the self-propelled velocity for the former, plus respective gradient terms, that finally give rise to the usual diffusion terms. Following Marchetti et al. [247] and for the sake of illustration, I decide to write this lengthy equation for the one-particle distribution in its totally explicit form as,

$$\partial_t\psi(\mathbf{r}_1, \theta_1; t) + V_0\mathbf{p}_1 \cdot \boldsymbol{\nabla}_1\psi = D\nabla^2\psi + D_r\partial_{\theta_1}^2\psi + \frac{1}{\gamma}\boldsymbol{\nabla}_1 \cdot \psi(\mathbf{r}_1, \theta_1; t)$$
$$\times \int_{\mathbf{r}_2, \theta_2} \boldsymbol{\nabla}_1 V(\mathbf{r}_1, \theta_1; \mathbf{r}_2, \theta_2)\psi(\mathbf{r}_2, \theta_2; t) + \frac{1}{\gamma_r}\partial_{\theta_1}\psi(\mathbf{r}_1, \theta_1; t)$$
$$\times \int_{\mathbf{r}_2, \theta_2} \partial_{\theta_1} V(\mathbf{r}_1, \theta_1; \mathbf{r}_2, \theta_2)\psi(\mathbf{r}_2, \theta_2; t). \tag{7.30}$$

The next step is to transform this kinetic equation into a set of equations for the moments of the one-particle distribution function. Given the particularly simple dependence of the $\mathbf{p}$ variable in terms of the elementary trigonometric function of the angle θ, a convenient trick consists ([29]) in using the angular Fourier transform of ψ, i.e. introducing $\psi_q(\mathbf{r};t) = \int \psi(\mathbf{r},\theta;t)e^{iq\theta}d\theta$. One can easily check that $\psi_0 = \rho$, the particle number density, $\psi_1 = w_x + iw_y$ with $\mathbf{w} = \rho\mathbf{p}$ for the polarization vector field, while ψ_2 will be related to the components of the nematic tensor field. Using the identity of the Fourier representation $2\pi\psi(\mathbf{r},\theta;t) = \sum_q \psi_q e^{-iq\theta}$, and retaining only linear terms in the gradients one obtains a hierarchy equations for the coupled components ψ_q,

$$\partial_t \psi_q + \frac{V_0}{2}\partial_x(\psi_{q+1} + \psi_{q-1}) + \frac{V_0}{2i}\partial_y(\psi_{q+1} - \psi_{q-1})$$
$$= -q^2 D_r \psi_q + \frac{iq\Gamma}{2\pi}\sum_{q'}\psi_{q'}V_{-q'}\psi_{k-q'} + O(\nabla^2), \tag{7.31}$$

where $V_{q'} = \int e^{iq'\theta}\sin\theta d\theta = i\pi(\delta_{q',1} - \delta_{q',-1})$. One further specializes this scheme to ψ_0 and ψ_1, assuming $\partial_t\psi_2 = 0$ as a fast relaxing variable. Moreover, one discards higher Fourier components, i.e. $\psi_q = 0$ for $q \geq 3$, and expresses ψ_2 as a function of ψ_0 and ψ_1, to finally obtain the pair of equations,

$$\partial_t\rho + V_0\nabla \cdot (\rho\mathbf{p}) = 0, \tag{7.32}$$

$$\partial_t\mathbf{p} + \lambda_1(\mathbf{p}\cdot\nabla)\mathbf{p} = [-\alpha(\rho) + \beta|\mathbf{p}|^2]\mathbf{p} - \mathbf{v_1}\cdot\nabla\rho + \frac{\lambda_3}{2}\nabla|\mathbf{p}|^2 + \lambda_2\mathbf{p}(\nabla\cdot\mathbf{p}). \tag{7.33}$$

Explicit expressions for the left coefficients, i.e. α, β, λ_i and $\mathbf{v_1}$ are given in [247]. Notice that this equation is the deterministic similar version of Eq. 7.6 at first order in gradients with the difference that the coefficient $\mathbf{v_1}$ now appears as a tensorial coefficient rather than as an scalar one.

7.2.2 Common Rationale: Phase-Separated Regimes

Since I have already commented in the previous section some of the most important results that follow from the analysis at an hydrodynamic level of the different categories of dry active systems, I will limit here to emphasize some particular ideas that the study of microscopic models have consolidated in the recent years. The first and likely most important idea emphasized by Chaté [58] is a paradigm shift that has resulted from a careful study of these systems: that of the phase coexistence between a disordered gas and an orientationally ordered liquid, rather than a direct disorder-order transition, as it was more or less explicitly advocated at the early days of the study of Vicsek-like models. This is nothing but a manifestation of the intrinsic conceptual basis of DADAM systems, i.e. moving particles carry themselves the orientational degrees of freedom. Thus, order and mass are trivially but crucially linked.

DADAM systems can be best analyzed in terms of phase diagrams organized around their two basic parameters, i.e. the particle density and the noise strength that is prescribed for particle interaction. In such a phase diagram one does not find a single line separating the disordered and ordered phases, but rather a peculiar construction of (doubled) binodal and spinodal lines. This is in contrast with what one learns, for instance, from the common phase separation scenarios from (partially) immiscible fluids. More in detail, two *binodals* B_{liq} and B_{gas}, spanning a wide region in between, separate the homogeneously disordered gas phase from the homogeneously ordered liquid state (see Fig. 1 in [58]). These binodals emerge from the origin and converge to a maximum noise strength at infinite density.

In the central region, a new configuration appears that consists in a system of *dense and ordered bands*. This supposes the presence of a central coexistence phase between the disordered gas and the orientationally ordered liquid. As a whole, this means that a symmetry is introduced in the system that is distinctively different from those endowing the respective homogeneous phases, either the gas or the liquid. In the Vicsek-polar class, bands extend transverse to the direction of order and propagate at a well-defined speed of the order of the elementary particle velocities. Conversely, order is established along the bands in the nematic class without a distinctive signature of sustained propagation.

Notice that this picture definitively precludes the existence of uniquely defined critical conditions at finite densities. As a matter of fact, one can describe this scenario as arising from the simultaneous presence of two transitions that are continuous (in the infinite size limit) but not critical [58]. Close to each binodal, inside the coexistence region, the nearby homogeneous phase is metastable. Deeper in the coexistence region in the area enclosed by the two *spinodal* lines unconditional phase separation takes place.

7.2.3 Specific Class-Dependent Features

Before finishing, I refer to specificities of each of the three previously identified DADAM classes. In most cases, these considerations suppose to add brief comments to specific questions already addressed in the previous section.

7.2.3.1 Traveling Bands in Polar Class

In numerical simulations (see Fig. 7.3), traveling bands eventually form a regular smectic A pattern (i.e an ordered array). At fixed parameter values for the density and noise strength, their number scales linearly with the system size in the direction of order i.e. transversal to the extended bands. At fixed system size their number also increase with the global density while they profile does not change much. Approaching the liquid binodal, their number increases so that they start to interact strongly and appear to never organize a regular smectic pattern. Hydrodynamic theories predict multiplicity of periodic

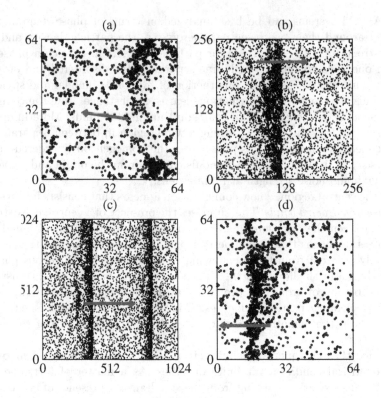

FIGURE 7.3
Collective motion of Vicsek-like particles. Panels (a) to (c): Traveling-band state with increasing system size. Panel (d) corresponds to a different class of noisy interaction, different from the purely angular referred to in the main text. Image and caption text adapted from Chaté et al. [59].

solutions, although the introduction of noise appears to break this degenerated set of solutions and reinstall pattern selection [58].

7.2.3.2   Unstable Nematic Bands

Dense bands observed in systems with nematic symmetry do not travel steadily as a net global motion is precluded. Instead, very large simulations in very large systems reveal a never ending scenario of band bending, splitting and merging giving rise to a a spatio-temporal disordered regime [277]. If the system is not too large a single band is observed, its width increasing with global density, while the gas density remains constant, supporting the idea of a phase-separation scenario. Hydrodynamic theories capture well this behavior as commented earlier on. The basic band, non-moving, solution is found to occupy a larger fraction of space as the global density increases.

7.2.4 Properties of the Liquid Ordered Phase

Toner and Tu original attention focused mostly on the ordered phase. They found that the two-point equal time correlation function for density inhomogeneities decay in the form $C(\mathbf{r}) = |\mathbf{r}_\perp|^{2\xi} f(r_\parallel/|\mathbf{r}_\perp|^\zeta)$, as commented earlier in relation to the question of giant number fluctuations . Symbols, denoted parallel and perpendicular. respectively refer to directions longitudinal and transverse to the direction of mean motion of the flock. Respective exponents ξ and ζ, as well as the scaling function were found with universal characteristics. The value of ξ controls the strength of orientational order. $\xi < 0$ is indicative of long-range order, and the order parameter takes a finite value in an infinite system. For $\xi = 0$, order is marginally long-range, and the order parameter decays algebraically with the system size. In the polar/Vicsek class of systems, it is accepted that $\xi < 0$ (I remind the reader that the value obtained by Toner and Tu was $\xi = -1/5$), and that the ordered phase of flocking displays long-range order in two-dimensions. For the active nematic class of systems $\xi = 0$ in two dimensions, while the scalar nematic order parameter varies with system size as $Q(L) \sim L^{-\sigma}$, with σ taking small values that increase linearly with the noise variance.

8

Appendix 1: Microswimming in Constrained and Disordered Environments

Although sometimes difficult to disentangle, I separate in two sections the discussion on the effects of swimming in constrained spaces from those that appear when motion takes place in disordered environments.

8.1 Microswimming under Constrained Motion

Needless to say that living, as well as artificial swimmers, often move across narrow passages. This is the aspect that I want to briefly comment on what follows. Just to emphasize the importance of this new aspect in the context of active microswimming, I quote the review paper published by Bechinger et al. [24] under the suggestive title of *Active particles in complex and crowded environments*. To address this topic, I refer first briefly to experimental observations on natural specimens since this is the context where this particular aspect has been mostly analyzed.

The most obvious realization, i.e. swimming in a channel, is what I consider in the first instance, both from experiments and modeling, to move later on into spherical and circular domains. A detailed experimental study was undertaken by Männik et al. [245] (references to previous works can be found in the cited paper). There, it was shown that *E. coli* and *B. subtilis* are still motile, with minor quantitative differences, when the width of the channel exceeds the specimens diameters only marginally. For smaller widths, the motility vanishes, but bacteria can still pass through these channels by growth and division. Another more recently investigated platform consists in connecting two chambers, as proposed by Paoluzzi et al. [295], and look for different self-sustained filling dynamics of the bacterial baths. Much more recently, bacterial hopping and trapping in porous media was reported by Bhattacharjee et al. [30].

From the modeling point of view, I mention a couple of works by Zhu et al. [459] and B. Liu et al. [233]. The first adopts a classical squirmer picture to conclude that in the case of swimming parallel to the tube axis, the locomotion speed is always reduced (respectively increased) for swimmers with

DOI: 10.1201/9781003302292-8

tangential (respectively normal) surface velocities. The second paper models the propulsion of a helical flagellum in a capillary tube, the conclusion being that, except for a small range of tube radii at the tightest confinements, the swimming speed at fixed rotation rate increases monotonically as the confinement becomes tighter. Still, using multiparticle collision dynamics, Münch et al. [266] studied an undulatory Taylor line (see also Sect. 9.3) swimming in a 2d microchannel and in a cubic lattice of obstacles. A somewhat different perspective was taken by Ledesma-Aguilar et al. [221], who addressed the effect of confinement from flexible boundaries on the motility of a model dipolar microswimmer. The main conclusion is that flexible boundaries are deformed by the velocity field of the swimmer in such a way that the motility of both extensile and contractile swimmers is enhanced.

Back to experimental works with microorganisms, confinement effects of bacteria in spherical domains reveal striking accumulation effects near boundaries, as evidenced by Vladescu et al. [422].[1] Experiments refer to the filling of an emulsion droplet by *E. coli*. The reported observation is that at low cell concentrations, the cell density peaks at the water-oil interface, while at increasing concentration, the bulk of each droplet fills up uniformly while the surface peak remains. Collective effects leading to vortical assemblies has been also reported in confined bacterial suspensions from experiments from Goldstein's group [443, 435, 242]. Moreover, large lattice arrangements of bacterial vortex lattices permit to identify ferromagnetic and antiferromagnetic distinctive features of ordered states in living matter as reported by Wioland et al. [434].

Changing a little bit the perspective, entrapment of swimming bacteria by convex walls (micro-fabricated pillars) of sufficiently low curvature, somehow ruling out steric effects, was reported by Sipos et al. [359]. Rectification of bacterial motion (asymmetric spatial accumulation) and trapping from static obstacles have been also observed experimentally by Galadja et al. [120], as well as for the flagellated algae by Kantsler et al. [190], and from models [427, 188].

In particular, Galajda et al. show that when a population of bacteria is exposed to a microfabricated wall of funnel-shaped openings, the random motion of bacteria through the openings is rectified by trapping of the swimming bacteria along the funnel wall. This leads to a buildup of the concentration of swimming cells on the narrow opening side of the funnel wall, but with no apparent concentration in the case of non-swimming cells. Similarly, it is shown that a series of such funnel walls functions as a multistage pump that can

---

[1] Observations on many biological swimmers show that this is the first generic phenomenon of microswimmers near surfaces. Importantly enough, this takes place without the need to invoke electrostatic or other sorts of attractive forces. Already in 1963, Rothschild observed an accumulation of sperm (pusher) cells on a glass cover slide [330]. Bacteria, have probably been the most studied living microswimmers in relation to bounding walls and surfaces. Classical observations of circular trajectories come from Berg's group [28, 112] that are eventually rectified into right-handed swimming when close to the floor plate (the other way around near the top plate), as reported by DiLuzio [84].

FIGURE 8.1
Bacterial ratchet motor. A nanofabricated asymmetric gear sedimented
at a liquid–air interface (48 μm external diameter, 10 μm thickness) rotates
clockwise at 1 rpm when immersed in an active bath of motile *E. coli*, visible in
the background. The circle points to a black spot on the gear that can be used
for visual angle tracking. Image and caption text adapted from DiLeonardo
et al. [318].

increase the concentration of motile bacteria exponentially with the number
of walls. With quite a different perspective, I mention at this point the work
by Nishiguchi et al. [281], reporting striking patterns of self-organization in
a concentrated suspension of motile bacteria dispersed in a two-dimensional
array of vertical pillars. The reported observation is that of a long range anti-
ferromagnetically organized lattice of bound vortices controlled by the pillar
spacing.

Considering other microorganisms, curvature-guided motility of flagellated
algae (*Chlamydomonas*) in geometric confinement was recently reported by
Ostapenko et al. [286], and crowding effects in a regular lattice of micro-pillars
by Bruny et al. [46]. The parallel question of microswimmers performing in
presence of (spatial) disorder will be more specifically addressed next. To finish
with these paragraphs devoted to living specimens and boundary effects, I
bring to the attention of the interested reader the detailed account dedicated
to constrained motion of microorganisms that can be found in the review by
Elgeti et al. [98].

A little bit apart from the strict context of confinement but similarly interesting, other studies consider the interaction of bacteria and moving surfaces.
More in particular, the issue of preparing bacterial baths with immersed microengines has attracted the attention of several groups. In this regard, I
mention attempts by Hiratsuka et al., Di Leonardo et al., and Sokolov et al.
[164, 8, 318, 365, 186] (see Fig. 8.1). Paralleling experimental observations,
a general theoretical treatment of simple engines driven by active fluids has
been very recently published by Pietzonka et al. [307].

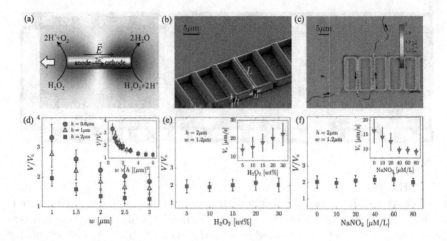

FIGURE 8.2

Bimetallic swimmers confined in linear channels. Upper panels correspond to schematics of the propulsion mechanism, the channel platform, and tracked swimming trajectories. The lower ones give more quantitative measures of the increased velocity and dependences with the chemical parameters of the system (colored image in original version). Image and caption text adapted from C. Liu et al. [235].

I leave hindered swimming of biological specimens to refer next to artificial realizations of constrained microswimming. A valuable experimental reference was published by C. Liu et al. [235], reporting the marked increase of the swimming velocity of bimetallic rods (see Sect. 3.1.1) with the degree of confinement (see Fig. 8.2). The results are interpreted in terms of electrostatic and electrohydrodynamic boundary effects that are explicitly worked out in a model whose numerical solution confirms the experimental observations. The idea underlying this result is that confinement helps to increase the strength of the self-generated electric field that drives the swimmer. Moreover, electroosmotic flows from the channel walls also contribute to speed up the swimmer.

However, practically published simultaneously, I mention another experimental and numerical paper bu Yu et al. [448] reporting an opposite conclusion for metallic/dielectric Janus particles. The authors proposed in this latter case that the motor slowed, possibly due to a tendency to hit and move along the wall as well as due to an implicit motor-wall interaction, given the diffusiophoretic mechanism that drives the swimmer in this situation, compared to the self-electrophoresis in the previously commented scenario. In any case, this tells us of the complexity of confined swimming and the many details that potentially may influence the precise swimming mode, and its corresponding response to geometric confinement. Finally, I quote a parallel realization based on the phase-separation driven microswimmer commented in Sect. 3.1.1.2, whose performance was studied in a patterned environment by Volpe et al. [425]. For a review of artificial swimming in microfluidic platforms a useful recent review has been published by Xiao et al. [446].

Directly related to constrained microswimming, let's finish this section turning our attention to the issue of *rectification* of swimming motion under particularly conditioned geometries. I already referred to the experimental studies of rectification of bacterial motion published by Galajda et al. [120] or for flagellated algae by Kantsler et al. [190].

Rectification of active Brownian motion in a ratchet-like spatial asymmetry has been theoretically demonstrated by P. K. Ghosh et al. [128], calling for experimental confirmation. Authors considered different compartment geometries, boundary collisional dynamics, and particle rotational diffusion, to conclude that ratcheting of Janus particles can be orders of magnitude stronger than for ordinary thermal potential ratchets. More recently, Stenhammar et al. [378] presented a computational proof-of-concept, showing that active rectification devices could be created directly from an unstructured "primordial soup" of light-controlled motile particles.

8.2 Microswimming under the Effects of Noise and Disorder

Similarly to considering the role of geometric constrainment, effects of noise (fluctuations) and disorder are potentially important as well when considering realistic scenarios of microswimming. On what remains of this appendix, I briefly comment on this particular aspect considering, again, living and artificial swimmers altogether.

As a matter of fact, this question has been addressed quite often in the context of active colloidal systems, singularly from a theoretical point of view or by means of numerical simulations, as commented at the end of this section, while experiments are much less abundant. Concerning living specimens, I have already referred in the previous section to motion of bacteria, or other microorganisms, in the presence of obstacles, but I am not aware at this moment of any specific study of the effects of disorder, either static or dynamic, on their swimming dynamics, neither at the level of single individuals, nor when looking for collective effects. Experiments with synthetic microswimmers are also scarce. The few experimental studies I am aware of are, however, very illustrative.

I start by referring to a paper by Morin et al. [263] using driven colloids (Quincke rotators) (see Fig. 8.3). The central question that motivated this study is whether or not flocks can propagate in disordered media. As commented earlier (see Sect. 3.2.2) the system consists of colloidal rollers that experience both hydrodynamic and electrostatic interactions which promote alignment of their translational velocities I remind the reader that when the roller packing fraction exceeds some critical value, polar interactions overcome rotational diffusion and macroscopic collective motion emerges [44]. In an homogeneous slab geometry, a long flock spontaneously forms and cruises through a dilute ensemble of rollers moving isotropically. In this typical platform, circular obstacles of radius a few micrometers were added by Morin et al. at various packing fractions ϕ. As the latter goes over some threshold ϕ_c, collisions of the rollers and obstacles suppress global orientational order, and any signature of macroscopic transport as well. As expected, denser fluids are more robust to disorder and ϕ_c is found to increase monotonically with the roller fraction.

Moreover, an order parameter for flocking is defined (roller current averaged over space and time) and, as expected, it is observed to decrease monotonically with ϕ, vanishing at ϕ_c. A deeper analysis of the flock structure is particularly rewarding, as three important features emerge: i) the flock length decreases and vanishes smoothly, ii) the maximal current amplitude exhibits a sharp drop and cancels discontinuously at ϕ_c, and iii) at ϕ_c flocks are intermittent. Authors conclude from these three observations that disorder appears to suppress flocking in the form of a first-order phase transition. The analysis is extended further to conclude that flocks propagate along river-like networks, with virtually no signatures of collective motion occurring in closed regions surrounded by flowing rivers. Sparsity of river networks increase with ϕ, and networks also become increasingly tortuous (see Fig. 8.3). An accompanying analytical study in the same paper provides some understanding of the principal observations reported.[2]

Practically simultaneously, and with reference to the same experimental system, the same group published an experimental and numerical study looking for diffusion, subdiffusion, and localization of (non-interacting) active colloids in random post lattices [262]. A smooth transition from diffusive to subdiffusive to localized dynamics upon increasing the obstacle density was first demonstrated, and the nature of these transitions further elucidated in terms of the different role played by colloid-obstacle interactions. Dynamic vortex glasses were recently reported by Chardac et al. [57].

Still with driven colloidal systems, Stoop et al. [379] reported the clogging and jamming of field tunable interacting colloids driven via a magnetic ratchet effect through a quenched disordered landscape of fixed obstacles (a random landscape of silica particles larger than the driven colloids). Authors claim their study provides the experimental observation of the "Faster is Slower effect" mediated by quenched disorder that occurs when increasing the particle speed.

---

[2]Very recently, flows of Quincke rollers in circular channels with obstacles displaying interesting chirality reversal effects were reported by B. Zhang et al. [450].

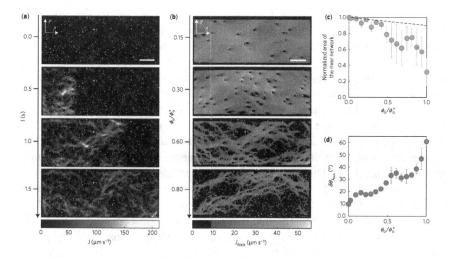

FIGURE 8.3
Flocking of rollers through quenched disorder. Left columns display the structures of river networks at different times represented in terms of the total and flocking currents respectively. Scale bar, 200 µm. Middle panels represent intensity of flow currents for different obstacle packing fractions. Right panels correspond to a measure of the sparsity of the network and the orientations fluctuations of the flock currents (colored image in original version). Image and caption text adapted from Morin et al. [263].

A similar experimental and numerical study corresponds to self-propelled Janus colloids moving atop a two-dimensional crystalline surface as reported by Choudhury et al. [66]. Flocking ferromagnetic colloids [187] in the form of moving vortices have been demonstrated to be easily manipulated from interactions with inert scatterers [202]. Moving to granular exercised matter, Kumar et al. [210] reported how millimetre-sized tapered rods, rendered motile by contact with an underlying vibrated surface, and interacting through a medium of spherical beads, undergo a phase transition to a state of spontaneous alignment of velocities and orientations above a threshold bead area fraction.

Another study, closer to the context of active colloids was published by Pinçe et al. [308]. The paper reports the long-time organization of a population of colloids in an active *E. coli* bacterial bath, where a controllable degree of disorder is introduced with an optical potential. Authors demonstrate that the presence of spatial disorder can alter the long-term dynamics in a colloidal active matter system, making it switch between gathering and dispersal of individuals. At equilibrium, colloidal particles always gather at the bottom of any attractive potential; however, under non-equilibrium driving forces in a bacterial bath, the colloids disperse if disorder is added to the potential. A somewhat related study, this time focusing on the effects of the topography of the environment regarding more efficient searching strategies of active particles was published by Volpe et al. [426]. Using simulations and theory, Jakuszeit et al. [179] recently studied the impact of boundary conditions on the diffusive transport of active particles in an obstacle lattice.

Mixtures of active and passive particles were analyzed by Stenhammar et al. [377] from the point of view of motility-induced phase separation. Later on, Alaimo et al. [3] revisited the problem from a microscopic field-theoretical approach looking for different dynamic phenomena, among them the enhanced crystallization due to the presence of active particles.

On the other hand, quite a number of theoretical, but mostly simulation studies, have addressed the influence of noise on the performing of model microswimmers. I will try to organize these different approaches chronologically. On what follows, I distinguish between temporal fluctuations, qualified generically as *noise*, from quenched (time-independent) random landscapes, that define what I have already referred to several times as (spatial) *disorder*.

The first work I want to mention is that of Chepizhko et al. [64]. It is devoted to the analysis of the effect of spatial heterogeneity on the collective motion of self-propelled particles in the context of the Vicsek-flocking model in 2d (see Sect. 7.1.1). The heterogeneity is modeled as a random distribution of either static or diffusive particles that the propelling particles avoid, while trying to align their movement. An optimal amplitude variance in the alignment rule is demonstrated to exist that maximizes collective motion. By increasing the number of obstacles, akin to augment the level of disorder, the existence of the maximum is preserved, although the value of the polar order parameter decreases. The existence of such optimal noise intensity values is a striking manifestation of a principle that was popular in the field of pattern forming systems at the end of the last century, and that was known as *noise-induced order* [332].

Authors also concluded that for weakly heterogeneous media (i.e. low obstacle densities), the transition to collective motion exhibits a unique critical point below which, the system exhibits long-range order, as in homogeneous media. Conversely, for strongly heterogeneous media (high obstacle densities), there are two critical points, with the system being disordered at both, large and low noise amplitudes, and exhibiting only quasi-long-range order in between these critical points. Effects on individual self-propelling particles (diffusion, subdiffusion and trapping) were considered in an accompanying paper by the same authors [65].

With a quite similar perspective, a different model, run-and-tumble rather than flocking, was considered by Reichhardt et al. [325].[3] These authors also observed an optimal noise value (here measured in terms of the length

---

[3] In the run-and-tumble particle model (RTP), in contrast to the active Brownian particle model, particles alternate running periods, during which self-propulsion keeps unaltered, with tumbling events, when the particle orientation is randomized. At the level of collective

persistence in the runs) for interacting and driven particles in a disordered environment. In the limit of zero noise, or infinite run length, even for a small number of obstacles jamming or clogging may appear.

Another perspective, closer to the idea of quenched disorder, was taken by Zeitz et al. [449]. A collection of non-interacting ABPs wandering in a (random) 2d *Lorentz gas* was analyzed from numerical simulations. A random Lorentz gas model supposes considering the obstacles as fixed, and randomly distributed according to a given area fraction. In the paper by Zeitz et al., steric interactions between ABPs and obstacles were explicitly introduced. As a reference situation, a test particle, either Brownian or ballistic, shows subdiffusive behavior below a critical density of obstacles, related to a continuum percolation of the underlying structure (see [449] and references therein), while above it particles get effectively trapped and long-range transport is suppressed. An active particle performs the same subdiffusive motion as ballistic and diffusive particles do, but it explores its environment faster, so that the long-time dynamics is reached earlier. At intermediate times, dynamics can be superdiffusive, according to the authors.

Recently, a study of a collection of self-propelled particles on a two-dimensional substrate in the presence of random quenched rotators was published by Das et al. [73]. These rotators act like obstacles which rotate the orientation of the self propelled particles by an angle determined by their intrinsic orientations. Results are compared with those of Chepizhko et al. [64] commented above.

As I did when commenting experimental realizations, let's finish with a reference to hindered flocking/swarming. Swarming transition in a 2d percolated lattice was addressed by Quint el al. [317]. Here the analyzed model consists of swarming agents, possessing both aligning and mutually avoiding repulsive interactions. Numerically a phase transition was found from a collectively moving swarm to a disordered gas-like state above a critical value of the topological or environmental disorder. A little bit later, flocking resistance to quenched disorder was revisited by Toner et al. in a couple of papers [410, 409]. Using the hydrodynamic-like approach of [412] authors reported that active polar systems are far more robust against quenched disorder than equilibrium ferromagnets. Specifically, long-range order with non-zero averaged velocity persists in the presence of quenched disorder even in spatial dimensions $d = 3$, contrary to equilibrium systems when only quasi-long-range order is possible; in $d = 2$, quasi-long-range order occurs, while the counterpart equilibrium systems, only display short-range order.

---

effects, the properties of RTP and ABP models are equivalent if one takes the inverse persistence time of the former (see Sect. 3.1.1.3) as the tumbling rate of the latter.

9

Appendix 2: Microswimming in Complex Fluids

When addressing the question of swimming in complex fluids one has to consider a two-way coupling between the swimming specimens and the fluid structure. Local mechanical stresses exerted by swimmers may alter their local environment, but at the same time the rheological properties of the complex fluid may modify the swimming gait and efficiency. Just as an example to stress the importance of this coupling remember that stresses in viscoelastic fluids are by definition viscous and elastic, and, as a consequence, time-dependent. As a consequence, kinematic reversibility may be broken down, and propulsion might be possible even for swimmers performing reciprocal motion, as commented below.

On what follows, I will briefly comment this subject from published experiments and theoretical treatments that encompass natural and artificial swimmers as well. As it will become clear, there is no universal answer to the principal question, i.e. whether motility is enhanced or hindered in complex fluids. The bottom line is that what really matters is the coupling between the details of the fluid structure and the characteristics of the swimming motion specific for each considered swimmer. A general overview on this particular subject was reviewed a few years ago by Patteson et al. [298]. A much more recent review paper with a more oriented fluid dynamics point of view, and dedicated to both natural and synthetic swimmers, has been published by Li et al. [225].

Just as a final introductory remark, let's add that attention will be here restricted to complex isotropic (i.e. non-ordered) fluids, since I already spotted in detail the use of Liquid Crystals as dispersing media for driven colloids in Chapter 3.

9.1 Motion of Microorganisms in Complex Fluids

One of the first papers reporting undulatory swimming in viscoelastic fluids was published by X. N. Shen et al. [357]. These authors analyzed the effects of fluid elasticity (solutions of the long flexible polymer carboxymethyl cellulose

DOI: 10.1201/9781003302292-9

with relaxation time of the order of 1 s) on the swimming behavior of the nematode *C. elegans* (see Fig. 9.1). Newtonian fluids of different viscosities were also employed for comparison, trying to effective disentangle pure viscous from viscoelastic effects. The referred specimen swims by generating traveling waves and the reported observations correspond to tracking its motion and measuring the associated velocity field. Compared with Newtonian solutions, it is found that fluid elasticity leads to slower propulsion and that, as a general trend, self-propulsion decreases as elastic stresses grow in magnitude in the fluid. Results were interpreted in terms of an enhanced resistance to flow near hyperbolic points for viscoelastic (polymeric) fluids. The chosen solutions were dilute enough to discard shear-thinning effects.

Moreover, when comparing with Newtonian fluids of similar viscosities it was found that the kinematic characteristics of motion (for instance

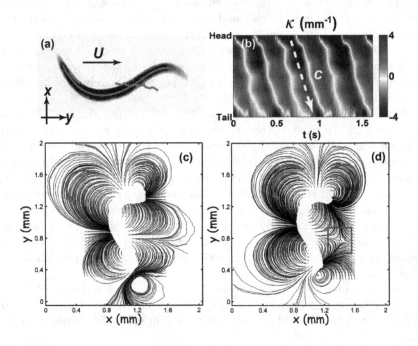

FIGURE 9.1
Swimming of *C. elegans* in different buffer solutions. (a) Sample snapshot with lines represent nematode's "skeleton" and its centroid path. (b) Corresponding contour plots of the nematode's bending curvature. (c) Streamlines computed from instantaneous velocity fields of Newtonian and (d) polymeric fluids. Arrows in panels (c) and (d) indicate flow direction and the box in (d) shows a hyperbolic point in the flow (colored image in original version). Image and caption text adapted from X. N. Shen et al. [357].

beating frequency and wave speed) are not influenced by viscoelasticity, but only depend on viscous effects. An observable where differences can be easily evidenced is, however, the swimming efficiency, defined here as the ratio of the swimming speed and the bending wave speed. For Newtonian fluids, efficiency increases with the viscosity, while for the polymeric fluids it is so at low viscosities but decreases beyond a crossover value at a **Deborah number** (De) of order unity. As a matter of fact, the Deborah number turns out to be a relevant dimensionless parameter of the problem, being defined as the ratio of the beating frequency and the inverse of the fluid relaxation time.

The case of semi-dilute solutions of shear-thinning xanthan gum was reported from the same group claiming that the velocity fields produced by the swimming nematode are modified without changing, however, the nematode's speed and beating kinematics [118]. Finally, when going to more concentrated solutions (entanglement of polymers and network formation), an enhancement in the nematode's swimming speed was found of approximately 65% in concentrated solutions compared to semi-dilute solutions [119].

More recently, the enhanced swimming speed of *C. elegans* was reported by Park et al. in [296] for a shear thinning polystyrene (PS) colloidal suspension, which is accompanied by marked alterations in the stroke form. To make the situation still a little bit more complex, the motion of bi-flagellated algal cells (*Chlamydomonas*) in viscoelastic (shear-thinning) fluids was studied with apparently contradictory results by Qin et al.. The beating frequency and the wave speed characterizing the cyclical bending were reported to be both enhanced by fluid elasticity. Despite these enhancements, the net swimming speed of the alga is hindered for fluids that are sufficiently elastic [315].

Changing to flagellated bacteria, Martinez et al. [251], following much earlier observations by Berg et al. [27], measured the swimming speed and the angular frequency of cell body rotation of motile *E. coli* as a function of polymer concentration in polyvinylpyrrolidone (PVP), and Ficoll solutions of different molecular weights. The measured concentration dependences of both observables for all but the highest-molecular-weight PVP were accounted for in terms of Newtonian hydrodynamics. However, non-Newtonian effects in the highest-molecular-weight PVP solution were evidenced. Calculations suggest that this is due to the fast-rotating flagella seeing a lower viscosity than the cell body. As a whole, experiments with swimming *E.coli* seem to indicate just a small increase in speed at low polymer concentrations, followed by a decrease at higher concentrations. A numerical study of this scenario was reported very recently by Zöttl et al. [463].

9.2 Artificial Microswimmers Performing in Complex Fluids

I start this subsection with a work a little bit out of the context of microswimming, but that may be useful to introduce the discussion that follows. I refer to the paper by B. Liu et al. [234]. The considered problem is that of the force-free swimming of a model helical flagellum in viscoelastic fluids. The helix is on the centimeter scale and it is plunged (and rotated) into a tank at a velocity adjusted to make the net hydrodynamic force vanish. Measures of the swimming speed as a function of helix rotation rate, helix geometry, and fluid properties are reported for highly viscous Newtonian fluids (high-molecular weight silicone oils), and for a complex (Boger) fluid (shear-rate independent viscosity) with a single relaxation time. When the relaxation time is short compared to the rotation period, the viscoelastic swimming speed is reported close to the viscous swimming speed. As the relaxation time increases, the former increases relative to the latter, reaching a peak when the relaxation time is comparable to the rotation period. As the relaxation time is further increased, the viscoelastic swimming speed decreases and eventually falls below the viscous swimming speed [234]. Mathematical modeling of this scenario was undertaken a couple of years later by Spagnolie et al. [372].

The first experimental realization of the principle of swimming of small (synthetic) rigid bodies by reciprocal motion at low Reynolds numbers was published by Keim et al. [194]. The considered prototype comprises two epoxy beads joined by a steel wire to render polar (asymmetric) and symmetric designs under two aligning electromagnets at constant current placed orthogonal to two driving magnets. The effects of inertia are absent due to high fluid viscosity ($\sim 4 \times 10^4$ cSt), resulting in Re $\leq 10^{-4}$, comparable to that of a swimming microorganism. The dimer is immersed in a container of either a Newtonian or a polymeric fluid. Evidence of purely elastic propulsion was shown for the polar dimer far away from boundaries at $De = 5.7$, and for the symmetric dimer near a wall at $De = 0.80$.

Under the same spirit, a study by Qiu et al. [316] reports a magnetically actuated symmetric 'micro-scallop', a single-hinge microswimmer, that can propel in shear thickening and shear thinning fluids by reciprocal motion at low Reynolds numbers. The clue of the experiment is the use of time-asymmetric stroke patterns. Excellent agreement between measurements, and both numerical and analytical theoretical predictions indicates that the net propulsion is caused by the modulation of the fluid viscosity upon varying the shear rate. Still, a time-periodic magnetic actuation on a hybrid magnetic head with a flexible tail was considered by Espinosa-Garcia et al. [102] who found that the elasticity of the fluid systematically enhances the locomotion speed of the swimmer, and that this enhancement increases with the Deborah number.

A rather different scenario, this time referring to the most prototypical example of autonomous artificial microswimming as represented by Janus particles, was addressed in an exhaustive study by Gomez-Solano et al. [142] (see Fig. 9.2).[1] The mechanism of motion was already described in Sect. 3.1.1.2,

---

[1]A more recent work from C. Bechinger's group was published by Narinder et al. [272]. In this case, the dynamics of activeparticles in a viscoelastic fluid under various geometric constraints, such as flat walls, spherical obstacles and cylindrical cavities, was analyzed.

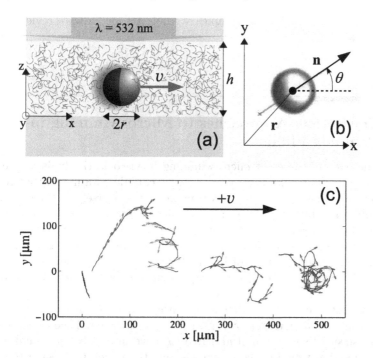

FIGURE 9.2
Self-propulsion of Janus particles in a viscoelastic fluid. Schematic illustration of the self-propulsion of Janus particles by light-induced demixing of a viscoelastic fluid. At the lower panel, examples of trajectories of the self-propelled particles with increasing velocity (from left to right). Image and caption text adapted from Gomez-Solano et al. [142].

and it is based on a light (thermal) actuation on a binary liquid mixture dispersing the Janus particle. To render the mixture viscoelastic a certain amount of polyacrylamide was added. The most important result of this study is the pronounced enhancement of the rotational diffusion coefficient. To be more precise, experiments were conducted in a regime of small values of the **Weissenberg number** We, defined as the product of the fluid relaxation time and the liquid rate of deformation ($Wi = V_0\tau/2R$). When comparing the Wi-dependent rotational diffusion coefficient with the reference value for a Newtonian fluid, the former was found to increase with increasing cruising speed by more than two orders of magnitude in a range of Wi extending up to 0.3. Interestingly enough, these anomalous effects were also observed under passive sedimentation. In the latter case, the translational diffusion coefficients (both parallel and perpendicular to the driving direction) were found to increase with Wi within the same range, although by a much smaller amount.

Actuation of nanopropellers in complex viscoelastic media were reviewed by Schamel et al. [347].

9.3   Theoretical Approaches to Microswimming in Complex Fluids

One of the earlier studies of microswimming in viscoelastic fluids is due to Chaudhury [61] more than forty years ago. He considered a model that turns out to be quite prototypical in this context: a transversely waving infinite (inextensible) flexible sheet (also named *Taylor sheet* after G. I. Taylor who first studied it in Newtonian fluids) in a viscoelastic liquid. It was found that the elastic property of the fluid augments self-propulsion (increases the induced velocity) in some range of Reynolds number and hampers it (reduces the induced velocity) in some other range under higher values of Reynolds number.

More recently, one of the first systematic studies of microswimming in complex fluids was undertaken by Lauga [214]. He considered the same model (a waving sheet of small amplitude) free to move in a polymeric fluid with a single relaxation time in a nonlinear viscoelastic model. Compared to self-propulsion in a Newtonian fluid occurring at a velocity V_N, the sheet swims with velocity $V/V_N = (1 + D_e^2 \eta_s/\eta)(1 + D_e^2)$ where η_s is the viscosity of the Newtonian solvent, η is the zero-shear-rate viscosity of the polymeric fluid, and D_e is the Deborah number for the wave motion. Notice that this result implies a slower velocity relative to that of a pure Newtonian fluid. Similar expressions were derived for the mechanical efficiency of the motion.[2]

Extending this approach, Fu et al. [113] considered the swimming velocity and hydrodynamic force exerted on an infinitely long cylinder with prescribed beating pattern in a nonlinear viscoelastic model. The reached conclusion is that viscoelasticity tends to decrease the swimming speed and that, in fact, changes in beating pattern due to viscoelastic effects can even reverse the swimming direction. A later paper from the same authors extend these calculations to examine how the scallop theorem is spoiled in a nonlinearly viscoelastic fluid at zero Reynolds number [114].

Following this series, I mention the paper by Teran et al. [396] again on the flapping sheet. The novelty is to consider fixed size swimmers rather than infinite models as in the previously mentioned works. Small-amplitude results for infinite sheets, that suggest that viscoelasticity impedes locomotion, are recovered, while the authors found that the opposite result applies when simulating

---

[2]The case of a cylindrical version of Taylor's swimming sheet in viscoelastic fluids was (experimentally) considered by Dasgupta et al. [74], and found that depending on the rheology, the speed can either increase or decrease relative to the speed in a Newtonian viscous fluid.

swimmers with large tail undulations, with both velocity and mechanical efficiency peaking for Deborah numbers near one. The complementary case of large undulations at the head, apparently more relevant to the swimming gait of *C. elegans*, was considered by Thomases et al. [402].

Changing to spherical swimmers, Zhu et al. [458] considered the eventual role of different hydrodynamic modes, pushers versus. pullers in relation to swimming behavior in viscoelastic ambients. It was found that, in all cases, the viscoelastic swimming speed is below the Newtonian one, with a minimum obtained for intermediate values of the Weissenberg number, We. An analysis of the flow field attributes the origin of this swimming degradation to non-Newtonian elongational stresses. The power required for swimming falls also systematically below the Newtonian power and it is always a decreasing function of We.

The study of active particles with prescribed surface velocities in non-Newtonian fluids was continued by Datt et al. [75], and considering spherical squirmers performing in polymer solutions by Qi et al. [314]. The interesting result in the former is that a Janus particle will always swim more slowly in a shear thinning fluid than in Newtonian fluid, assuming a fixed slip velocity. In the second situation, a drastic enhancement of the rotational diffusion by more than an order of magnitude in the presence of activity is obtained, evoking the results by Gomez-Solano et al. [142]. According to these authors, the amplification is a consequence of two effects: a decrease of the amount of adsorbed polymers by active motion, and an asymmetric encounter with polymers on the squirmer surface, which yields an additional torque and random noise for the rotational motion.

A very recent paper brought a new perspective of microswimming in complex fluids. More in particular, Miles et al. [258] investigated the dynamics of a dilute suspension of hydrodynamically interacting motile or non-motile stress-generating swimmers as they invade a surrounding viscous fluid. Colonies of aligned pusher particles are shown to elongate in the direction of particle orientation. At the same time, they undergo a cascade of transverse concentration instabilities, governed at small times by an equation that also describes the Saffman-Taylor instability in a Hele-Shaw cell, or the Rayleigh-Taylor instability in a two-dimensional flow through a porous medium. Thin sheets of aligned pusher particles are always unstable, while sheets of aligned puller particles can either be stable (non-motile particles), or unstable (motile particles) with a growth rate that is non monotonic in the force dipole strength.

10

Appendix 3: Motility Assays

A rewarding strategy at the early days of the in-vitro studies of self-organizing patterns assembled from filamentary proteins consisted in the use of the so-called motility assays [204]. Although in the main text I have primarily referred to tubulin (i.e. microtubules)-based systems, I will start by reviewing the in fact more numerous realizations of motility assays based on its analogue, i.e. the actin protein, powered (normally) by myosin motors. In the next section, I will briefly refer to gliding microtubules, under the action of kinesin proteins.

10.1 Motility Assays Based on the Actin System

Standard actin/myosin motility assays involve nitrocellulose covered-slips secured to a glass slide with enough space for injecting solutions. A saturating concentration of myosin is flowed through the cell, adhering to the nitrocellulose, and in the presence of ATP (Adenosine Triphosphate) interact with fluorescent labeled actin filaments, and propel them across the surface. These actin filaments are visualized with an inverted epifluorescence microscope, and their trajectories can be easily analyzed (see Fig. 10.1).

Schaller et al. [345] reported the formation of polar patterns of actin filaments driven by non-processive motor proteins (heavy meromyosin HMM). The basic control parameter in Schaller's experiments is the filament density (see Fig. 10.2). Below a critical value ρ_c, a disordered phase was observed. Filaments of a typical length of 10 µm perform persistent random walk without any directional preference. The average length of the actin filaments was controlled by adding the regulator gelsolin before actin polymerization. The speed of the entrained filaments, at values of a few microns per second, is set by motor proteins and ATP concentration.

At the transition density ρ_c, an ordered phase assembles characterized by a polymorphism of different polar nematic patterns coherently moving. At the smallest densities, yet above ρ_c, a swarming phase is observed, while at the highest densities stable bands, corresponding to density waves, propagate system-wide. Clusters, continuously losing and recruiting new filaments, move independently in an erratic manner, with sizes ranging from 20 to 500 µm. Their stability is only affected by collisions with other clusters or with

DOI: 10.1201/9781003302292-10

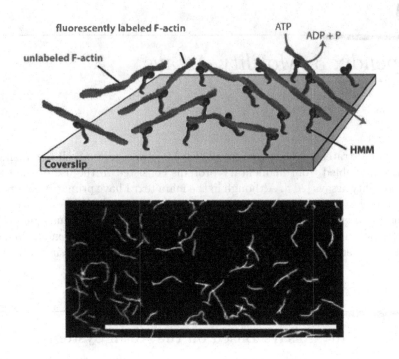

FIGURE 10.1
Schematics of a high-density motility assay based on immobilized myosin motors and gliding actin filaments. (Top) The molecular motor HMM is immobilized on a cover-slip and the filament motion is visualized by the use of fluorescent labeled reporter filaments. (Bottom) For low filament densities, a disordered phase is found. The individual filaments perform persistent random walks without any specific directional preferences. Encounters between filaments lead to crossing events with only slight reorientations. Image and caption text adapted from Schaller et al. [345].

confining boundaries. Swirls or spirals were also occasionally observed for all the densities above ρ_c.

Experiments were supplemented with agent-based simulations. Agents were considered as finite-length filaments, each performing a persistent random walk, and interacting each other through steric repulsions and weak local alignment directions. Hydrodynamic interactions were discarded. In any case, the symmetry-breaking nature of the local alignment interactions turns out to be crucial.

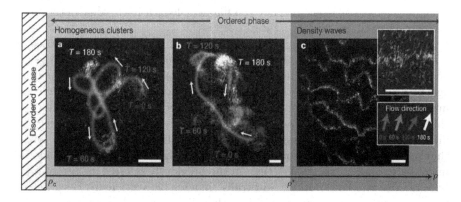

FIGURE 10.2
Motility patterns as a function of the filament density. (a) Above a certain critical density, ρ_c, in an intermediate-density regime, the disordered phase is unstable and small polar nematic clusters of coherently moving filaments start to form. (b) At higher densities, these clusters become larger but remain homogeneous. (c) Above a threshold density, ρ^*, in the high-density regime, persistent density fluctuations lead to the formation of wave-like structures (colored image jn original version). Image and caption text adapted from Schaller et al. [345].

To better understand the stability of the patterns observed in the motility assays just mentioned, and immediately following the 2010 Nature paper [345], Schaller et al. published additional results stressing the role of the hydrodynamic interactions that are self-induced by the coherently moving filaments [344]. The existence of a depletion zone between approaching clusters prevents coarsening and could explain the somewhat surprising stability of individual clusters. To enable quantitative comparison with the theories of collective motion, local density fluctuations were evaluated in a following paper, and confirmed to behave anomalously in conditions where the overall density can be considered homogeneous [342].

Still from A. R. Bausch's group, another aspect that was experimentally investigated was the polar versus nematic nature of the collisions between filaments. For low filament densities the binary collision statistics was extracted and two main conclusions were reported by R. Suzuki et al. [387] (see also [386]). First, the degree of alignment caused by a single collision event is normally weak, and the resulting alignment exhibits neither perfect nematic nor polar symmetries. Instead, depending on the collision angle, there is a weak tendency to favor either alignment or anti-alignment of the filaments. Second, and by varying the filament length, it was concluded that it is the total number of simultaneous collisions and not the filament length itself what determines the transition to ordered structures.

In a more recent paper published by Huber et al. [167], the emergence of a dynamic coexistence of ordered states with fluctuating nematic and polar symmetry in an actomyosin motility assay was demonstrated when a depleting agent (PEG) was added to the standard preparation.[1] In this way, the local interaction between filaments can be easily tuned. When small amounts of this depletant were added, and despite the rather minute changes caused in interaction characteristics slightly favoring nematic alignment, it was found that polar flocks no longer form. Instead, the moving filaments, less than one micrometer long, much smaller than in the previously referred assays, quickly self-organize into a network of "ant trails" moving bidirectionally. Because the filaments move along these tracks in either direction with equal probability, the overall order is nematic.

As a matter of fact, more (biologically) complex actomyosin motility assays, incorporating this time cross-linking proteins had been published from the same group in a couple of earlier papers [346, 343]. The declared aim in the first of the mentioned papers, using fascin as a cross-linker, was to investigate the eventual existence of what authors refer to as *frozen active states*, i.e. quiescent or absorbing states with frozen fluctuations. A frozen active state would combine self-organization and growth processes. In particular under the conditions reported in the first study, fascin is known to assemble polar filaments. At low concentrations of the cross-linking molecule and under high filament density, the observed patterns largely resemble those previously identified, i.e. coherently moving structures in the form of clusters or density waves. However, above a critical concentration of the added fascin, the patterns of self-organization change drastically as manifested by constantly rotating clusters that appear either as closed or open structures (see Fig. 10.3).

10.2   Motility Assays Based on the Tubulin System

Motility assays, practically simultaneous to those previously reported for the actomyosin system but this time involving microtubules were published by L. Liu et al. [236], and a little bit later by Sumino et al. [383]. In the first paper, authors claim that MTs, appearing as rigid rods (persistence length around 1 mm) and with lengths 5–15 μm, assembled, though not-strictly in a two-dimensional layer, when concentrated at relatively high densities and sheared by kinesins. The first time a motility assay, strictly two-dimensional,

---

[1]I remind the reader that the use of PEG as a depleting agent was crucial to the development of active gels/active nematics based on microtubules and kinesins.

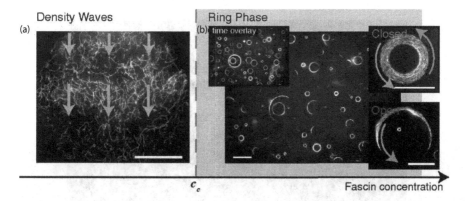

FIGURE 10.3
Phase behavior in an actomyosin motility assay incorporating fascin as a cross-linking protein. (a) Under low fascin concentration conventional propagating bands are observed. (b) Increasing cross-linking, either closed or open constantly rotating rings are observed (colored image in original version). Image and caption text adapted from Schaller et al. [346].

was demonstrated to organize patterns of microtubules driven by kinesin motors was reported by Inoue et al. [173]. The trick used by Inoue et al. consisted again in the use of a depleting agent (this time methylcellulose MC was employed) to regulate the interaction of MTs.

A much more systematic experimental study, accompanied by a theoretical model, was reported in the paper by Sumino et al.[383]. Microtubules are propelled this time by surface-bound dynein.[2] Dynein motors were grafted at a glass surfaces at densities around a few thousands of molecules per square micrometer, while flushed with MTs around 10 μm in length.

Following the addition of ATP, many aligning collisions were observed, leading to the formation of streams, with MTs moving in both directions. These streams started to meander over the surface, and later degenerated into the formation of a vortex lattice, eventually organizing a hexagonal, non perfect, arrangement. Shapes of the vortices changed with the course of time, although their size, around 400 μm stayed practically constant (see Fig. 10.4). Vortices consist of a sparse core and a dense peripheral annulus, inside which MTs move both clockwise and counter-clockwise, sliding past each other in a clear nematic arrangement. Microtubules were never trapped in a single vortex, but rather they circulated inside one vortex for some time before

---

[2]I remind the reader that this is an alternative typical motor for MTs, different from the most usual kinesin.

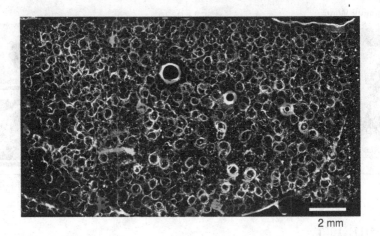

2 mm

FIGURE 10.4
Emergence of a lattice of vortices in a motility assay with micro-tubules. Large-scale lattice of vortices that can be observed everywhere on the surface of the flow cell. Three air bubbles in the flow cell can be seen distinctively owing to their greater size and thicker edges (colored image in original version). Image and caption text adapted from Sumino et al. [383].

moving to a neighboring or towards a more distant one. Long-range hydro-dynamic effects were observed to be negligible, compared with alignment or anti-alignment effects.

11

Appendix 4: Active Nematic Concepts in the Context of Cell Tissues

Chapter 4 was entirely devoted to review experimental developments on active fluids that can be viewed as minimal reconstitutions of cell subsystems, fundamentally the cell cytoskeleton. Beyond this level one can go a little bit higher in the hierarchy of cell biophysics and refer to cell assemblies. In fact, it is not difficult to imagine the cells themselves as active units embedded into a larger structure which is the tissue ensemble. The latter encodes, albeit in a very complex manner, the structure of active systems. In this respect, it is well-known that cells exert active stresses and promote self-organized motion patterns not unlike what I have extensively reviewed within the context of assemblies of filamentary motor proteins. From this point of view, it is well-accepted that cell tissues display modes of dynamic self-organization that bear close similarities with active nematics.

There are many and very important aspects of the tissue dynamics, singularly those associated with cell proliferation or migration, that fall out of the scope of this text. Rather, I am going to briefly comment only several papers that clearly convey the idea that some dynamic modes observed at a cell tissue level may be easily and conveniently interpreted in terms of experimental observables and modeling approaches totally similar to what has become conventional for active fluids. A review paper that clearly frames this sort of perspective and that can be a guide to the scenarios introduced later on in this appendix was published recently by Saw et al. [341].

11.1 Textures, Flows, and Defects in Cell Tissues

One of the main characteristics of nematic textures, either in conventional liquid crystals or in active realizations, and that has been largely discussed in Chapters 4, and 6, is the presence of defects. As commented in the main text, defects are, however, slightly reinterpreted in active nematics, where they are viewed more as regions void of active components than as strict orientational singularities common to classic liquid crystals (see Sect. 2.2.2).

DOI: 10.1201/9781003302292-11

It is worth stressing that this latter perspective is precisely fully recovered when referring to defects in the context of cell tissues.

The recognition of the role of topological defects in cell tissues is rooted in the very early observations by Elsdale at al. [100] relative to the orientational order of fibroblast cells. With a similar perspective, it is worth mentioning the experiments conducted by Bonhoeffer at al. more than twenty yeard later [40], who demonstrated that the orientation maps of cells in the cat visual cortex are characterized by patches of cells, encircling center points, termed 'pinwheel' structures. The pinwheels play an important role in the organization of neurons in the visual cortex and in the response to visual stimuli. The analogy to liquid crystals showed that these structures are indeed related to $\pm 1/2$ topological defects, and dedicated theories based on the simple orientational dynamics of passive nematics faithfully reproduced the organizational patterns of pinwheels observed in the experiments [438, 222].

A little bit more focused into our own perspective of active fluids, pioneering experiments by Gruler et al. [145] and Kemkemer et al. [195] described the analogy between nematic liquid crystals and assemblies of amoeboid cells. These authors showed that nematic ordering and topological defects emerge in dense ensembles of melanocytes (cells located in skin, eye, inner ear, bones and heart), fibroblasts (cells important for tissue maintenance and wound healing), osteoblasts (cells that synthesize bone), and lipocytes (fat cells) (see Fig. 11.1). By estimating the orientational elastic constants from defect structures in melanocyte cells, authors concluded that resistance to bend deformations is stronger than to splay ($K_{splay} < K_{bend}$), and it was additionally demonstrated that, in analogy to conventional liquid crystals, the nematic director can be guided by creating parallel scratches on the cell substrate.

Much more recent is the work by Duclos et al. [95]. Duclos et al. report that elongated, weakly interacting, apolar, fibroblast cells cultured at confluence align together, forming large domains (correlation length of the order of 500 μm) where they are perfectly ordered. Cells are initially very motile and the monolayer is characterized by anomalous density fluctuations, a signature typical of active rod-shaped particles as commented in Chapter 7. As the cell density increases because of proliferation, the cells align with each other forming large oriented domains while, at the same time, the cellular movements and the density fluctuations freeze. Topological defects that are characteristic of nematic phases remain trapped at long times, thereby preventing the development of infinite domains. When confined within adhesive stripes of given widths (from 30 μm to 1.5 mm) cells spontaneously align with the domain edges. This orientation then propagates toward the pattern center. For widths smaller than the orientation correlation length, cells perfectly align in the direction of the stripe.

Circular confinement at small radii was considered for the same kind of cells in a later paper by some of the previous authors [94]. The important addressed question there was the role of substrate friction opposing cells activity. Results shown demonstrate that at the chosen conditions of the reported

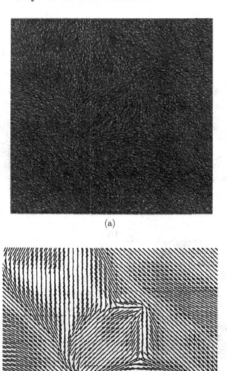

(a)

(b)

FIGURE 11.1
Nematic arrangements of amoeboid cells. (a) Human melanocytes exposed to a glass surface display nematic characteristics with the presence of topological defects. (b) Calculated mean cell orientation in panel (a). Image and caption text adapted from Kemkemer et al. [195].

experiments, activity is effectively damped by friction, and that the interaction between defects is controlled by the system's elastic nematic energy. As a matter of fact, a large initial number of defects at confluence resulting from local alignment annihilate with the course of time, without observing creation of new defects via the splay instability of contractile systems, demonstrating prevalence of friction damping. Finally, the system evolves towards two identical $+1/2$ disclinations facing each other (see Fig. 11.2). The reduced positions of these defects, once scaled with the radius of the pattern, are independent of

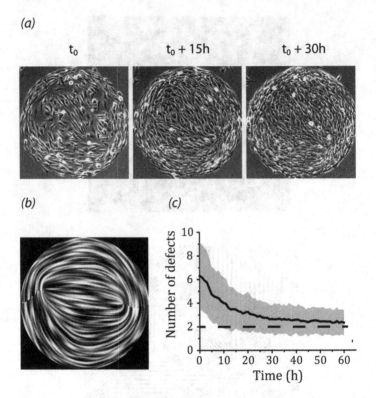

FIGURE 11.2
Effect of confinement of spindle-shaped cells in a disk. (a) Coarsening
of defects with time until they reach a stable position (Radius of the disk is
350 µm). (b) Identification of the left two +1/2 defects. (c) Defects decrease
with the course of time until a pair of positive ones remain to satisfy the total
topological charge of +1. Image and caption text adapted from Duclos et al.
[94].

the size of the disk, the cells' activity or even the cell type, and follow the principles of conventional nematic liquid crystals. Working with smaller circular
confinement conditions, and with a different class of specimens, here epithelial (extensile) cells that develop very strong cell–cell adhesions, Deforet et al.
[80] had earlier demonstrated the importance of three-dimensional effects as
peripheral cell cords develop at the domains edges by differential extrusion.

Still considering cells under geometric confinement, there is a later article
from Silberzan's group that connects with the experiments mentioned earlier
by Hardoüin et al. [156] (see Sect. 4.4.2), and with the theoretical studies by
Voituriez et al. [423] (see Sect. 6.2) on channeled active fluids. The paper by

Duclos et al. [93] reports the onset of flows for elongated cells constrained to channels.

More in particular, an ensemble of spindle-shaped cells plated in a well-defined stripe is shown to spontaneously develop a shear flow whose character-istics depend on the width of the stripe. On wide stripes, the cells self-organize in a nematic phase with a director at a well-defined angle with the stripe's direction, and develop a shear flow close to the stripe's edges. However, on stripes narrower than a critical width, the cells perfectly align with the stripe's direction and the net flow vanishes.

If I have referred several times in the main text to the topic of active turbulence as a hallmark of active matter it is not strange that this issue reappears here in the particular context of cell assemblies.[1] This is the ques-tion addressed by Blanch-Mercader et al. [37]. Their study refers to human bronchial epithelial cells (HBEC) demonstrated to be extensile according to the flow patterns associated to the positive defects (see Fig. 11.3).

Contrary to other epithelial cell lines, HBECs are weakly cohesive at the initial stages, meaning that when cells fully cover the substrate, they are still highly motile and exhibit long-range collective movements. The system gradually slows down due to cell proliferation, approaching asymptotically a jammed state after time lapse of the order of 60 h, in which cells hardly move beyond their own sizes. As a matter of fact, extracted values of the kinetic energy and enstrophy show exponentially decrease with time over more than three decades. This feature is consistent with more than a decade increase in the shear viscosity, the authors attribute both to incipient jamming and maturation of the cell-cell contacts. Different from the case commented earlier in relation to the paper by Duclos et al. [94], where large friction largely prevented defect formation, the different kind of cell tissue investigated here, show repeatedly episodes of equilibrated defect creation and annihilation.

A complete analysis of the defect landscape and flow patterns is proposed in the original reference [37]. By analyzing a large population of vortices, these authors show that their area follows an exponential law with a characteristic length scale and their rotational frequency is size independent, both being characteristic features of the chaotic dynamics of active nematic suspensions, as I have commented earlier in relation to experiments with microtubule-based active nematics ([149]; see Sect. 4.2.1). Moreover, the associated active length scale is extracted as well. As a general characteristics already commented in relation to cell tissues, nematic defects are found at the interface between domains with a total number that remains constant due to the dynamical balance of nucleation and annihilation events.

The role of defects in cell tissues is far more intriguing as exemplified by a couple of simultaneously published papers. Saw et al. [340], working with different types of epithelia cells, found a universal correlation between

---

[1]Topological turbulence in the membrane of a living cell was analyzed very recently in a paper published by T. H. Tan et al. [393].

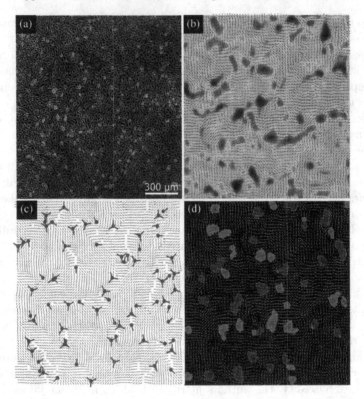

FIGURE 11.3
Nematodynamics of a monolayer of human bronchial epithelial cells.
(a) Phase contrast image. (b) Normalized vorticity map with the particle image velocimetry flow field illustrating the collective behavior. (c) Cell orientation map with marked defects. (d) Map of the Okubo-Weiss field. Scale bar is 300 μm (colored image in original version). Image and caption text adapted from Blanc-Mercader et al. [37].

extrusion sites (places of cell death and further expelling from the cell monolayer) and the positions of nematic defects in the cell orientation field (see Fig. 11.4). The results confirm the active nematic nature of epithelia, and demonstrate that defect-induced isotropic stresses are the primary precursors of the entire biochemical cue involved in cell death and extrusion. Importantly, the defect-driven extrusion mechanism depends on inter-cellular junctions, because the weakening of cell-cell interactions increases both the number of defects and extrusion rates. Saw et al. further demonstrate the ability to control extrusion hot spots by geometrically inducing defects through microcontact

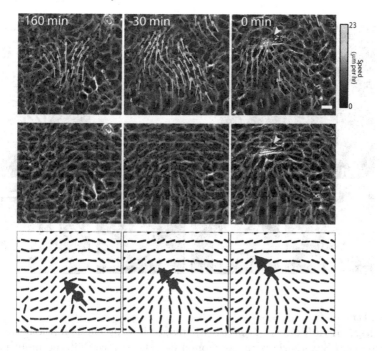

FIGURE 11.4
Correlation between defect position and cell extrusion in epithelia.
Top panels: Phase-contrast images showing monolayer dynamics before extrusion (arrowhead) at t = 0 min, overlaid with velocity field vectors. Extrusion is preceded by large-scale flows that bear the characteristic comet-like symmetry of +1/2 defects. Length of vectors is proportional to their magnitude. Corresponding images overlaid with lines (in mid panels), and represented as black lines (bottom panels) showing average local orientation of cells (colored image in original version). Image and caption text adapted from Saw et al. [340].

printing of patterned monolayers. Continuous active nematic modeling predicts the flow field and the (motile) defect dynamics.

With another kind of cells (neural progenitor which are multipotent stem cells that give rise to cells in the central nervous system) Kawaguchi et al. [192] reported cell alignment patterns and distinctive characteristics of positive and negative half-integer defects. More precisely, at low densities, cells moved randomly in an amoeba-like fashion. However, at high density cells elongate and align their shapes with one another, gliding at relatively high velocities. Individual cells may reverse stochastically motion along their axis, yet they are capable of forming an aligned pattern up to length scales similar to that of the migratory stream observed in the adult stage. Moreover, rapid cell

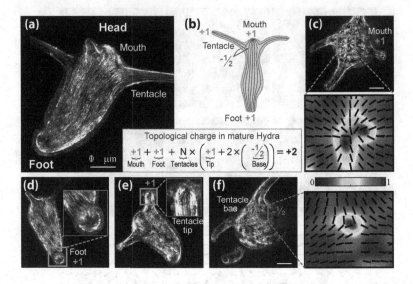

FIGURE 11.5
Topological defects in *Hydra*. (a) Image showing the nematic actin organization. (b) Schematic with positioned defects localized at the mouth (+1),
foot (+1), and tentacles (+1) tip and two (−1/2) at the base. Total defect
charge is constrained to be +2. (c)-(f) Images of actin fiber organization showing the defects at the head (c), apex of the foot (d) and the tip (e) and base
(f), respectively, of a tentacle. All shown images are 2d projections extracted
from 3d spinning-disk confocal images. Scale bar is 100 μm (colored image
in original version). Image and caption adapted from Maroudas-Sacks et al.
[250].

accumulation at +1/2 defects is identified, concurrent with the formation of
three-dimensional mounds, and depletion around the negative counterparts.
A generic mechanism for the instability in cell density around the defects that
arises from the interplay between the anisotropic friction and the active force
field is proposed, although the formation of mounds is still unclear according
to the authors.

Normally the previous considerations correspond to observations made on
more or less elongated cells of different sorts. However, the possibility to observe nematic features emerging from assemblies of isotropic cells has been
theoretically addressed as well by Mueller et al. [264]. In this paper, a minimal model of cellular monolayers based on cell deformation and force trans-
mission at the cell-cell interface is presented. Authors argue that this model
can explain the formation of topological defects while it captures the flow field
and stress patterns around them. Moreover, authors emphasize the role of a
feedback mechanism between shape deformation and active driving.

Stressing the importance of topological defects in the context of cell biophysics, I finish this appendix with a reference to a recently appeared paper published by Maroudas-Sacks et al. [250] stressing the role of these singularities in *Hydra* morphogenesis (see Fig. 11.5).

Hydra regeneration from tissue pieces proceeds through the formation of spheroids that assemble a nematic ordering of actin fibers. Similarly to what was reported for the cortical enclosed nematic flows analyzed by Keber et al. [193] in Sect. 4.4.1, this assembling leads to the formation of defects with a total accumulated charge of +2. Two +1 defects are located at the head and the base of the foot, thus defining the body axis of the animal. Further development leads to the formation of tentacles, again associated with defects of charge +1 at the tip, and two −1/2 at the basis.

Bibliography

[1] R. Aditi Simha and S. Ramaswamy. Hydrodynamic fluctuations and instabilities in ordered suspensions of self-propelled particles. *Physical Review Letters*, 89(5):058101, 2002.

[2] A. Ahmadi, T. B. Liverpool, and M. C. Marchetti. Nematic and polar order in active filament solutions. *Physical Review E*, 72(6):060901, 2005.

[3] F. Alaimo and A. Voight. Microscopic field-theoretical approach for mixtures of active and passive particles. *Physical Review E*, 98(3):032605, 2018.

[4] R. Alert, J. Casademunt, and J. F. Joanny. Active Turbulence. *Annual Review of Condensed Matter Physics*, 13(1):143–170, 2022.

[5] R. Alert, J.-F. Joanny, and J. Casademunt. Universal scaling of active nematic turbulence. *Nature Physics*, 16:682–688, 2020.

[6] J. Alvarado, M. Sheinman, A. Sharma, F. C. Mackintosh, and G. H. Koenderink. Molecular motors robustly drive active gels to a critically connected state. *Nature Physics*, 9(9):591–597, 2013.

[7] J. Anderson. Colloid Transport By Interfacial Forces. *Annual Review of Fluid Mechanics*, 21(1):61–99, 1989.

[8] L. Angelani, R. Di Leonardo, and G. Ruocco. Self-starting micromotors in a bacterial bath. *Physical Review Letters*, 102(4):048104, 2009.

[9] I. S. Aranson, A. Snezkho, J. S. Olafsen, and J. S. Urbach. Comment on " Long-Lived Giant Number Fluctuations in a Swarming Granular Nematic. *Science*, 320(May):612, 2008.

[10] I. S. Aranson, A. Sokolov, J. O. Kessler, and R. E. Goldstein. Model for dynamical coherence in thin films of self-propelled microorganisms. *Physical Review E*, 75(4):040901, 2007.

[11] H. Aref. Stirring by chaotic advection. *Journal of Fluid Mechanics*, 143:1–21, 1984.

[12] S. Asakura and F. Oosawa. On interaction between two bodies immersed in a solution of macromolecules. *The Journal of Chemical Physics*, 22(7):1255–1256, 1954.

[13] A. Aubret, M. Youssef, S. Sacanna, and J. Palacci. Targeted assembly and synchronization of self-spinning microgears. *Nature Physics*, 14(11):1114–1118, 2018.

[14] K. Avila, D. Moxey, A. De Lozar, M. Avila, D. Barkley, and B. Hof. The onset of turbulence in pipe flow. *Science*, 333(6039):192–196, 2011.

[15] F. Backouche, L. Haviv, D. Groswasser, and A. Bernheim-Groswasser. Active gels: Dynamics of patterning and self-organization. *Physical Biology*, 3(4):264–273, 2006.

[16] D. Banerjee, A. Souslov, A. G. Abanov, and V. Vitelli. Odd viscosity in chiral active fluids. *Nature Communications*, 8(1):1–12, 2017.

[17] L. Barberis and F. Peruani. Large-Scale Patterns in a Minimal Cognitive Flocking Model: Incidental Leaders, Nematic Patterns, and Aggregates. *Physical Review Letters*, 117(24):248001, 2016.

[18] A. Baskaran and M. C. Marchetti. Enhanced diffusion and ordering of self-propelled rods. *Physical Review Letters*, 101(26):268101, 2008.

[19] A. Baskaran and M. C. Marchetti. Statistical mechanics and hydrodynamics of bacterial suspensions. *Proceedings of the National Academy of Sciences*, 106(37):15567–15572, 2009.

[20] A. Baskaran and M. C. Marchetti. Nonequilibrium statistical mechanics of self-propelled hard rods. *Journal of Statistical Mechanics*, page 04019, 2010.

[21] A. Basu, J. F. Joanny, F. Jülicher, and J. Prost. Thermal and non-thermal fluctuations in active polar gels. *European Physical Journal E*, 27(2):149–160, 2008.

[22] T. Bäuerle, A. Fischer, T. Speck, and C. Bechinger. Self-organization of active particles by quorum sensing rules. *Nature Communications*, 9:3232, 2018.

[23] T. Bäuerle, R. C. Löffler, and C. Bechinger. Formation of stable and responsive collective states in suspensions of active colloids. *Nature Communications*, 11(1):1–9, 2020.

[24] C. Bechinger, R. Di Leonardo, H. Löwen, C. Reichhardt, G. Volpe, and G. Volpe. Active particles in complex and crowded environments. *Reviews of Modern Physics*, 88(4):045006, 2016.

[25] H. C. Berg. *E. coli in motion*. Springer-Verlag, 2004.

[26] H. C. Berg and D. A. Brown. Chemotaxis in escherichia coli analysed by three-dimensional tracking. *Nature*, 239:500–504, 1972.

[27] H. C. Berg and L. Turner. Movement of microorganisms in viscous environments [12], 1979.

[28] H. C. Berg and L. Turner. Chemotaxis of bacteria in glass-capillary arrays. *scherichia coli*, motility, microchanel plate, and lighht-scattering. *Biophysical Journal*, 58:919–930, 1990.

[29] E. Bertin, M. Droz, and G. Grégoire. Boltzmann and hydrodynamic description for self-propelled particles. *Physical Review E*, 74(2):022101, 2006.

[30] T. Bhattacharjee and S. S. Datta. Bacterial hopping and trapping in porous media. *Nature Communications*, 10(10):2075, 2019.

[31] J. Bialké, H. Löwen, and T. Speck. Microscopic theory for the phase separation of self-propelled repulsive disks Microscopic theory for the phase separation of self-propelled. *Europhysics Letters*, 103:30008, 2013.

[32] J. Bialké, J. T. Siebert, H. Löwen, and T. Speck. Negative Interfacial Tension in Phase-Separated Active Brownian Particles. *Physical Review Letters*, 115(9):098301, 2015.

[33] J. Bialké, T. Speck, and H. Löwen. Crystallization in a dense suspension of self-propelled particles. *Physical Review Letters*, 108(16):168301, 2012.

[34] J. R. Blake. A shperical envelope approach to ciliary propulsion. *Journal of Fluid Mechanics*, 46:199–208, 1971.

[35] C. Blanc, D. Coursault, and Lacaze E. Ordering nano- and microparticles assemblies with liquid crystalsl. *Liquid Crystals Reviews*, 1(2):83–109, 2013.

[36] C. Blanc and M. Kleman. Tiling the plane with noncongruent toric focal conic domains. *Physical Review E*, 62(5):6739–6748, 2000.

[37] C. Blanch-Mercader, V. Yashunsky, S. Garcia, G. Duclos, L. Giomi, and P. Silberzan. Turbulent dynamics of epithelial cell cultures. *Physical Review Letters*, 120(20):208101, 2018.

[38] L. Blanchoin, R. Boujemaa-Paterski, C. Sykes, and J. Plastino. Actin dynamics, architecture, and mechanics in cell motility. *Physiological Reviews*, 94(1):235–263, 2014.

[39] M. L. Blow, S. P. Thampi, and J. M. Yeomans. Biphasic, lyotropic, active nematics. *Physical Review Letters*, 113(24):248303, 2014.

[40] T. Bonhoeffer and A Grinvald. Iso-orientation domains in cat visual cortex are arranged in pinwheel-like patterns. *Nature*, 353:429, 1991.

[41] M. Bourgoin, R. Kervil, C. Cottin-Bizonne, F. Raynal, R. Volk, and C. Ybert. Kolmogorovian Active Turbulence of a Sparse Assembly of Interacting Marangoni Surfers. *Physical Review X*, 10(2):21065, 2020.

[42] V. Bratanov, F. Jenko, and E. Frey. New class of turbulence in active fluids. *Proceedings of the National Academy of Sciences*, 112(49):15048–15053, 2015.

[43] A. Bricard, J. B. Caussin, D. Das, C. Savoie, V. Chikkadi, K. Shitara, O. Chepizhko, F. Peruani, D. Saintillan, and D. Bartolo. Emergent vortices in populations of colloidal rollers. *Nature Communications*, 6:7470, 2015.

[44] A. Bricard, J. B. Caussin, N. Desreumaux, O. Dauchot, and D. Bartolo. Emergence of macroscopic directed motion in populations of motile colloids. *Nature*, 503(7474):95–98, 2013.

[45] A. M. Brooks, S. Sabrina, and K. J. M. Bishop. Shape-directed dynamics of active colloids powered by induced-charge electrophoresis. *Proceedings of the National Academy of Sciences*, pages E1090–E1099, 2018.

[46] M. Brun-Cosme-Bruny, E. Bertin, B. Coasne, P. Peyla, and S Rafaï. Effective diffusivity of microswimmers in a crowded environment. *Journal of Chemical Physics*, 150(February):104901, 2019.

[47] L. Bruno, V. Levi, and M. Brunstein. Transition to superdiffusive behavior in intracellular actin-based transport. *Physical Review E*, 80(1):011912, 2009.

[48] I. Buttinoni, J. Bialké, F. Kümmel, H. Löwen, C. Bechinger, and T. Speck. Dynamical clustering and phase separation in suspensions of self-propelled colloidal particles. *Physical Review Letters*, 110(23):238301, 2013.

[49] I. Buttinoni, G. Volpe, F. Kümmel, G. Volpe, and C. Bechinger. Active Brownian motion tunable by light. *Journal of Physics Condensed Matter*, 24:284129, 2012.

[50] A. C. Callan-Jones and F. Jülicher. Hydrodynamics of active permeating gels. *New Journal of Physics*, 13, 2011.

[51] G. Carleo, I. Cirac, K. Cranmer, L. Daudet, M. Schuld, N. Tishby, L. Vogt-Maranto, and L. Zdeborová. Machine learning and the physical sciences. *Reviews of Modern Physics*, 91(4):45002, 2019.

[52] M. E. Cates and J. Tailleur. When are active Brownian particles and run-and-tumble particles equivalent? Consequences for motility-induced phase separation. *Europhysics Letters*, 101(2):20010, 2013.

[53] M. E. Cates and J. Tailleur. Motility-Induced Phase Separation. *Annual Review of Condensed Matter Physics*, 6(1):219–244, 2015.

[54] J.-B. Caussin, A. Solon, A. Peshkov, H. Chaté, T. Dauxois, J. Tailleur, V. Vitelli, and D. Bartolo. Emergent Spatial Structures in Flocking Models : A Dynamical System Insight. *Physical Review Letters*, 112(14):148102, 2014.

[55] S. Chandragiri, A. Doostmohammadi, J. M. Yeomans, and S. P. Thampi. Flow States and Transitions of an Active Nematic in a Three-Dimensional Channel. *Physical Review Letters*, 125(14):148002, 2020.

[56] P. Chandrakar, M. Varghese, S. A. Aghvami, A. Baskaran, Z. Dogic, and G. Duclos. Confinement Controls the Bend Instability of Three-Dimensional Active Liquid Crystals. *Physical Review Letters*, 125(25):257801, 2020.

[57] A. Chardac, S. Shankar, M. Cristina Marchetti, and D. Bartolo. Emergence of dynamic vortex glasses in disordered polar active fluids. *Proceedings of the National Academy of Sciences of the United States of America*, 118(10), 2021.

[58] H. Chaté. Dry Aligning Dilute Active Matter. *Annual Review of Condensed Matter Physics*, 11(1):189–212, 2020.

[59] H. Chaté, F Ginelli, G. Guillaume, and F. Raynaud. Collective motion of self-propelled particles interacting without cohesion. *Physical Review E*, 77(4):046113, 2008.

[60] H. Chaté, F. Ginelli, and R. Montagne. Simple model for active nematics: Quasi-long-range order and giant fluctuations. *Physical Review Letters*, 96(18):180602, 2006.

[61] T. K. Chaudhury. On swimming in a visco-elastic liquid. *Journal of Fluid Mechanics*, 95:189–197, 1979.

[62] K. Chen, O. J. Gebhardt, R. Devendra, G. Drazer, R. D. Kamien, H. Reich, and R. L. Leheny. Colloidal transport within nematic liquid crystals with array of obstacles. *Soft Matter*, 14:83–91, 2017.

[63] S. Chen, P. Gao, and T. Gao. Dynamics and structure of an apolar active suspension in an annulus. *Journal of Fluid Mechanics*, 835:393–405, 2018.

[64] O. Chepizhko, E. G. Altmann, and F. Peruani. Optimal Noise Maximizes Collective Motion in Heterogeneous Media. *Physical Review Letters*, 110(23):238101, 2013.

[65] O. Chepizhko and F. Peruani. Diffusion, subdiffusion, and trapping of active particles in heterogeneous media. *Physical Review Letters*, 111(16):160604, 2013.

[66] U. Choudhury, A. V. Straube, P. Fischer, J. G. Gibbs, and F. Höfling. Active colloidal propulsion over a crystalline surface. *New Journal of Physics*, 19:125010, 2017.

[67] L. H. Cisneros, C. Cortez, C. Dombrowski, R. E. Goldstein, and J. O. Kessler. Fluid dynamics of self-propelled microorganisms from individuals to concentrated populations. *Experimental Fluids*, 43:737–753, 2007.

[68] J. Colen, M. Han, R. Zhang, S. A. Redford, L. M. Lemma, L. Morgang, P. V. Ruijgrokh, R. Adkins, Z. Bryant, Z. Dogic, M. L. Gardel, J. J. de Pablo, and V. Vitelli. Machine learning active-nematic hydrodynamics. *Proceedings of the National Academy of Sciences*, 118:e2016708118, 2021.

[69] S. Čopar, J. Aplinc, Ž. Kos, S. Žumer, and M. Ravnik. Topology of Three-Dimensional Active Nematic Turbulence Confined to Droplets. *Physical Review X*, 9(3):031051, 2019.

[70] M. C. Cross and P. C. Hohenberg. Pattern formation outside of equilibrium. *Review of Modern Physics*, 65:851–1112, 1993.

[71] A. Daddi-moussa ider and A. M. Menzel. Dynamics of a simple model microswimmer in an anisotropic fluid: Implications for alignment behavior and active transport in a nematic liquid crystal. *Physical Review Fluids*, 3:094102, 2018.

[72] N. C. Darnton, L. Turner, S. Rojevsky, and H. C. Berg. Dynamics of bacterial swarming. *Biophysical Journal*, 98(10):2082–2090, 2010.

[73] R. Das, M. Kumar, and S. Mishra. Polar flock in the presence of random quenched rotators. *Physical Review E*, 98(6):060602, 2018.

[74] M. Dasgupta, B. Liu, H. C. Fu, M. Berhanu, K. S. Breuer, T. R. Powers, and A. Kudrolli. Speed of a swimming sheet in Newtonian and viscoelastic fluids. *Physical Review E*, 87(1):013015, 2013.

[75] C. Datt, G. Natale, S. G. Hatzikiriakos, and G. J. Elfring. An active particle in a complex fluid. *Journal of Fluid Mechanics*, 823:675–688, 2017.

[76] M. De Corato, X. Arqué, T. Patiño, M. Arroyo, S. Sánchez, and I. Pagonabarraga. Self-Propulsion of Active Colloids via Ion Release: Theory and Experiments. *Physical Review Letters*, 124(10):108001, 2020.

[77] P. G. de Gennes and J. Prost. *The Physics of Liquid Crystals*. Clarendon, Oxford, 1993.

[78] S. R. de Groot and P. Mazur. *Non-Equilibrium Thermodynamics*. Dover, New York, 1984.

[79] S. J. DeCamp, G. S. Redner, A. Baskaran, M. F. Hagan, and Z. Dogic. Orientational order of motile defects in active nematics. *Nature Materials*, 14(11):1110–1115, 2015.

[80] M. Deforet, V. Hakim, H. G. Yevick, G. Duclos, and P. Silberzan. Emergence of collective modes and tri-dimensional structures from epithelial confinement. *Nature Communications*, 5(May):3747, 2014.

[81] J. Deseigne, O. Dauchot, and H. Chaté. Collective Motion of Vibrated Polar Disks. *Physical Review Letters*, 105(9):098001, 2010.

[82] M. A. Despósito, C. Pallavicini, V. Levi, and L. Bruno. Active transport in complex media : Relationship between persistence and superdiffusion. *Physica A*, 390(6):1026–1032, 2011.

[83] K. Dietrich, N. Jaensson, I. Buttinoni, G. Volpe, and L. Isa. Microscale Marangoni Surfers. *Physical Review Letters*, 125(9):98001, 2020.

[84] W. R. DiLuzio, L. Turner, M. Mayer, P. Garstecki, D. B. Weibel, H. C. Berg, and G. M. Whitesides. Escherichia coli swim on the right-hand side. *Nature*, 435(7046):1271–1274, 2005.

[85] Z. Dogic, P. Sharma, and M. J. Zakhary. Hypercomplex liquid crystals. *Annual Review of Condensed Matter Physics*, 5(1):137–157, 2014.

[86] C. Dombrowski, L. Cisneros, S. Chatkaew, R. E. Goldstein, and J. O. Kessler. Self-concentration and large-scale coherence in bacterial dynamics. *Physical Review Letters*, 93(9):098103, 2004.

[87] A. Doostmohammadi, J. Ignés-Mullol, J. M. Yeomans, and F. Sagués. Active nematics. *Nature Communications*, 9:3246, 2018.

[88] A. Doostmohammadi, T. N. Shendruk, K. Thijssen, and J. M. Yeomans. Onset of meso-scale turbulence in active nematics. *Nature Communications*, 8:15326, 2017.

[89] K. Drescher, J. Dunkel, L. H. Cisneros, S. Ganguly, and R. E. Goldstein. Fluid dynamics and noise in bacterial cell-cell and cell-surface scattering. *Proceedings of the National Academy of Sciences*, 108(27):10940–10945, 2011.

[90] K. Drescher, R. E. Goldstein, N. Michel, M. Polin, and I. Tuval. Direct measurement of the flow field around swimming microorganisms. *Physical Review Letters*, 105(16):168101, 2010.

[91] R. Dreyfus, J. Baudry, M. L. Roper, M. Fermigier, H. A. Stone, and J. Bibette. Microscopic artificial swimmers. *Nature*, 437(7060):862–865, 2005.

[92] G. Duclos, R. Adkins, D. Banerjee, M. S. E. Peterson, M. Varghese, I. Kolvin, A. Baskaran, R. A. Pelcovits, T. R. Powers, A. Bascaran, F. Toschi, M. F. Hagan, S. J. Streitchan, V. Vitelli, D. A. Beller, and Z. Dogic. Topological structure and dynamics of three-dimensional active nematics. *Science*, 367(March):1120–1124, 2020.

[93] G. Duclos, C. Blanch-Mercader, V. Yashunsky, G. Salbreux, J. F. Joanny, J. Prost, and P. Silberzan. Spontaneous shear flow in confined cellular nematics. *Nature Physics*, 14(7):728–732, 2018.

[94] G. Duclos, C. Erlenkämper, J. F. Joanny, and P. Silberzan. Topological defects in confined populations of spindle-shaped cells. *Nature Physics*, 13(1):58–62, 2017.

[95] G. Duclos, S. Garcia, H. G. Yevick, and P. Silberzan. Perfect nematic order in confined monolayers of spindle-shaped cells. *Soft Matter*, 10(14):2346–2353, 2014.

[96] J. Dunkel, S. Heidenreich, K. Drescher, H. H. Wensink, M. Bär, and R. E. Goldstein. Fluid dynamics of bacterial turbulence. *Physical Review Letters*, 110(22):228102, 2013.

[97] B. J. Edwards, A. N. Beris, and M. Grmela. Generalized constitutive equation for polymeric liquid crystals Part 1. Model formulation using the Hamiltonian (poisson bracket) formulation. *Journal of Non-Newtonian Fluid Mechanics*, 35(1):51–72, 1990.

[98] J. Elgeti, R. G. Winkler, and G. Gompper. Physics of microswimmers - Single particle motion and collective behavior: A review. *Reports on Progress in Physics*, 78(5):056601, 2015.

[99] P. W. Ellis, D. J. G. Pearce, Y-W. Chang, G. Goldsztein, L. Giomi, and A. Fernandez-Nieves. Curvature-induced defect unbinding and dynamics in active nematic toroids. *Nature Physics*, 14(1):85–90, 2018.

[100] T. R. Elsdale. Parallel orientation of fibroblasts in vitro. *Experimental Cell Research*, 51(2-3):439–450, 1968.

[101] M. Enculescu and H. Stark. Active colloidal suspensions exhibit polar order under gravity. *Physical Review Letters*, 107(5):058301, 2011.

[102] J. Espinosa-Garcia, E. Lauga, and R. Zenit. Fluid elasticity increases the locomotion of flexible swimmers. *Physics of Fluids*, 25(3):031701, 2013.

[103] T. F. F. Farage, P. Krinninger, and J. M. Brader. Effective interactions in active Brownian suspensions. *Physical Review E*, 91(4):042310, 2015.

[104] C. Ferreiro-Córdova, J. Toner, H. Löwen, and H. H. Wensink. Long-time anomalous swimmer diffusion in smectic liquid crystals. *Physical Review E*, 97(6):062606, 2018.

[105] Y. Fily, A. Baskaran, and M. F. Hagan. Dynamics of self-propelled particles under strong confinement. *Soft Matter*, 10(30):5609–5617, 2014.

[106] Y. Fily and M. C. Marchetti. Athermal phase separation of self-propelled particles with no alignment. *Physical Review Letters*, 108(23):235702, 2012.

[107] E. Fodor, C. Nardini, M. E. Cates, J. Tailleur, P. Visco, and F. Van Wijland. How Far from Equilibrium Is Active Matter? *Physical Review Letters*, 117(3):038103, 2016.

[108] S. Fournier-Bidoz, A. C. Arsenault, I. Manners, and G. A. Ozin. Synthetic self-propelled nanorotors. *Chemical Communications*, 441:441–443, 2005.

[109] B. M. Friedrich and F. Jülicher. Chemotaxis of sperm cells. *Proceedings of the National Academy of Sciences*, 104(33):13256–13261, 2007.

[110] B. M. Friedrich, I. H. Riedel-Kruse, J. Howard, and F. Jülicher. High-precision tracking of sperm swimming fine structure provides strong test of resistive force theory. *Journal of Experimental Biology*, 213(8):1226–1234, 2010.

[111] U. Frisch. *Turbulence. The Legacy of A. N. Kolmogorov.* Cambridge University Press, 1995.

[112] P. D. Frymier, R. M. Ford, H. C. Berg, and P. T. Cummings. Three-dimensional tracking of motile bacteria near a solid planar surface. *Proceedings of the National Academy of Sciences*, 92(13):6195–6199, 1995.

[113] H. C. Fu, T. R. Powers, and C. W. Wolgemuth. Theory of Swimming Filaments in Viscoelastic Media ψ. *Physical Review Letters*, 99(25):258101, 2007.

[114] H. C. Fu, C. W. Wolgemuth, and T. R. Powers. Swimming speeds of filaments in nonlinearly viscoelastic fluids. *Physics of Fluids*, 21(3):033102, 2009.

[115] S. Fürthauer, M. Neef, S. W. Grill, K. Kruse, and F. Jülicher. The Taylor-Couette motor: Spontaneous flows of active polar fluids between two coaxial cylinders. *New Journal of Physics*, 14:023001, 2012.

[116] S. Fürthauer, M. Strempel, S. W. Grill, and F. Jülicher. Active chiral fluids. *European Physical Journal E*, 35(9):89, 2012.

[117] D. A. Gagnon, C. Dessi, J. P. Berezney, D. T.-N. Chen, R. Boros, Z. Dogic, and D. L. Blair. Shear-induced gelation of self-yielding active networks. *Physical Review Letters*, 125(17):178003, 2020.

[118] D. A. Gagnon, N. C. Keim, and P. E. Arratia. Undulatory swimming in shear-thinning fluids : experiments with Caenorhabditis elegans. *Journal of Fluid Mechanics*, 758:R3:1–11, 2014.

[119] D.A. Gagnon, X. N. Shen, and P. E. Arratia. Undulatory swimming in fluids with polymer networks. *Europhysics Letters*, 104:14004, 2013.

[120] P. Galajda, J. Keymer, P. Chaikin, and R. Austin. A wall of funnels concentrates swimming bacteria. *Journal of Bacteriology*, 189(23):8704–8707, 2007.

[121] S. Gangwal, O. J. Cayre, M. Z. Bazant, and O. D. Velev. Induced-charge electrophoresis of metallodielectric particles. *Physical Review Letters*, 100(5):058302, 2008.

[122] T. Gao, M. D. Betterton, A. S. Jhang, and M. J. Shelley. Analytical structure, dynamics, and coarse graining of a kinetic model of an active fluid. *Physical Review Fluids*, 2(9):093302, 2017.

[123] T. Gao and Z. Li. Self-Driven Droplet Powered by Active Nematics. *Physical Review Letters*, 119(10):108002, 2017.

[124] W. Gao, A. Pei, X. Feng, C. Hennessy, and J. Wang. Organized self-assembly of janus micromotors with hydrophobic hemispheres. *Journal of the American Chemical Society*, 135(3):998–1001, 2013.

[125] W. Gao and J. Wang. The environmental impact of micro/nanomachines: A review. *ACS Nano*, 8(4):3170–3180, 2014.

[126] D. Geyer, A. Morin, and D. Bartolo. Sounds and hydrodynamics of polar active fluids. *Nature Materials*, 17(9):789–793, 2018.

[127] A. Ghosh and P. Fischer. Controlled propulsion of artificial magnetic nanostructured propellers. *Nano Letters*, 9(6):2243–2245, 2009.

[128] P. K. Ghosh, V. R. Misko, F. Marchesoni, and F. Nori. Self-propelled janus particles in a ratchet: Numerical simulations. *Physical Review Letters*, 110(26):268301, 2013.

[129] F. Ginelli, F. Peruani, M. Bär, and H. Chaté. Large-scale collective properties of self-propelled rods. *Physical Review Letters*, 104(18):184502, 2010.

[130] F. Ginot, I. Theurkauff, F. Detcheverry, C. Ybert, and C. Cottin-Bizonne. Aggregation-fragmentation and individual dynamics of active clusters. *Nature Communications*, 9(1):696, 2018.

[131] F. Ginot, I. Theurkauff, D. Levis, C. Ybert, L. Bocquet, L. Berthier, and C. Cottin-Bizonne. Nonequilibrium equation of state in suspensions of active colloids. *Physical Review X*, 5(1):011004, 2015.

[132] L. Giomi. Geometry and topology of Turbulence in active nematics. *Physical Review X*, 5(3):031003, 2015.

[133] L. Giomi, M. J. Bowick, X. Ma, and M. C. Marchetti. Defect annihilation and proliferation in active Nematics. *Physical Review Letters*, 110(22):228101, 2013.

[134] L. Giomi, M. J. Bowick, X. Ma, and M. C. Marchetti. Erratum: Defect annihilation and proliferation in active nematics (Physical Review Letters (2013) 110 (228101)). *Physical Review Letters*, 111(20):209901, 2013.

[135] L. Giomi, M. J. Bowick, P. Mishra, R. Sknepnek, and M. C. Marchetti. Defect dynamics in active nematics. *Philosophical Transactions of the Royal Society A: Mathematical, Physical and Engineering Sciences*, 372(2029):20130365, 2014.

[136] L. Giomi and A. DeSimone. Spontaneous division and motility in active nematic droplets. *Physical Review Letters*, 112(14):147802, 2014.

[137] L. Giomi, Ž. Kos, M. Ravnik, and A. Sengupta. Cross-talk between topological defects in different fields revealed by nematic microfluidics. *Proceedings of the National Academy of Sciences of the United States of America*, 114(29):E5771–E5777, 2017.

[138] L. Giomi, L. Mahadevan, B. Chakraborty, and M. F. Hagan. Excitable patterns in active nematics. *Physical Review Letters*, 106(21):218101, 2011.

[139] L. Giomi, M. C. Marchetti, and T. B. Liverpool. Complex spontaneous flows and concentration banding in active polar films. *Physical Review Letters*, 101(19):198101, 2008.

[140] R. Golestanian, T. B. Liverpool, and A. Ajdari. Propulsion of a molecular machine by asymmetric distribution of reaction products. *Physical Review Letters*, 94(22):220801, 2005.

[141] R. Golestanian, T. B. Liverpool, and A. Ajdari. Designing phoretic micro- and nano-swimmers. *New Journal of Physics*, 9:126, 2007.

[142] J. R. Gomez-Solano, A. Blokhuis, and C. Bechinger. Dynamics of Self-Propelled Janus Particles in Viscoelastic Fluids. *Physical Review Letters*, 116(13):138301, 2016.

[143] G. Grégoire and H. Chaté. Onset of Collective and Cohesive Motion. *Physical Review Letters*, 92(2):025702, 2004.

[144] R. Großmann, P. Romanczuk, M. Bär, and L. Schimansky-Geier. Vortex arrays and mesoscale turbulence of self-propelled particles. *Physical Review Letters*, 113(25):258104, 2014.

[145] H. Gruler, U. Dewald, and M. Eberhardt. Nematic liquid crystals formed by living amoeboid cells. *European Physical Journal B*, 11(1):187–192, 1999.

[146] P. Guillamat, J. Hardoüin, B. M. Prat, J. Ignés-Mullol, and F. Sagués. Control of active turbulence through addressable soft interfaces. *Journal of Physics Condensed Matter*, 29(50):504003, 2017.

[147] P. Guillamat, J. Ignés-Mullol, and F. Sagués. Control of active liquid crystals with a magnetic field. *Proceedings of the National Academy of Sciences*, 113(20):5498–5502, 2016.

[148] P. Guillamat, J. Ignés-Mullol, and F. Sagués. Control of active nematics with passive liquid crystals. *Molecular Crystals and Liquid Crystals*, 646(1):226–234, 2017.

[149] P. Guillamat, J. Ignés-Mullol, and F. Sagués. Taming active turbulence with patterned soft interfaces. *Nature Communications*, 8(1):564, 2017.

[150] P. Guillamat, J. Ignés-Mullol, S. Shankar, M. C. Marchetti, and F. Sagués. Probing the shear viscosity of an active nematic film. *Physical Review E*, 94(6):060602, 2016.

[151] P. Guillamat, Ž. Kos, J. Hardoüin, J. Ignés-Mullol, M. Ravnik, and F. Sagués. Active nematic emulsions. *Science Advances*, 4:eaao1470, 2018.

[152] Kraichnan R. H. and Montgomery D. Two-dimensional turbulence. *Reports on Progress in Physics*, 43:547–619, 1980.

[153] M. Han, J. Yan, S. Granick, and E. Luijten. Effective temperature concept evaluated in an active colloid mixture. *Proceedings of the National Academy of Sciences*, 114(29):7513–7518, 2017.

[154] J. Happel and H. Brenner. *Low Reynolds Number Hydrodynamics: With Special Applications to Particulate Media*. Kluwer Academic Publishers, 2nd ed., 2012.

[155] J. Hardoüin, P. Guillamat, F. Sagués, and J. Ignés-Mullol. Dynamics of Ring Disclinations Driven by Active Nematic Shells. *Frontiers in Physics*, 7(October):165, 2019.

[156] J. Hardoüin, R. Hughes, A. Doostmohammadi, J. Laurent, T. Lopez-Leon, J. M. Yeomans, J. Ignés-Mullol, and F. Sagués. Reconfigurable flows and defect landscape of confined active nematics. *Communications Physics*, 2(1):1–9, 2019.

[157] J. Hardoüin, J. Laurent, T. Lopez-Leon, J. Ignés-Mullol, and F. Sagués. Active microfluidic transport in two-dimensional handlebodies. *Soft Matter*, 2020.

[158] E. J. Hemingway, P. Mishra, M. C. Marchetti, and S. M. Fielding. Correlation lengths in hydrodynamic models of active nematics. *Soft Matter*, 12(38):7943–7952, 2016.

[159] G. Henkin, S. J. DeCamp, D. T. N. Chen, T. Sanchez, and Z. Dogic. Tunable dynamics of microtubule-based active isotropic gels. *Philosophical Transactions of the Royal Society A: Mathematical, Physical and Engineering Sciences*, 372:20140142, 2014.

[160] S. Herminghaus, C. C. Maass, C. Krüger, S. Thutupalli, L. Goehring, and C. Bahr. Interfacial mechanisms in active emulsions. *Soft Matter*, 10:7008–7022, 2014.

[161] S. Hernàndez-Navarro, P. Tierno, J. A. Farrera, J. Ignés-Mullol, and F. Sagués. Reconfigurable swarms of nematic colloids controlled by photoactivated surface patterns. *Angewandte Chemie International Edicion England*, 53(40):10696–700, 2014.

[162] S. Hernàndez-Navarro, P. Tierno, J. Ignés-Mullol, and F. Sagués. Ac electrophoresis of microdroplets in anisotropic liquids: transport, assembling and reaction. *Soft Matter*, 9(33):7999–8004, 2013.

[163] S. Hernàndez-Navarro, P. Tierno, J. Ignés-Mullol, and F. Sagués. Liquid-crystal enabled electrophoresis: Scenarios for driving and reconfigurable assembling of colloids. *European Physics Journal Special Topics*, 224:1263–1273, 2015.

[164] Y. Hiratsuka, M. Miyata, T. Tada, and T. Q. P. Uyeda. A microrotary motor powered by bacteria. *Proceedings of the National Academy of Sciences*, 103(37):13618–13623, 2006.

[165] Y. Hong, N. M. K. Blackman, N. D. Kopp, A. Sen, and D. Velegol. Chemotaxis of nonbiological colloidal rods. *Physical Review Letters*, 99(17):178103, 2007.

[166] J. R. Howse, R. A. L. Jones, A. J. Ryan, T. Gough, R. Vafabakhsh, and R Golestanian. Self-Motile Colloidal Particles: From Directed Propulsion to Random Walk. *Physical Review Letters*, 99(4):048102, 2007.

[167] L. Huber, R. Suzuki, E. Frey, and A. R. Bausch. Emergence of coexisting ordered states in active matter systems. *Science*, 258(July):255–258, 2018.

[168] R. J. Hunter. *Foundations of Colloid Science*. Oxford University Press, 2000.

[169] M. Ibele, T. E. Mallouk, and A. Sen. Schooling behavior of light-powered autonomous micromotors in water. *Angewandte Chemie - International Edition*, 48(18):3308–3312, 2009.

[170] Y. Ibrahim, R. Golestanian, and T. B. Liverpool. Multiple phoretic mechanisms in the self-propulsion of a Pt-insulator Janus swimmer. *Journal of Fluid Mechanics*, 828:318–352, 2017.

[171] J. Ignés-Mullol and F. Sagués. Active, self-motile, and driven emulsions. *Current Opinion in Colloid and Interface Science*, 49:16–26, 2020.

[172] P. Illien, R. Golestanian, and A. Sen. "Fuelled" motion: Phoretic motility and collective behaviour of active colloids. *Chemical Society Reviews*, 46(18):5508–5518, 2017.

[173] D. Inoue, B. Mahmot, A. M. Rashedul Kabir, T. I. Farhana, K. Tokuraku, K. Sada, A. Konagaya, and A. Kakugo. Depletion force induced collective motion of microtubules driven by kinesin. *Nanoscale*, 7(43), 2015.

[174] T. Ishikawa and T. J. Pedley. Coherent structures in monolayers of swimming particles. *Physical Review Letters*, 100(8):088103, 2008.

[175] T. Ishikawa, M. P. Simmonds, and T. J. Pedley. Hydrodynamic interaction of two swimming model micro-organisms. *Journal of Fluid Mechanics*, 568:119–160, 2006.

[176] T. Ishikawa, N. Yoshida, H. Ueno, M. Wiedeman, Y. Imai, and T. Yamaguchi. Energy transport in a concentrated suspension of bacteria. *Physical Review Letters*, 107(2):028102, 2011.

[177] R. F. Ismagilov, A. Schwartz, N. Bowden, and G. M. Whitesides. Autonomous movement and self-assembly. *Angewandte Chemie - International Edition*, 41(4):652–654, 2002.

[178] Z. Izri, M. N. Van Der Linden, S. Michelin, and O. Dauchot. Self-propulsion of pure water droplets by spontaneous Marangoni-stress-driven motion. *Physical Review Letters*, 113(24):248302, 2014.

[179] T. Jakuszeit, O. A. Croze, and S. Bell. Diffusion of active particles in a complex environment : Role of surface scattering. *Physical Review E*, 99(1):012610, 2019.

[180] M. James, W. J.T. Bos, and M. Wilczek. Turbulence and turbulent pattern formation in a minimal model for active fluids. *Physical Review Fluids*, 3(6):061101, 2018.

[181] H. R. Jiang, N. Yoshinaga, and M. Sano. Active motion of a Janus particle by self-thermophoresis in a defocused laser beam. *Physical Review Letters*, 105(26):268302, 2010.

[182] J. F. Joanny, F. Jülicher, K. Kruse, and J. Prost. Hydrodynamic theory for multi-component active polar gels. *New Journal of Physics*, 9:422, 2007.

[183] J. F. Joanny and S. Ramaswamy. A drop of active matter. *Journal of Fluid Mechanics*, 705:46–57, 2012.

[184] F. Jülicher, K. Kruse, J. Prost, and J. F. Joanny. Active behavior of the Cytoskeleton. *Physics Reports*, 449:3–28, 2007.

[185] G. Junot, G. Briand, R. Ledesma-Alonso, and O. Dauchot. Active versus Passive Hard Disks against a Membrane: Mechanical Pressure and Instability. *Physical Review Letters*, 119(2):028002, 2017.

[186] A. Kaiser, A. Peshkov, A. Sokolov, B. Ten Hagen, H. Löwen, and I. S. Aranson. Transport powered by bacterial turbulence. *Physical Review Letters*, 112(15):158101, 2014.

[187] A. Kaiser, A. Snezhko, and I. S. Aranson. Flocking ferromagnetic colloids. *Science Advances*, 3:e1601469, 2017.

[188] A. Kaiser, H. H. Wensink, and H. Löwen. How to capture active particles. *Physical Review Letters*, 108(26):268307, 2012.

[189] R. D. Kamien. The geometry of soft materials: a primer. *Reviews of Modern Physics*, 74(4):953–971, 2002.

[190] V. Kantsler, J. Dunkel, M. Polin, and R. E. Goldstein. Ciliary contact interactions dominate surface scattering of swimming eukaryotes. *Proceedings of the National Academy of Sciences*, 110(4):1187–1192, 2013.

[191] H. Karani, G. E. Pradillo, and P. M. Vlahovska. Tuning the Random Walk of Active Colloids: From Individual Run-And-Tumble to Dynamic Clustering. *Physical Review Letters*, 123(20):208002, 2019.

[192] K. Kawaguchi, R. Kageyama, and M. Sano. Topological defects control collective dynamics in neural progenitor cell cultures. *Nature*, 545(7654):327–331, 2017.

[193] F. Keber, E. Loiseau, T. Sanchez, S. DeCamp, L. Giomi, M. Bowick, M. C. Marchetti, Z. Dogic, and A. Bausch. Topology and Dynamics of Active Nematic Vesicles. *Science*, 345(6201):1135–1139, 2015.

[194] N. C. Keim, M. Garcia, and P. E. Arratia. Fluid elasticity can enable propulsion at low Reynolds number. *Physics of Fluids*, 24(8):081703, 2012.

[195] R. Kemkemer, D. Kling, D. Kaufmann, and H. Gruler. Elastic properties of nematoid arrangements formed by amoeboid cells. *European Physical Journal E*, 1:215–225, 2000.

[196] D. Khoromskaia and G. P. Alexander. Motility of active fluid drops on surfaces. *Physical Review E*, 92(6):062311, 2015.

[197] M. Kleman and O. D. Lavrentovich. *Soft Matter Physics: An Introduction*. Springer, New York, 2003.

[198] C.-M. Koch and M. Wilczek. The role of advective inertia in active nematic turbulence. *Physical Review Letters*, 127(26):268005, 2021.

[199] G. H. Koenderink, Z. Dogic, F. Nakamura, P. M. Bendix, F. C. Mackintosh, J. H. Hartwig, T. P. Stossel, and D. A. Weitz. An active biopolymer network controlled by molecular motors. *Proceedings of the National Academy of Sciences*, 106(36):15192, 2009.

[200] S. Köhler, V. Schaller, and A. R. Bausch. Structure formation in active networks. *Nature Materials*, 10(6):462–468, 2011.

[201] G. Kokot, S. Das, R. G. Winkler, G. Gompper, I. S. Aranson, and A. Snezhko. Active turbulence in a gas of self-assembled spinners. *Proceedings of the National Academy of Sciences*, 114(49):12870–12875, 2017.

[202] G. Kokot and A. Snezhko. Manipulation of emergent vortices in swarms of magnetic rollers. *Nature Communications*, 9:2344, 2018.

[203] P. Kraikivski, R. Lipowsky, and J. Kierfeld. Enhanced ordering of interacting filaments by molecular motors. *Physical Review Letters*, 96(25):258103, 2006.

[204] S. J. Kron and J. A. Spudich. Fluorescent actin filaments move on myosin fixed to a glass surface. *Proceedings of the National Academy of Sciences*, 83(17):6272–6276, 1986.

[205] C. Krüger, G. Klös, C. Bahr, and C. C. Maass. Curling Liquid Crystal Microswimmers: A Cascade of Spontaneous Symmetry Breaking. *Physical Review Letters*, 117(4):048003, 2016.

[206] K. Kruse, J. F., F. Jülicher, J. Prost, and K. Sekimoto. Generic theory of active polar gels: A paradigm for cytoskeletal dynamics. *European Physical Journal E*, 16(1):5–16, 2005.

[207] K. Kruse, J. F. Joanny, F. Jülicher, J. Prost, and K. Sekimoto. Asters, Vortices, and Rotating Spirals in Active Gels of Polar Filaments. *Physical Review Letters*, 92(7):078101, 2004.

[208] K. Kruse, J. F. Joanny, F. Jülicher, J. Prost, and K. Sekimoto. Erratum: Asters, vortices, and rotating spirals in active gels of polar filaments (Physical Review Letters (2004) 92 (078101)). *Physical Review Letters*, 93(9):099902, 2004.

[209] A. Kudrolli, G. Lumay, D. Volfson, and L. S. Tsimring. Swarming and Swirling in Self-Propelled Polar Granular Rods. *Physical Review Letters*, 100(5):058001, 2008.

[210] N. Kumar, H. Soni, S. Ramaswamy, and A. K. Sood. Flocking at a distance in active granular matter. *Nature Communications*, 5:4688, 2014.

[211] N. Kumar, R. Zhang, J. J. De Pablo, and M. L. Gardel. Tunable structure and dynamics of active liquid crystals. *Science Advances*, 4:eaat7779, 2018.

[212] F. Kümmel, B. ten Hagen, R. Wittkowski, I. Buttinoni, R. Eichhorn, G. Volpe, H. Löwen, and C. Bechinger. Circular motion of asymmetric self-propelling particles. *Physical Review Letters*, 110(19):198302, 2013.

[213] C. P. Lapointe, T. G. Mason, and I. I. Smalyukh. Shape-controlled colloidal interactions in nematic liquid crystals. *Science*, 326(5956):1083–1086, 2009.

[214] E. Lauga. Propulsion in a viscoelastic fluid. *Physics of Fluids*, 19(8):083104, 2007.

[215] E. Lauga and T. R. Powers. The hydrodynamics of swimming microorganisms. *Reports on Progress in Physics*, 72(9):096601, 2009.

[216] F. A. Lavergne, H. Wendehenne, T. Bäuerle, and C. Bechinger. Group formation and cohesion of active particles with visual perception – dependent motility. *Science*, 364:70–74, 2019.

[217] O. D. Lavrentovich. Transport of particles in liquid crystals. *Soft Matter*, 19:1264–1283, 2014.

[218] O. D. Lavrentovich. Active colloids in liquid crystals. *Current Opinion in Colloid & Interface Science*, 21:97–109, 2016.

[219] O. D. Lavrentovich, I. Lazo, and O. P. Pishnyak. Nonlinear electrophoresis of dielectric and metal spheres in a nematic liquid crystal. *Nature*, 467(7318):947–50, 2010.

[220] I. Lazo, C. Peng, J. Xiang, S. V. Shiyanovskii, and O. D. Lavrentovich. Liquid crystal-enabled electro-osmosis through spatial charge separation in distorted regions as a novel mechanism of electrokinetics. *Nat. Commun.*, 5:5033, 2014.

[221] R. Ledesma-Aguilar and J. M. Yeomans. Enhanced Motility of a Microswimmer in Rigid and Elastic Confinement. *Physical Review Letters*, 111(13):138101, 2013.

[222] H. Y. Lee, M. Yahyanejad, and M. Kardar. Symmetry considerations and development of pinwheels in visual maps. *Proceedings of the National Academy of Sciences of the United States of America*, 100(26):16036–16040, 2003.

[223] L. M. Lemma, S. J. DeCamp, Z. You, L. Giomi, and Z. Dogic. Statistical properties of autonomous flows in 2D active nematics. *Soft Matter*, 15(15):3264–3272, 2019.

[224] L. M. Lemma, M. M. Norton, A. M. Tayar, S. J. DeCamp, S. A. Aghvami, S. Fraden, M. F. Hagan, and Z Dogic. Multiscale Microtubule Dynamics in Active Nematics. *Physical Review Letters*, 127(14):148001, 2021.

[225] G. Li, E. Lauga, and A. M. Ardekani. Journal of Non-Newtonian Fluid Mechanics Microswimming in viscoelastic fluids. *Journal of Non-Newtonian Fluid Mechanics*, 297(April):104655, 2021.

[226] H. Li, X. Shi, M. Huang, X. Chen, M. Xiao, C. Liu, H. Chaté, and H. P. Zhang. Data-driven quantitative modeling of bacterial active nematics. *Proceedings of the National Academy of Sciences*, 116(3):777–785, 2019.

[227] B. Liebchen and H. Löwen. Synthetic Chemotaxis and Collective Behavior in Active Matter. *Accounts of Chemical Research*, 51(12):2982–2990, 2018.

[228] B. Liebchen, D. Marenduzzo, I. Pagonabarraga, and M. E. Cates. Clustering and Pattern Formation in Chemorepulsive Active Colloids. *Physical Review Letters*, 115(25):258301, 2015.

[229] M. J. Lighthill. On the squirming motion of nearly spherical deformable bodies through liquids at very small reynolds numbers. *Communications on Pure and Applied Mathematics*, 5(2):109–118, 1952.

[230] M. Linkmann, G. Boffetta, M. C. Marchetti, and B. Eckhardt. Phase Transition to Large Scale Coherent Structures in Two-Dimensional Active Matter Turbulence. *Physical Review Letters*, 122(21):214503, 2019.

[231] J. S. Lintuvuori, A. Würger, and K. Stratford. Hydrodynamics Defines the Stable Swimming Direction of Spherical Squirmers in a Nematic Liquid Crystal. *Physical Review Letters*, 119(6):068001, 2017.

[232] T. W. Lion and R. J. Allen. Osmosis with active solutes. *Europhysics Letters*, 106:34003, 2014.

[233] B. Liu, K. S. Breuer, and T. R. Powers. Propulsion by a helical flagellum in a capillary tube. *Physics of Fluids*, 26(January):011701, 2014.

[234] B. Liu, T. R. Powers, and K. S. Breuer. Force-free swimming of a model helical flagellum in viscoelastic fluids. *Proceedings of the National Academy of Sciences*, 108(49):19516, 2011.

[235] C. Liu, C. Zhou, W. Wang, and H. P. Zhang. Bimetallic Microswimmers Speed Up in Confining Channels. *Physical Review Letters*, 117(19):198001, 2016.

[236] L. Liu, E. Tüzel, and J. L. Ross. Loop formation of microtubules during gliding at high density. *Journal of Physics Condensed Matter*, 23(37), 2011.

[237] T. B. Liverpool and M. C. Marchetti. Rheology of active filament solutions. *Physical Review Letters*, 97(26):268101, 2006.

[238] D. Loi, S. Mossa, and L. F. Cugliandolo. Effective temperature of active matter. *Physical Review E*, 77(5):051111, 2008.

[239] T. Lopez-Leon, V. Koning, K. B.S. Devaiah, V. Vitelli, and A. Fernandez-Nieves. Frustrated nematic order in spherical geometries. *Nature Physics*, 7(5):391–394, 2011.

[240] D. K. Lubensky and R. E. Goldstein. Hydrodynamics of monolayer domains at the air-water interface. *Physics of Fluids*, 8(4):843–854, 1996.

[241] T. C. Lubensky and J. Prost. Orientational order and vesicle shape. *Journal de Physique II*, 2:371–382, 1992.

[242] E. Lushi, H. Wioland, and R. E. Goldstein. Fluid flows created by swimming bacteria drive self-organization in confined suspensions. *Proceedings of the National Academy of Sciences*, 111(27):9733–9738, 2014.

[243] C. C. Maass, C. Krüger, S. Herminghaus, and C. Bahr. Swimming Droplets. *Annual Review of Condensed Matter Physics*, 7(1):171–193, 2016.

[244] S. A. Mallory, A. Šarić, C. Valeriani, and A. Cacciuto. Anomalous thermomechanical properties of a self-propelled colloidal fluid. *Physical Review E*, 89(5):052303, 2014.

[245] J. Männik, R. Driessen, P. Galajda, J. E. Keymer, and C. Dekker. Bacterial growth and motility in sub-micron constrictions. *Proceedings of the National Academy of Sciences*, 106:14861, 2009.

[246] M. C. Marchetti, Y. Fily, S. Henkes, A. Patch, and D. Yllanes. Current Opinion in Colloid & Interface Science Minimal model of active colloids highlights the role of mechanical interactions in controlling the emergent behavior of active matter. *Current Opinion in Colloid & Interface Science*, 21:34–43, 2016.

[247] M. C. Marchetti, J. F. Joanny, S. Ramaswamy, T. B. Liverpool, J. Prost, M. Rao, and R. Aditi Simha. Hydrodynamics of soft active matter. *Review of Modern Physics*, 85(3):1143–1189, 2013.

[248] D. Marenduzzo, E. Orlandini, M. E. Cates, and J. M. Yeomans. Steady-state hydrodynamic instabilities of active liquid crystals: Hybrid lattice Boltzmann simulations. *Physical Review E*, 76(3):031921, 2007.

[249] D. Marenduzzo, E. Orlandini, and J. M. Yeomans. Hydrodynamics and rheology of active liquid crystals: A numerical investigation. *Physical Review Letters*, 98(11):118102, 2007.

[250] Y. Maroudas-Sacks, L. Garion, L. Shani-Zerbib, A. Livshits, E. Braun, and K. Keren. Topological defects in the nematic order of actin fibers as organization centers of Hydra morphogenesis. *Nature Physics*, 17:251–259, 2021.

[251] V. A. Martinez, J. Schwarz-Linek, M. Reufer, L. G. Wilson, and A. N. Morozov. Flagellated bacterial motility in polymer solutions. *Proceedings of the National Academy of Sciences*, 111(50):17771–17776, 2014.

[252] B. Martínez-Prat, R. Alert, F. Meng, J. Ignés-Mullol, J. F. Joanny, J. Casademunt, R. Golestanian, and F. Sagués. Scaling regimes of active turbulence with external dissipation. *Physical Review X*, 11:031065, 2021.

[253] B. Martínez-Prat, J. Ignés-Mullol, J. Casademunt, and F. Sagués. Selection mechanism at the onset of active turbulence. *Nature Physics*, 15(4):362–366, 2019.

[254] Y. Mei, A. A. Solovev, S. Sánchez, and O. G. Schmidt. Rolled-up nanotech on polymers: from basic perception to self-propelled catalytic microengines. *Chemical Society Reviews*, 40(5):2109–2119, 2011.

[255] N. H. Mendelson, A. Bourque, K. Wilkening, K. R Anderson, and J. C Watkins. Organized Cell Swimming Motions in *Bacillus subtilis* Colonies: Patterns of short lived whirls and jets. *Journal of Bacteriology*, 181(2):600–609, 1999.

[256] L. Metselaar, J. M. Yeomans, and A. Doostmohammadi. Topology and Morphology of Self-Deforming Active Shells. *Physical Review Letters*, 123(20):208001, 2019.

[257] S. Michelin and E. Lauga. Phoretic self-propulsion at finite Péclet numbers. *Journal of Fluid Mechanics*, 747:572–604, 2014.

[258] C. J. Miles, A. A. Evans, M. J. Shelley, and S. E. Spagnolie. Active matter invasion of a viscous fluid: Unstable sheets and a no-flow theorem. *Physical Review Letters*, 122(9):98002, 2019.

[259] S. Mishra and S. Ramaswamy. Active nematics are intrinsically phase separated. *Physical Review Letters*, 97(9):090602, 2006.

[260] D. Mizuno, C. Tardin, C. F. Schmidt, and F. C. MacKintosh. Nonequilibrium Mechanics of Active Cytoskeletal Networks. *Science*, 315(January):370–374, 2007.

[261] A. Morin and D. Bartolo. Flowing Active Liquids in a Pipe : Hysteretic Response of Polar Flocks to External Fields. *Physical Review X*, 8(2):021037, 2018.

[262] A. Morin, D. L. Cardozo, V. Chikkadi, and D. Bartolo. Diffusion, subdiffusion, and localization of active colloids in random post lattices. *Physical Review E*, 96(4):042611, 2017.

[263] A. Morin, N. Desreumaux, J. B. Caussin, and D. Bartolo. Distortion and destruction of colloidal flocks in disordered environments. *Nature Physics*, 13(1):63–67, 2017.

[264] R. Mueller, J. M. Yeomans, and A. Doostmohammadi. Emergence of Active Nematic Behavior in Monolayers of Isotropic Cells. *Physical Review Letters*, 122(4):48004, 2019.

[265] A. Müller-Deku, J. C. M. Meiring, K. Loy, Y. Kraus, C. Heise, R. Bingham, K. I. Jansen, X. Qu, F. Bartolini, L. C. Kapitein, A. Akhmanova, J. Ahlfeld, D. Trauner, and O. Thorn-Seshold. Photoswitchable paclitaxel-based microtubule stabilisers allow optical control over the microtubule cytoskeleton. *Nature communications*, 11(1):4640, 2020.

[266] J. L. Münch, D. Alizadehrad, S. B. Babu, and H. Stark. Taylor line swimming in microchannels and cubic lattices of obstacles. *Soft Matter*, 12:7350–7363, 2016.

[267] I. Mušević. *Liquid Crystal Colloids*. Springer, 2017.

[268] I. Mušević, M. Škarabot, U. Tkalec, M. Ravnik, and S. Žumer. Two-Dimensional Nematic Colloidal Crystals Self-Assembled by Topological Defects. *Science*, 313(5789):954–959, 2006.

[269] A. Najafi and R. Golestanian. Simple swimmer at low Reynolds number: Three linked spheres. *Physical Review E*, 69(6):062901, 2004.

[270] S. Nakata, Y. Iguchi, S. Ose, M. Kuboyama, T. Ishii, and K. Yoshikawa. Self-Rotation of a Camphor Scraping on Water: New Insight into the Old Problem. *Langmuir*, 13(16):4454–4458, 1997.

[271] V. Narayan, S. Ramaswamy, and N. Menon. Long-Lived Giant Number Fluctuations. *Science*, 317(July):105–109, 2007.

[272] N Narinder, J. R. Gomez-Solano, and C. Bechinger. Active particles in geometrically confined viscoelastic fluids. *New Journal of Physics*, 21:093058, 2019.

[273] F. J. Nédélec, T. Surrey, A. C. Maggs, and S. Leibler. Self-organization of microtubules and motors. *Nature*, 389(6648):305–308, 1997.

[274] D. Needleman and Z. Dogic. Active matter at the interface between materials science and cell biology. *Nature Reviews Materials*, 2:17048, 2017.

[275] B. J. Nelson, I. K. Kaliakatsos, and J. J. Abbott. Microrobots for Minimally Invasive Medicine. *Annual Review of Biomedical Engineering*, 12(1):55–85, 2010.

[276] D. R. Nelson. Toward a Tetravalent Chemistry of Colloids. *Nano Letters*, 2(10):1125–1129, 2002.

[277] S. Ngo, A. Peshkov, I. S. Aranson, E. Bertin, F. Ginelli, and H. Chaté. Large-scale chaos and fluctuations in active nematics. *Physical Review Letters*, 113(3):038302, 2014.

[278] N. H. P. Nguyen, D. Klotsa, M. Engel, and S. C. Glotzer. Emergent collective phenomena in a mixture of hard shapes through active rotation. *Physical Review Letters*, 112(7):075701, 2014.

[279] R. Ni, M. A. Cohen Stuart, and P. G. Bolhuis. Tunable long range forces mediated by self-propelled colloidal hard spheres. *Physical Review Letters*, 114(1):018302, 2015.

[280] N. Nikola, A. P. Solon, Y. Kafri, M. Kardar, J. Tailleur, and R. Voituriez. Active Particles with Soft and Curved Walls: Equation of State, Ratchets, and Instabilities. *Physical Review Letters*, 117(9):098001, 2016.

[281] D. Nishiguchi, I. S. Aranson, A. Snezhko, and A. Sokolov. Engineering bacterial vortex lattice via direct laser lithography. *Nature Communications*, 9(1):4486, 2018.

[282] D. Nishiguchi and M. Sano. Mesoscopic turbulence and local order in Janus particles self-propelling under an ac electric field. *Physical Review E*, 92(5):052309, 2015.

[283] M. M. Norton, A. Baskaran, A. Opathalage, B. Langeslay, S. Fraden, A. Baskaran, and M. F. Hagan. Insensitivity of active nematic liquid crystal dynamics to topological constraints. *Physical Review E*, 97(1):012702, 2018.

[284] M. M. Norton, P. Grover, M. F. Hagan, and S. Fraden. Optimal Control of Active Nematics. *Physical Review Letters*, 125(17):178005, 2020.

[285] A. Opathalage, M. M. Norton, M. P. N. Juniper, B. Langeslay, S. A. Aghvami, S. Fraden, and Z. Dogic. Self-organized dynamics and the transition to turbulence of confined active nematics. *Proceedings of the National Academy of Sciences*, 116(11):4788–4797, 2019.

[286] T. Ostapenko, F. J. Schwarzendahl, T. J. Böddeker, C. T. Kreis, J. Cammann, M. G. Mazza, and O. Bäumchen. Curvature-Guided Motility of Microalgae in Geometric Confinement. *Physical Review Letters*, 120(6):68002, 2018.

[287] P. Oswald and P. Pieranski. *Nematic and Cholesteric Liquid Crystals: Concepts and Physical Properties Illustrated by Experiments*. Taylor and Francis, Boca Raton, 2005.

[288] G. A. Ozin, I. Manners, S. Fournier-Bidoz, and A. Arsenault. Dream nanomachines. *Advanced Materials*, 17(24):3011–3018, 2005.

[289] J. M. Pagès, J. Ignés-Mullol, and F. Sagués. Anomalous diffusion of motile colloids dispersed in liquid crystals. *Physical Review Letters*, 122(19):198001, 2019.

[290] J. M. Pagès, A. V. Straube, P. Tierno, J. Ignés-Mullol, and F. Sagués. Inhomogeneous assembly of driven nematic colloids. *Soft Matter*, 15:312–320, 2019.

[291] J. Palacci, C. Cottin-Bizonne, C. Ybert, and L. Bocquet. Sedimentation and effective temperature of active colloidal suspensions. *Physical Review Letters*, 105(8):088304, 2010.

[292] J. Palacci, S. Sacanna, A. P. Steinberg, D. J Pine, and P. M. Chaikin. Living crystals of light-activated colloidal surfers. *Science*, 339(6122):936–939, 2013.

[293] S. Paladugu, C. Conklin, J. Viñals, and O. D Lavrentovich. Nonlinear Electrophoresis of Colloids Controlled by Anisotropic Conductivity and Permittivity of Liquid-Crystalline Electrolyte. *Physical Review Applied*, 7:034033, 2017.

[294] S. Palagi, A. G. Mark, S. Y. Reigh, K. Melde, T. Qiu, H. Zeng, C. Parmeggiani, D. Martella, A. Sanchez-Castillo, N. Kapernaum, F. Giesselmann, D. S. Wiersma, E. Lauga, and P. Fischer. Structured light enables biomimetic swimming and versatile locomotion of photoresponsive soft microrobots. *Nature Materials*, 15(6):647–653, 2016.

[295] M. Paoluzzi, R. Di Leonardo, and L. Angelani. Self-Sustained Density Oscillations of Swimming Bacteria Confined in Microchambers. *Physical Review Letters*, 115(18):188303, 2015.

[296] J-S. Park, D. Kim, J. H. Shin, and D. A Weitz. Efficient nematode swimming in a shear thinning colloidal suspension. *Soft Matter*, 12:1892–1897, 2016.

[297] D. Patra, S. Sengupta, W. Duan, H. Zhang, R. Pavlick, and A. Sen. Intelligent, self-powered, drug delivery systems. *Nanoscale*, 5(4):1273–1283, 2013.

[298] A. E. Patteson, A. Gopinath, and P. E. Arratia. Active colloids in complex fluids. *Current Opinion in Colloid and Interface Science*, 21:86–96, 2016.

[299] W. F. Paxton, K. C. Kistler, C. C. Olmeda, A. Sen, S. K. St. Angelo, Y. Cao, T. E. Mallouk, P. E. Lammert, and V. H. Crespi. Catalytic nanomotors: Autonomous movement of striped nanorods. *Journal of the American Chemical Society*, 126(41):13424–13431, 2004.

[300] W. F. Paxton, S. Sundararajan, T. E. Mallouk, and A. Sen. Minireviews Microscopic Machines Chemical Locomotion. *Angewandte Chemie - International Edition*, 45:5420–5429, 2006.

[301] D. J. G. Pearce, P. W. Ellis, A. Fernandez-Nieves, and L. Giomi. Geometrical Control of Active Turbulence in Curved Topographies. *Physical Review Letters*, 122(16):168002, 2019.

[302] D. J. G. Pearce, J. Nambisan, P. W. Ellis, A. Fernandez-Nieves, and L. Giomi. Orientational Correlations in Active and Passive Nematic Defects. *Physical Review Letters*, 127(19):197801, 2021.

[303] C. Peng, Y. Guo, C. Conklin, J. Viñals, S. V. Shiyanovskii, Q. Wei, and O. D. Lavrentovich. Liquid crystals with patterned molecular orientation as an electrolytic active medium. *Physical Review E*, 92(5):052502, 2015.

[304] Y. Peng, Z. Liu, and X. Cheng. Imaging the emergence of bacterial turbulence: Phase diagram and transition kinetics. *Science Advances*, 7(17), 2021.

[305] F. Peruani, A. Deutsch, and M. Bär. Nonequilibrium clustering of self-propelled rods. *Physical Review E*, 74(3):030904, 2006.

[306] F. Peruani, J. Starruß, V. Jakovljevic, L. Søgaard-Andersen, A. Deutsch, and M. Bär. Collective motion and nonequilibrium cluster formation in colonies of gliding bacteria. *Physical Review Letters*, 108(9):098102, 2012.

[307] P. Pietzonka, E. Fodor, C. Lohrmann, M. E. Cates, and U. Seifert. Autonomous Engines Driven by Active Matter: Energetics and Design Principles. *Physical Review X*, 9(4):41032, 2019.

[308] E. Pinçe, S. K. P. Velu, A. Callegari, P. Elahi, S. Gigan, G. Volpe, and G. Volpe. Disorder-mediated crowd control in an active matter system ˇ. *Nature Communications*, 7:10907, 2016.

[309] L. M. Pismen. Dynamics of defects in an active nematic layer. *Physical Review E*, 88(5):050502, 2013.

[310] L. M. Pismen and F. Sagués. Viscous dissipation and dynamics of defects in an active nematic interface. *European Physical Journal E*, 40(10):1–8, 2017.

[311] O. Pohl and H. Stark. Dynamic clustering and chemotactic collapse of self-phoretic active particles. *Physical Review Letters*, 112(23):238303, 2014.

[312] M. Polin, I. Tuval, K. Drescher, J. P. Gollub, and R. E. Goldstein. Chlamydomonas Swims with Two Gears in a Eukaryotic Version of Run-and-Tumble Locomotion. *Science*, 325:487–490, 2009.

[313] E. M. Purcell. Life at low Reynolds number. *American Journal of Physics*, 45:3–11, 1977.

[314] K. Qi, E. Westphal, G. Gompper, and R. G. Winkler. Enhanced Rotational Motion of Spherical Squirmer in Polymer Solutions. *Physical Review Letters*, 124(6):68001, 2020.

[315] B. Qin, A. Gopinath, J. Yang, J. P. Gollub, and P. E. Arratia. Flagellar Kinematics and Swimming of Algal Cells in Viscoelastic Fluids. *Scientific Reports*, 5:9190, 2015.

[316] T. Qiu, T-Ch. Lee, A. G Mark, K. I. Morozov, R. Mu, O. Mierka, S. Turek, A. M. Leshansky, and P. Fischer. Swimming by reciprocal motion at low Reynolds number. *Nature Communications*, 5:5119, 2014.

[317] D. A. Quint and A. Gopinathan. Topologically induced swarming phase transition on a 2D percolated lattice. *Physical Biology*, 12:046008, 2015.

[318] Di Leonardo R., L. Angelani, D. Dell'Arciprete, G. Ruocco, V. Iebba, S. Schippa, M. P. Conte, F. Mecarini, F. De Angelis, and E. Di Fabrizio. A bacterial ratchet motor. *Proceedings of the National Academy of Sciences*, 107(21):9541–9545, 2010.

[319] S. Ramaswamy. The Mechanics and Statistics of Active Matter. *Annual Review of Condensed Matter Physics*, 1:323–343, 2010.

[320] S. Ramaswamy, R. Aditi Simha, and J. Toner. Active nematics on a substrate: Giant number fluctuations and long-time tails. *Europhysics Letters*, 62(2):196–202, 2003.

[321] K. J. Rao, F. Li, L. Meng, H. Zheng, F. Cai, and W. Wang. A Force to Be Reckoned With : A Review of Synthetic Microswimmers Powered by Ultrasound. *Small*, 11(24):2836–2846, 2015.

[322] M. Ravnik and J. M. Yeomans. Confined active nematic flow in cylindrical capillaries. *Physical Review Letters*, 110(2):026001, 2013.

[323] M. Ravnik and S. Žumer. Landau-de Gennes modelling of nematic liquid crystal colloids. *Liquid Crystals*, 36(10-11):1201–1214, 2009.

[324] G. S. Redner, M. F. Hagan, and A. Baskaran. Structure and dynamics of a phase-separating active colloidal fluid. *Physical Review Letters*, 110(5):055701, 2013.

[325] C. Reichhardt and C. J. Olson-Reichhardt. Active matter transport and jamming on disordered landscapes. *Physical Review E*, 90(1):012701, 2014.

[326] H. Reinken, S. H. L. Klapp, M. Bär, and S. Heinderich. Derivation of a hydrodynamic theory for mesoscale dynamics in microswimmer suspensions. *Physical Review E*, 97(2):022613, 2018.

[327] D. P. Rivas, T. N. Shendruk, R. R. Henry, D. H. Reich, and R. L. Leheny. Driven topological transitions in active nematic films. *Soft Matter*, 16(40):9331–9338, 2020.

[328] P. Romanczuk, M. Bär, W. Ebeling, B. Lindner, and L. Schimansky-Geier. Active Brownian Particles. From Individual to Collective Stochastic Dynamics. *European Physics Journal Special Topics*, 162:1–162, 2012.

[329] T. D. Ross, H. J. Lee, Z. Qu, R. A. Banks, R. Phillips, and M. Thomson. Controlling organization and forces in active matter through optically defined boundaries. *Nature*, 572(7768):224–229, 2019.

[330] Rothschild. Non-random distribution of bull spermatozoa in a drop of sperm suspension. *Nature*, 198(4886):1221–1222, 1963.

[331] L. J. Ruske and J. M. Yeomans. Morphology of Active Deformable 3D Droplets. *Physical Review X*, 11(2):21001, 2021.

[332] F. Sagués, J. M. Sancho, and J. García-Ojalvo. Spatiotemporal order out of noise. *Reviews of Modern Physics*, 79(3):829–882, 2007.

[333] S. Saha, R. Golestanian, and S. Ramaswamy. Clusters, asters, and collective oscillations in chemotactic colloids. *Physical Review E*, 89(6):062316, 2014.

[334] D. K. Sahu, S. Kole, S. Ramaswamy, and S. Dhara. Omnidirectional transport and navigation of Janus particles through a nematic liquid crystal film. *Physical Review Research*, 2:032009, 2020.

[335] D. Saintillan and M. J. Shelley. Instabilities, pattern formation, and mixing in active suspensions. *Physics of Fluids*, 20(12):123304, 2008.

[336] S. Samin and R. Van Roij. Self-Propulsion Mechanism of Active Janus Particles in Near-Critical Binary Mixtures. *Physical Review Letters*, 115(18):188305, 2015.

[337] T. Sanchez, D. T. N. Chen, S. J. Decamp, M. Heymann, and Z. Dogic. Spontaneous motion in hierarchically assembled active matter. *Nature*, 491(7424):431–434, 2012.

[338] T. Sanchez, D. Welch, D. Nicastro, and Z. Dogic. Cilia-like beating of active microtubule bundles. *Science*, 333(January):456, 2011.

[339] S. Sankararaman and S. Ramaswamy. Instabilities and waves in thin films of living fluids. *Physical Review Letters*, 102(11):118107, 2009.

[340] T. B. Saw, A. Doostmohammadi, V. Nier, L. Kocgozlu, S. Thampi, Y. Toyama, P. Marcq, C. T. Lim, J. M. Yeomans, and B. Ladoux. Topological defects in epithelia govern cell death and extrusion. *Nature*, 544(7649):212–216, 2017.

[341] T. B. Saw, W. Xi, B. Ladoux, and C. T. Lim. Biological Tissues as Active Nematic Liquid Crystals. *Advanced Materials*, 30(47):1802579, 2018.

[342] V. Schaller and A. R Bausch. Topological defects and density fl uctuations in collectively moving systems. *Proceedings of the National Academy of Sciences*, 110(12):4488–4493, 2013.

[343] V. Schaller, K. M. Schmoller, E. Karakose, B. Hammerich, M. Maier, and A. R. Bausch. Crosslinking proteins modulate the self-organization of driven systems. *Soft Matter*, 9:7229–7233, 2013.

[344] V. Schaller, C. Weber, E. Frey, and A. R. Bausch. Polar pattern formation: Hydrodynamic coupling of driven filaments. *Soft Matter*, 7(7):3213–3218, 2011.

[345] V. Schaller, C. Weber, C. Semmrich, E. Frey, and A. R. Bausch. Polar patterns of driven filaments. *Nature*, 467(7311):73–77, 2010.

[346] V. Schaller, C. A. Weber, B. Hammerich, E. Frey, and A. R. Bausch. Frozen steady states in active systems. *Proceedings of the National Academy of Sciences*, 108(48):19183–19188, 2011.

[347] D. Schamel, A. G. Mark, J. G. Gibbs, C. Miksch, K. I Morozov, A. M. Leshansky, and P. Fischer. Nanopropellers and Their Actuation in Complex Viscoelastic Media. *ACS Nano*, 8(9):8794–8801, 2014.

[348] M. Schmitt and H. Stark. Swimming active droplet: A theoretical analysis. *Europhysics Letters*, 101(4):44008, 2013.

[349] M. Schmitt and H. Stark. Marangoni flow at droplet interfaces: Three-dimensional solution and applications. *Physics of Fluids*, 28:012106, 2016.

[350] J. Schwarz-Linek, C. Valeriani, A. Cacciuto, M. E. Cates, D. Marenduzzo, A. N. Morozov, and W. C. K. Poon. Phase separation and rotor self-assembly in active particle suspensions. *Proceedings of the National Academy of Sciences*, 109(11):4052–4057, 2012.

[351] F. Schweitzer. *Brownian Agents and Active Particles: Collective Dynamics in the Natural and Social Sciences*. Springer-Verlag, Heidelberg, 2007.

[352] A. Senoussi, S. Kashida, R. Voituriez, J. C. Galas, A. Maitra, and A. Estevez-Torres. Tunable corrugated patterns in an active nematic sheet. *Proceedings of the National Academy of Sciences*, 116(45):22464–22470, 2019.

[353] B. Senyuk, Q. Liu, S. He, R. D. Kamien, R. B. Kusner, T. C. Lubensky, and I. I. Smalyukh. Topological colloids. *Nature*, 493(7431):200–205, 2013.

[354] M. R. Shaebani, A. Wysocki, R. G. Winkler, G. Gompper, and H. Rieger. Computational models for active matter. *Nature Reviews Physics*, 2(4):181–199, 2020.

[355] S. Shankar, S. Ramaswamy, M. C. Marchetti, and M. J. Bowick. Defect Unbinding in Active Nematics. *Physical Review Letters*, 121(10):108002, 2018.

[356] M. J. Shelley. The Dynamics of Microtubule/Motor-Protein Assemblies in Biology and Physics. *Annual Review of Fluid Mechanics*, 48(1):487–506, 2016.

[357] X. N. Shen and P. E. Arratia. Undulatory Swimming in Viscoelastic Fluids. *Physical Review Letters*, 106(20):208101, 2011.

[358] T. N. Shendruk, A. Doostmohammadi, K. Thijssen, and J. M. Yeomans. Dancing disclinations in confined active nematics. *Soft Matter*, 13(21):3853–3862, 2017.

[359] O. Sipos, K. Nagy, R. Di Leonardo, and P. Galajda. Hydrodynamic Trapping of Swimming Bacteria by Convex Walls. *Physical Review Letters*, 114(25):258104, 2015.

[360] J. Slomka and J. Dunkel. Spontaneous mirror-symmetry breaking induces inverse energy cascade in 3D active fluid. *Proceedings of the National Academy of Sciences*, 114(9):2119–2124, 2017.

[361] A. Snezhko and I. S. Aranson. Magnetic manipulation of self-assembled colloidal asters. *Nature Materials*, 10(9):698–703, 2011.

[362] A. Snezhko, I. S. Aranson, and W. K. Kwok. Surface wave assisted self-assembly of multidomain magnetic structures. *Physical Review Letters*, 96(7):078701, 2006.

[363] A. Snezhko, M. Belkin, I. S. Aranson, and W. K. Kwok. Self-assembled magnetic surface swimmers. *Physical Review Letters*, 102(11):118103, 2009.

[364] M. Soares, M. Depken, B. Stuhrmann, M. Korsten, F. C. Mackintosh, and G. H. Koenderink. Active multistage coarsening of actin networks driven by myosin motors. *Proceedings of the National Academy of Sciences*, 108(23):9408–9413, 2011.

[365] A. Sokolov, M. M. Apodaca, B. A. Grzybowski, and I. S. Aranson. Swimming bacteria power microscopic gears. *Proceedings of the National Academy of Sciences*, 107(3):969–974, 2010.

[366] A. Sokolov and I. S. Aranson. Physical properties of collective motion in suspensions of bacteria. *Physical Review Letters*, 109(24):248109, 2012.

[367] A. Sokolov, I. S. Aranson, J. O. Kessler, and R. E. Goldstein. Concentration dependence of the collective dynamics of swimming bacteria. *Physical Review Letters*, 98(15):158102, 2007.

[368] A. P. Solon, Y. Fily, A. Baskaran, M. E. Cates, Y. Kafri, M. Kardar, and J. Tailleur. Pressure is not a state function for generic active fluids. *Nature Physics*, 11(8):673–678, 2015.

[369] A. P. Solon, J. Stenhammar, R. Wittkowski, M. Kardar, Y. Kafri, M. E. Cates, and J. Tailleur. Pressure and phase equilibria in interacting active Brownian spheres. *Physical Review Letters*, 114(19):198301, 2015.

[370] A. A. Solovev, Y. Mei, E. Bermúdez Ureña, G. Huang, and O. G. Schmidt. Catalytic microtubular jet engines self-propelled by accumulated gas bubbles. *Small*, 5(14):1688–1692, 2009.

[371] V. Soni, E. S. Bililign, S. Magkiriadou, S. Sacanna, D. Bartolo, M. J. Shelley, and W. T. M. Irvine. The odd free surface flows of a colloidal chiral fluid. *Nature Physics*, 15(11):1188–1194, 2019.

[372] S. E. Spagnolie, B. Liu, and T. R. Powers. Locomotion of Helical Bodies in Viscoelastic Fluids : Enhanced Swimming at Large Helical Amplitudes. *Physical Review Letters*, 111(6):068101, 2013.

[373] T. Speck, J. Bialké, A. M. Menzel, and H. Löwen. Effective Cahn-Hilliard Equation for the Phase Separation of Active Brownian Particles. *Physical Review Letters*, 112(21):218304, 2014.

[374] T. M. Squires and M. Z. Bazant. Induced-charge electro-osmosis. *Journal of Fluid Mechanics*, 509:217–252, 2004.

[375] T. Stearns and M. Kirschner. In vitro reconstitution of centrosome assembly and function: The central role of γ-tubulin. *Cell*, 76(4):623–637, 1994.

[376] J. Stenhammar, A. Tiribocchi, R. J. Allen, D. Marenduzzo, and M. E. Cates. Continuum theory of phase separation kinetics for active brownian particles. *Physical Review Letters*, 111(14):145702, 2013.

[377] J. Stenhammar, R. Wittkowski, D. Marenduzzo, and M. E. Cates. Activity-induced phase separation and self-assembly in mixtures of active and passive particles. *Physical Review Letters*, 114(1):018301, 2015.

[378] J. Stenhammar, R. Wittkowski, D. Marenduzzo, and M. E. Cates. Light-induced self-assembly of active rectification devices. *Science Advances*, 2:e1501850, 2016.

[379] R. L. Stoop and P. Tierno. Clogging and jamming of colloidal monolayers driven across disordered landscapes. *Communications Physics*, 1(1):68, 2018.

[380] A. V. Straube, J. M. Pagès, A. Ortiz-Ambriz, P. Tierno, J. Ignés-Mullol, and F. Sagués. Assembly and transport of nematic colloidal swarms above photo-patterned defects and surfaces. *New Journal of Physics*, 20(7):075006, 2018.

[381] T. Strübing, A. Khosravanizadeh, A. Vilfan, E. Bodenschatz, R. Golestanian, and I. Guido. Wrinkling Instability in 3D Active Nematics. *Nano Letters*, 20:6281–6288, 2020.

[382] Y. Sumino, N. Magome, T. Hamada, and K. Yoshikawa. Self-running droplet: Emergence of regular motion from nonequilibrium noise. *Physical Review Letters*, 94(6):068301, 2005.

[383] Y. Sumino, K. H. Nagai, Y. Shitaka, D. Tanaka, K. Yoshikawa, H. Chaté, and K. Oiwa. Large-scale vortex lattice emerging from collectively moving microtubules. *Nature*, 483(7390):448–452, 2012.

[384] T. Surrey, F. Nédélec, S. Leibler, and E. Karsenti. Physical properties determining self-organization of motors and microtubules. *Science*, 292(5519):1167–1171, 2001.

[385] K. Suzuki, M. Miyazaki, J. Takagi, T. Itabashi, and S. Ishiwata. Spatial confinement of active microtubule networks induces large-scale rotational cytoplasmic flow. *Proceedings of the National Academy of Sciences of the United States of America*, 114(11):2922–2927, 2017.

[386] R. Suzuki and A. R. Bausch. The emergence and transient behavious of collective motion in active filament systems. *Nature Communications*, 8:41, 2017.

[387] R. Suzuki, C. A. Weber, E. Frey, and A. R. Bausch. Polar pattern formation in driven filament systems requires non-binary particle collisions. *Nature Physics*, 11(10):839–844, 2015.

[388] J. Tailleur and M. E. Cates. Statistical mechanics of interacting run-and-tumble bacteria. *Physical Review Letters*, 100(21):218103, 2008.

[389] S. C. Takatori and J. F. Brady. Towards a thermodynamics of active matter. *Physical Review E*, 91(3):032117, 2015.

[390] S. C. Takatori and J. F. Brady. Forces, stresses and the (thermo?) dynamics of active matter. *Current Opinion in Colloid and Interface Science*, 21:24–33, 2016.

[391] S. C. Takatori, W. Yan, and J. F. Brady. Swim pressure: Stress generation in active matter. *Physical Review Letters*, 113(2):028103, 2014.

[392] A. J. Tan, E. Roberts, S. A. Smith, U. A. Olvera, J. Arteaga, S. Fortini, K. A. Mitchell, and L. S. Hirst. Topological chaos in active nematics. *Nature Physics*, 15(10):1033–1039, 2019.

[393] T. H. Tan, J. Liu, P. W. Miller, M. Tekant, J. Dunkel, and N. Fakhri. Topological turbulence in the membrane of a living cell. *Nature Physics*, 16(6):657–662, 2020.

[394] A. M. Tayar, M. F. Hagan, and Z. Dogic. Active liquid crystals powered by force-sensing DNA-motor clusters. *Proceedings of the National Academy of Sciences of the United States of America*, 118(30):1–10, 2021.

[395] B. Ten Hagen, R. Wittkowski, and H. Löwen. Brownian dynamics of a self-propelled particle in shear flow. *Physical Review E*, 84(3):031105, 2011.

[396] J. Teran, L. Fauci, and M. J. Shelley. Viscoelastic Fluid Response Can Increase the Speed and Efficiency of a Free Swimmer. *Physical Review Letters*, 104(3):038101, 2010.

[397] S. P. Thampi, R. Golestanian, and J. M. Yeomans. Velocity correlations in an active nematic. *Physical Review Letters*, 111(11):118101, 2013.

[398] S. P. Thampi, R. Golestanian, and Julia M. Yeomans. Instabilities and topological defects in active nematics. *EPL*, 105(1):18001, 2014.

[399] S. P. Thampi and J. M. Yeomans. Active turbulence in active nematics. *European Physical Journal: Special Topics*, 225(4):651–662, 2016.

[400] I. Theurkauff, C. Cottin-Bizonne, J. Palacci, C. Ybert, and L. Bocquet. Dynamic clustering in active colloidal suspensions with chemical signaling. *Physical Review Letters*, 108(26):268303, 2012.

[401] K. Thijssen, D. A. Khaladj, S. A. Aghvami, M. A. Gharbi, S. Fraden, J. M. Yeomans, L. S. Hirst, and T. N. Shendruk. Submersed micropatterned structures control active nematic flow, topology, and concentration. *Proceedings of the National Academy of Sciences of the United States of America*, 118(38):1–10, 2021.

[402] B. Thomases and R. D. Guy. Mechanisms of Elastic Enhancement and Hindrance for Finite-Length Undulatory Swimmers in Viscoelastic Fluids. *Physical Review Letters*, 113(9):098102, 2014.

[403] S. Thutupalli, R. Seemann, and S. Herminghaus. Swarming behavior of simple model squirmers. *New Journal of Physics*, 13:073021, 2011.

[404] P. Tierno, R. Golestanian, I. Pagonabarraga, and F. Sagués. Controlled swimming in confined fluids of magnetically actuated colloidal rotors. *Physical Review Letters*, 101(21):218304, 2008.

[405] A. Tiribocchi, R. Wittkowski, D. Marenduzzo, and M. E. Cates. Active Model H: Scalar Active Matter in a Momentum-Conserving Fluid. *Physical Review Letters*, 115(18):188302, 2015.

[406] E. Tjhung, D. Marenduzzo, and M. E. Cates. Spontaneous symmetry breaking in active droplets provides a generic route to motility. *Proceedings of the National Academy of Sciences*, 109(31):12381–12386, 2012.

[407] E. Tjhung, C. Nardini, and M. E. Cates. Cluster Phases and Bubbly Phase Separation in Active Fluids: Reversal of the Ostwald Process. *Physical Review X*, 8(3):31080, 2018.

[408] E. Tjhung, A. Tiribocchi, D. Marenduzzo, and M. E. Cates. A minimal physical model captures the shapes of crawling cells. *Nature Communications*, 6:5420, 2015.

[409] J. Toner, N. Guttenberg, and Y. Tu. Hydrodynamic theory of flocking in the presence of quenched disorder. *Physical Review E*, 98(6):062604, 2018.

[410] J. Toner, N. Guttenberg, and Y. Tu. Swarming in the Dirt: Ordered Flocks with Quenched Disorder. *Physical Review Letters*, 121(24):248002, 2018.

[411] J. Toner, H. Löwen, and H. H. Wensink. Following fluctuating signs: Anomalous active superdiffusion of swimmers in anisotropic media. *Physical Review E*, 93(6):062610, 2016.

[412] J. Toner and Y. Tu. Long-Range Order in a Two-Dimensional Dynamical XY Model: How Birds Fly Together. *Physical Review Letters*, 75(23):4326–4329, 1995.

[413] J. Toner and Y. Tu. Flocks, herds, and schools: A quantitative theory of flocking. *Physical Review E*, 58(4):4828–4858, 1998.

[414] J. Toner, Y. Tu, and S. Ramaswamy. Hydrodynamics and phases of flocks. *Annals of Physics*, 318:170–244, 2005.

[415] G. Tóth, C. Denniston, and J. M. Yeomans. Hydrodynamics of Topological Defects in Nematic Liquid Crystals. *Physical Review Letters*, 88(10):105504, 2002.

[416] T. Turiv, J. Krieger, G. Babakhanova, H. Yu, S. V. Shiyanovskii, Q. H. Wei, M. H. Kim, and O. D. Lavrentovich. Topology control of human fibroblast cells monolayer by liquid crystal elastomer. *Science Advances*, 6(20):1–11, 2020.

[417] R. Urrutia, M. A. Mcniven, J. P. Albanesi, D. B. Murphy, and B. Kachar. Purified kinesin promotes vesicle motility and induces active sliding between microtubules in vitro. *Proceedings of the National Academy of Sciences*, 88(15):6701–6705, 1991.

[418] J. Urzay, A. Doostmohammadi, and J. M. Yeomans. Multi-scale statistics of turbulence motorized by active matter. *Journal of Fluid Mechanics*, 822:762–773, 2017.

[419] S. Van Teeffelen and H. Löwen. Dynamics of a Brownian circle swimmer. *Physical Review E*, 78(2):020101, 2008.

[420] M. Varghese, A. Baskaran, M. F. Hagan, and A. Baskaran. Confinement-Induced Self-Pumping in 3D Active Fluids. *Physical Review Letters*, 125(26):268003, 2020.

[421] T. Vicsek, A. Czirok, E. Ben-Jacob, I. Cohen, and O. Shochet. Novel type of phase transition in a system of self-riven particles. *Physical Review Letters*, 75(6):1226–1229, 1995.

[422] I. D. Vladescu, E. J. Marsden, J. Schwarz-Linek, V. A. Martinez, J. Arlt, A. N. Morozov, D. Marenduzzo, M. E. Cates, and W. C. K. Poon. Filling an Emulsion Drop with Motile Bacteria. *Physical Review Letters*, 113(26):268101, 2014.

[423] R. Voituriez, J. F. Joanny, and J. Prost. Spontaneous flow transition in active polar gels. *Europhysics Letters*, 70(3):404–410, 2005.

[424] R. Voituriez, J. F. Joanny, and J. Prost. Generic phase diagram of active polar films. *Physical Review Letters*, 96(2):028102, 2006.

[425] G. Volpe, I. Buttinoni, D. Vogt, H.-J. Kuemmerer, and C. Bechinger. Microswimmers in Patterned Environments. *Soft Matter*, 7:8810–8815, 2011.

[426] G. Volpe and G. Volpe. The topography of the environment alters the optimal search strategy for active particles. *Proceedings of the National Academy of Sciences*, 114:11350, 2017.

[427] M. B. Wan, C. J. Olson Reichhardt, Z. Nussinov, and C. Reichhardt. Rectification of swimming bacteria and self-driven particle systems by arrays of asymmetric barriers. *Physical Review Letters*, 101(1):018102, 2008.

[428] W. Wang, W. Duan, S. Ahmed, A. Sen, and T. E. Mallouk. From One to Many: Dynamic Assembly and Collective Behavior of Self-Propelled Colloidal Motors. *Accounts of Chemical Research*, 48:1938–1946, 2015.

[429] W. Wang, W. Duan, A. Sen, and T. E. Mallouk. Catalytically powered dynamic assembly of rod-shaped nanomotors and passive tracer particles. *Proceedings of the National Academy of Sciences*, 110(44):17744–17749, 2013.

[430] K. L. Weirich, K. Dasbiswas, T. A. Witten, S. Vaikuntanathan, and M. L. Gardel. Self-organizing motors divide active liquid droplets. *Proceedings of the National Academy of Sciences of the United States of America*, 166(23):11125–11130, 2019.

[431] H. H. Wensink, J. Dunkel, S. Heidenreich, K. Drescher, R. E. Goldstein, H. Löwen, and J. M. Yeomans. Meso-scale turbulence in living fluids. *Proceedings of the National Academy of Sciences*, 109(36):14308–14313, 2012.

[432] B. J. Williams, S. V. Anand, J. Rajagopalan, and M. T. A. Saif. A self-propelled biohybrid swimmer at low Reynolds number. *Nature Communications*, 5:3081, 2014.

[433] R. G. Winkler, A. Wysocki, and G. Gompper. Virial pressure in systems of active Brownian particles. *Soft Matter*, 11:6680–6691, 2015.

[434] H. Wioland, F. G. Woodhouse, J. Dunkel, and R. E. Goldstein. Ferromagnetic and antiferromagnetic order in bacterial vortex lattices. *Nature Physics*, 12(4):341–346, 2016.

[435] H. Wioland, F. G. Woodhouse, J. Dunkel, J. O. Kessler, and R. E. Goldstein. Confinement stabilizes a bacterial suspension into a spiral vortex. *Physical Review Letters*, 110(26):268102, 2013.

[436] R. Wittkowski and H. Löwen. Self-propelled Brownian spinning top: Dynamics of a biaxial swimmer at low Reynolds numbers. *Physical Review E*, 85(2):021406, 2012.

[437] R. Wittkowski, A. Tiribocchi, J. Stenhammar, R. J. Allen, D. Marenduzzo, and M. E. Cates. Scalar ϕ^4 field theory for active-particle phase separation. *Nature Communications*, 5:4351, 2014.

[438] F Wolf and T Geisel. Spontaneous pinwheel annihilation during visual development. *Nature*, 395:73, 1998.

[439] C. W. Wolgemuth. Collective swimming and the dynamics of bacterial turbulence. *Biophysical Journal*, 95(4):1564–1574, 2008.

[440] A. J. M. Wollman, C. Sanchez-Cano, H. M. J. Carstairs, R. A. Cross, and A. J. Turberfield. Transport and self-organization across different length scales powered by motor proteins and programmed by DNA. *Nature Nanotechnology*, 9(1):44–47, 2014.

[441] T. A. Wood, J. S. Lintuvuori, A. B. Schofield, D. Marenduzzo, and W. C. K. Poon. A self-quenched defect glass in a colloid-nematic liquid crystal composite. *Science*, 334(6052):79–83, 2011.

[442] F. G. Woodhouse and J. Dunkel. Active matter logic for autonomous microfluidics. *Nature Communications*, 8:15169, 2017.

[443] F. G. Woodhouse and R. E. Goldstein. Spontaneous circulation of confined active suspensions. *Physical Review Letters*, 109(16):168105, 2012.

[444] K. T. Wu, J. B. Hishamunda, D. T. N. Chen, S. J. DeCamp, Y. W. Chang, A. Fernández-Nieves, S. Fraden, and Z. Dogic. Transition from turbulent to coherent flows in confined three-dimensional active fluids. *Science*, 355(6331), 2017.

[445] A. Würger. Self-Diffusiophoresis of Janus Particles in Near-Critical Mixtures. *Physical Review Letters*, 115(18):188304, 2015.

[446] Z. Xiao, M. Wei, and W. Wang. A Review of Micromotors in Confinements : Pores , Channels , Grooves , Steps , Interfaces , Chains , and Swimming in the Bulk. *ACS Applied Materials & Interfaces*, 11:6667–6684, 2019.

[447] Z. Xu, M. Chen, H. Lee, S. Feng, J. Y. Park, S. Lee, and J. T. Kim. X-ray-Powered Micromotors. *ACS Applied Materials & Interfaces*, 11:15727–15732, 2019.

[448] H. Yu, A. Kopach, V. R. Misko, A. A. Vasylenko, and D. Makarov. Confined Catalytic Janus Swimmers in a Crowded Channel : Geometry-Driven Rectification Transients and Directional Locking. *Small*, 12(42):5882–5890, 2016.

[449] M. Zeitz, K. Wolff, and H. Stark. Active Brownian particles moving in a random Lorentz gas. *European Physical Journal E*, 40:23, 2017.

[450] B. Zhang, B. Hilton, C. Short, A. Souslov, and A. Snezhko. Oscillatory chiral flows in confined active fluids with obstacles. *Physical Review Research*, 2(4):43225, 2020.

[451] H. P. Zhang, A. Be'er, E.-L. Florin, and H. L. Swinney. Collective motion and density fluctuations in bacterial colonies. *Proceedings of the National Academy of Sciences*, 107(31):13626–13630, 2010.

[452] J. Zhang, E. Luijten, B. A. Grzybowski, and S. Granick. Active colloids with collective mobility status and research opportunities. *Chemical Society Reviews*, 46(18):5551–5569, 2017.

[453] R. Zhang, N. Kumar, J. L. Ross, M. L. Gardel, and J. J. De Pablo. Interplay of structure, elasticity, and dynamics in actin-based nematic materials. *Proceedings of the National Academy of Sciences*, 115(2):E124–E133, 2017.

[454] R. Zhang, S. A. Redford, P. V. Ruijgrok, N. Kumar, A. Mozaffari, S. Zemsky, A. R. Dinner, V. Vitelli, Z. Bryant, M. L. Gardel, and J. J. de Pablo. Spatiotemporal control of liquid crystal structure and dynamics through activity patterning. *Nature Materials*, 20(6):875–882, 2021.

[455] R. Zhang, Y. Zhou, M. Rahimi, and J. J. De Pablo. Dynamic structure of active nematic shells. *Nature Communications*, 7:13483, 2016.

[456] L. Zhao, L. Yao, D. Golovaty, J. Ignés-Mullol, F. Sagués, and M. Carme Calderer. Stability analysis of flow of active extensile fibers in confined domains. *Chaos*, 30:113105, 2020.

[457] Z. Zhou, C. Joshi, R. Liu, M. M. Norton, L. Lemma, Z. Dogic, M. F. Hagan, S. Fraden, and P. Hong. Machine learning forecasting of active nematics. *Soft Matter*, 17(3):738–747, 2021.

[458] L. Zhu, E. Lauga, and L. Brandt. Self-propulsion in viscoelastic fluids : Pushers vs . pullers. *Physics of Fluids*, 24(5):051902, 2012.

[459] L. Zhu, E. Lauga, and L. Brandt. Low-Reynolds-number swimming in a capillary tube. *Journal of Fluid Mechanics*, 726:285–311, 2013.

[460] A. Zöttl and H. Stark. Nonlinear dynamics of a microswimmer in Poiseuille flow. *Physical Review Letters*, 108(21):218104, 2012.

[461] A. Zöttl and H. Stark. Hydrodynamics determines collective motion and phase behavior of active colloids in quasi-two-dimensional confinement. *Physical Review Letters*, 112(11):118101, 2014.

[462] A. Zöttl and H. Stark. Emergent behavior in active colloids. *Journal of Physics Condensed Matter*, 28(25):253001, 2016.

[463] A. Zöttl and J. M. Yeomans. Enhanced bacterial swimming speeds in macromolecular polymer solutions. *Nature Physics*, 15(6):554, 2019.

Index

Printed in the United States
by Baker & Taylor Publisher Services